Regulatory Support for Off-Grid Renewable Electricity

Edited by
Ngozi Chinwa Ole,
Eduardo G. Pereira,
Peter Kayode Oniemola, and
Gustavo Kaercher Loureiro

Routledge
Taylor & Francis Group
LONDON AND NEW YORK
earthscan
from Routledge

First published 2023
by Routledge
4 Park Square, Milton Park, Abingdon, Oxon OX14 4RN

and by Routledge
605 Third Avenue, New York, NY 10158

Routledge is an imprint of the Taylor & Francis Group, an informa business

British Library Cataloguing-in-Publication Data
A catalogue record for this book is available from the British Library

ISBN: 978-1-032-01294-0 (hbk)
ISBN: 978-1-032-01295-7 (pbk)
ISBN: 978-1-003-17808-8 (ebk)

DOI: 10.4324/9781003178088

Typeset in Times New Roman
by Deanta Global Publishing Services, Chennai, India

To the infinite intelligence

Contents

List of contributors ix

1 **General Introduction: Conceptualising Off-Grid Renewable Electricity in Global Energy Transition** 1
NGOZI CHINWA OLE

PART 1
Sustainable Electricity and Off-grid Renewable Energy: Critical Issues and Challenges 15

2 **International Legal Framework for Sustainable Energy** 17
PETER KAYODE ONIEMOLA AND ADEJUMOKE J AKINBUSOYE

3 **Sustainable Development and Off-Grid Renewable Electricity: Current Status and Challenges** 36
CHITZI C. OGBUMGBADA, RASAKI STEPHEN DAUDA, AND EDUARDO G. PEREIRA

4 **The International Climate Change Regime and Off-Grid Renewable Electricity** 52
NGOZI CHINWA OLE, EMPIRE HECHIME NYEKWERE, COSMOS NIKE NWEDU, KINGSLEY OSINACHI ONU, AND ADESOLA OMOTOLA

PART 2
Africa 71

5 **Kenya** 73
KINGSLEY OSINACHI ONU AND SAMUEL PIYANILE LOMOLE

6 **Nigeria** 93
ETI BEST HERBERT AND PETER KAYODE ONIEMOLA

7 Uganda 110
OLUGBENGA OKE-SAMUEL, KAMORU T. LAWAL, AND TAJUDEEN SANNI

PART 3
Asia 133

8 China 135
NGOZI CHINWA OLE, OPEYEMI OMOTUYI, AND IMAM MULYANA

9 Philippines 154
NGOZI CHINWA OLE, DICKSON E. OMUKORO, AND ZIYANA NAZEEMUDEEN MOHAMED

10 Indonesia 173
ANTON LATIEF AND RANDY HENDRIKA

PART 4
North and South America 201

11 Brazil 203
CLARISSA EMANUELA LEÃO LIMA, LUIZ GUSTAVO KAERCHER LOUREIRO, AND EDUARDO G. PEREIRA

12 Mexico 222
MARÍA SERNA, NGOZI CHINWA OLE, AND IZUOMA EGERUOH-ADINDU

13 Trinidad and Tobago 241
ALICIA ELIAS-ROBERTS

PART 5
Towards a Supportive Regulatory Framework for Off-Grid Renewable Electricity 257

14 Reflections and Conclusion 259
GUSTAVO KAERCHER LOURIERO AND EDUARDO G. PEREIRA

Index 273

Contributors

Adejumoke J. Akinbusoye is a legal practitioner and a researcher with interests in the field of environment, development, energy, and climate change law. She obtained her LLB from the University of Lagos (2010) and her LLM in Environmental Law and Policy from the Centre for Energy, Petroleum, Mineral Law and Policy (CEPMLP), University of Dundee (2014). She was called to the Nigerian Bar in the year 2012. She is currently concluding her doctoral studies at the Department of Petroleum, Energy, Economics and Law, University of Ibadan. She teaches International Trade Law, Mineral and Water Law, and Petroleum and Energy Law at the University of Lagos, Nigeria. Her experience in legal practice spans litigation, corporate commercial practice, corporate governance, and legal advocacy. She is a member of the Nigerian Bar Association, the American Society of International Law, and The Institute of Chartered Secretaries Administrators of Nigeria.

Rasaki Stephen Dauda is an Associate Professor of Economics, Ag. Head, Department of Economics and Head, Water Economics and Social Issues Unit, African Centre of Excellence for Water and Environmental Research (ACEWATER), Redeemer's University, Ede, Osun State, Nigeria. He holds a PhD in Economics from the University of Ibadan, Nigeria. His research focuses on economic development, covering issues such as poverty, inequality, health, and human resource development. Dauda is an Alumnus of the African Economic Research Consortium, Kenya; a World Bank Scholar; TETFUND Grantee, Laureate of the Council for the Development of Social Science Research in Africa (CODESRIA), Senegal; a Grantee of the Association of African Universities (AAU), Ghana; Best Researcher of the Year, 2021, Redeemer's University; among others. He was one of the facilitators that estimated the Demographic Dividend (DD) for Kaduna State under the UNDP in December 2019. He consults for the International Business School, The Hague, Netherlands; Centro Studi di Politica Internazionale (CeSPI), Rome, Italy on Mondopoli Project, and Chartered Institute of Bankers (CIBN), Nigeria. He is a reviewer for several local and international journals, some of which are published by Springer, Routledge, Taylor & Francis, and Wiley & Sons. He has several publications in both national and international journals, some of which are indexed in Web of Science and Scopus.

Izuoma Egeruoh-Adindu graduated with an LLB, Second Class Upper Division, from the Imo State University, Owerri, Nigeria, in 2009. She was enrolled to practice as a barrister and solicitor of the Supreme Court of Nigeria in 2010 after passing her Bar examinations (BL) at the Nigerian Law School with Second Class Honours Upper Division. She proceeded to the University of Jos, Nigeria, for postgraduate studies where she obtained her Master of Laws (LLM) specialising in Environmental Law. She is currently a doctoral candidate (specialising in Environmental Law and Policy) at the same university. Egeruoh-Adindu joined the services of the Nigerian Institute of Advanced Legal Studies (NIALS) as a Research Fellow in 2010 with special research interests in environmental, natural resources, and energy law, where she is widely published. She is a Fellow of the Oxford School of Climate Change (2022). She currently serves as the Acting Coordinator for the Center for Continuing Legal Education – NIALS, Secretary to the NIALS *Journal of Business Law*, and an editorial assistant to the NIALS *Environmental Law Journal*, one of the flagship journals of the Institute. Izuoma is currently a Senior Research Fellow and the Acting Coordinator of Continuing Legal Education NIALS. She is a Lecturer and Content Developer in the NIALS postgraduate school, Facilitator, and Examiner in the Faculty of Law, National Open University of Nigeria, where she teaches Environmental Law and Company Law.

Alicia Elias-Roberts is an academic who teaches Oil and Gas Law and International Environmental Law at The University of the West Indies (UWI). She graduated from the University of Guyana, where she obtained an LLB. She is also the recipient of a Masters of Law (BCL) from the University of Oxford in the UK and an LLM in Energy, Environment and Natural Resource Law from the University of Houston in Texas, USA, where she was an OAS/LASPAU scholar. Alicia is also the recipient of a United Nations fellowship in International Law. She is currently a PhD student at Queen's University, Canada, where her research is focused on Offshore Energy Development. She was awarded several scholarships to support her PhD studies. Alicia is the former Deputy Dean of the Faculty of Law at the UWI St Augustine Campus in Trinidad and Tobago and the former Head of the Department of Law at the University of Guyana. She has over 20 years of experience as an energy and environmental law legal consultant. She was previously a Legal Treaty Consultant with the Caribbean Community (CARICOM) Secretariat and has done legal consultancies for UNAIDS, ILO, PAHO, and various governments in the Commonwealth Caribbean. She has provided expert legal advice in conservation and biodiversity law, multilateral environmental agreements, maritime law, oil and gas law, treaty law, and procurement law, to name a few. She is an attorney-at-law, admitted to practice in New York, USA, Trinidad and Tobago, and Guyana.

Randy Hendrika is an Associate and Founding Member of Latief & Co (LCO) Law Firm, based in Jakarta, Indonesia. He has extensive practical experience in advising and representing clients in oil and gas deals, among others in matters concerning the acquisition or termination of Production Sharing Contracts, the construction and utilisation of facilities, etc. His experience includes negotiating and preparing of Farm-in/out Agreement, Facility Sharing Agreements, Tie-in Agreements, etc. He also advises major Indonesian companies in banking and finance, M&A, foreign direct investment, and general corporate compliance projects.

Eti Best Herbert is a Research Fellow in the College of Law, Bowen University, Iwo, Osun State, Nigeria. He holds an LLB from the University of Uyo, Nigeria, and an LLM from the University of Ibadan, Nigeria. He is also a barrister and solicitor of the Supreme Court of Nigeria. He is a team member of the Environment and Water Regulation Unit, African Centre for Water and Environmental Research (ACEWATER), Osun State.

Anton Latief is the Founding Partner of Latief & Co (LCO) in Jakarta, Indonesia. Prior to establishing LCO, he successfully established the Energy Practice Group of one of Indonesia's prominent law firms and had the opportunity to lead the Indonesia Legal Department of a French oil supermajor after being assigned to their Gas and Power division in Paris, France. With more than 20 years of experience, he has been involved in the development of major Indonesian and international upstream oil and gas projects, such as the Yamal LNG (Russia), Shah Deniz – Phase 2 (Azerbaijan), the Mahakam PSC (Indonesia), and Abadi LNG Project (Indonesia). He also has substantial experience in assisting major international oil companies in the negotiation and finalisation of M&A transactions, acquisition or termination of PSCs, joint venture agreements, project financing-related agreements, construction and use of facilities agreements, and the natural gas, LPG, and LNG sales and purchases agreements. He is ranked in 2021, 2022 and 2023 Chambers & Partners Global and the Asia Pacific.

Kamoru Lawal is the principal partner in K.T. LAWAL & CO., a commercial law firm in Ibadan, Nigeria. He obtained an MPhil in Energy and Environmental Law (with Distinction) from the University of Adelaide, Australia, and an LLB (First Class Hons) from the University of Ibadan, Nigeria. Kamoru has over a decade of experience in commercial litigation and advisory and has appeared before all levels of court in Nigeria. Kamoru researches topics related to energy (electricity regulation, oil and gas, and renewable energy), environment, mining, tax, and corporate law. He has published articles and book chapters on tax, renewable energy, oil & gas, and human rights.

Clarissa Emanuela Leão Lima is a lawyer and master's candidate at the Graduate Program in Energy at the University of São Paulo (PPGE/USP). She is a Researcher in the Advocacy Group at the Research Centre for Greenhouse Gas Innovation (RCGI). Clarissa holds a bachelor's in law with emphasis in economics from Fundação Getulio Vargas (RJ), Brazil. She is a Consultant for the Inter-American Development Bank through PPP Americas 2021.

Samuel Piyanile Lomole is the Managing Partner of Piyanile and Co Advocates, Juba, South Sudan. He holds a master's in Oil and Gas Law from the University of Aberdeen, Scotland. He holds a Bachelor of Law certificate from Moi University, Kenya. Samuel is enrolled to practice law in Kenya and South Sudan as an advocate. His areas of practice include, but are not limited to, renewable energy law, oil and gas law, commercial transactions, contract review and drafting, etc. Prior to establishing Piyanile & Co. Advocates, Lomole worked at Robson Harris & Co. Advocates (Kenya) and later at South Sudan Associated Advocates (South Sudan). In addition to working at the firm, Samuel lectures at the University of Juba, College of Law; he is also a consultant in energy law at the National Petroleum and Gas Commission and serves on the technical committee established by the Board of Nile Petroleum Corporation (the national oil company). Lomole fluently speaks and writes English and Swahili, and fluently speaks local Arabic and Bari (the local language in South Sudan).

Gustavo Kaercher Loureiro is currently a Senior Researcher at Center for Regulatory and Infrastructure Studies – CERI – Fundação Getúlio Vargas – FGV/RJ. Gustavo holds a PhD in Public Law from the Federal University of Rio Grande do Sul (2007). The latter was followed by a post-doctorate (fellowship) on the International Centre of Economic Research – ICE, Turin, Italy (2013). He is a former Associate Professor of Administrative and Constitutional Law at the Federal University of Brasília – UnB, Brazil (2007–2014). Loureiro has authored several books and papers on infrastructure regulation and Brazilian public law such as *Electric Energy Law Manual. Quartier Latin: São Paulo, 2021* (book); *Studies on the Economical-Financial Regime of Concession Contracts. Quartier Latin: São Paulo, 2020* (book); *Institutions of Energy Law. Quartier Latin: São Paulo, 2019* (book); 'Is There Really Constitutional Basis on the Economical-Financial Balance of the Concessions? For a Flexible Model of the Economical Regime of the Concessions of Public Service' (paper); 'The Origins and the Commitments of the Figures of Economical-financial Balance in the Concession of Public Services and Just Compensation of the Employed Capital on the Activities of Public Utility' (paper); 'Unpredictability, Economical-Financial Balance and the Prince Factor in the Concessions of Public Services' (paper).

Ziyana Nazeemudeen Mohamed is a Teaching Fellow at the SOAS, University of London, and a former lecturer at the University of Stirling, Scotland, and the Faculty of Law, University of Colombo, Sri Lanka. She recently completed a research project at the University of Aberdeen, Scotland which was funded by the British Council under the grant of South Asia Science Programme for the project titled 'Reproductive Health Care and Policy Concerns : Regulation of Surrogacy Arrangements made in Sri Lanka and Lessons Learned from the United Kingdom'. Her research focus on Human Rights, family law and family aspect on private international law with special interests in Asia.

Imam Mulyana holds an LLB and an LLM from the Universitas Padjadjaran, Indonesia, and earned his doctorate from the University of Aberdeen, Scotland. Mulyana has been a Lecturer and Researcher at the Department of International Law of the Universitas Padjadjaran since 2006. He also serves as an editor for several premium law journals in Indonesia. Mulyana's current research agenda focuses on the legal theory of state sovereignty over natural resources, environmental democracy, renewable energy, and public participation in environmental decision-making. Imam is also a member of the International Law Association.

Cosmos Nike Nwedu teaches and researches law at the Faculty of Law, Federal University Ndufu-Alike Ikwo, Nigeria. He received his LLM in Climate Change and Energy Law and Policy from the Centre for Energy, Petroleum and Mineral Law and Policy (CEPMLP), University of Dundee, UK; BL from Nigerian Law School, Abuja; and LLB (Hons) from Ebonyi State University, Nigeria. His research focus brings interdisciplinarity to the interface of climate change and energy law and policy, including oil, gas, and renewable energy. He is a member of the Young International Council for Commercial Arbitration (Young ICCA), Young International Arbitration Group (YIAG), Energy Institute (EI), UK, Nigerian Association of Law Teachers (NALTs), and Nigerian Bar Association (NBA).

Empire Hechime Nyekwere is a law teacher and researcher at Bowen University, Iwo, Nigeria. In the latter capacity, he teaches environmental law and energy law. He has also supervised various research projects that border on the subject matter of energy and environmental law. He is a member of the environment and water regulation unit of the African Center of Excellence for Water and Environmental Research (ACEWATER), Nigeria. He is widely published with his publications featuring in various international and national journals, including the *Groningen Journal of International Law* and the *Journal of Law, Policy and Globalisation.*

Chitzi C. Ogbumgbada is a Lecturer in Law at the University of Huddersfield, United Kingdom. He is a PhD candidate at the School of Law, University of Strathclyde, United Kingdom. He is also a member of the Strathclyde Centre for Environmental Law and Governance, and the African Natural Resources and Energy Law Network. Ogbumgbada is a recipient of the John and Anne Benson Prize 2020/21 and is currently the editor-in-chief of the *Strathclyde Law Review.*

Olugbenga Oke-Samuel is an Associate Professor of Law and current Dean of the Faculty of Law, Adekunle Ajasin University, Akungba Akoko, Nigeria. He was formerly a senior lecturer at Kampala International University, Uganda (2018–2021). He is a barrister and solicitor of the Supreme Court of Nigeria. Oke-Samuel has a doctorate (PhD/LLD) in Environmental Law from the University of Zululand, South Africa. His research areas include environmental law, renewable energy law, Nigerian legal system, international law, justice education, clinical legal education, and African law. He is the author of *Legislation and Control of Gas Emission in Nigeria and South Africa,* co-author of *Freedom of Information Manual* (NULAI Nigeria, 2016), and editor of *Trends in Nigerian Law* (Nigeria, 2007). A passionate environmental lawyer, he is the Executive Director of the Centre for Environment and Justice in Africa (CEJAFRICA). Oke-Samuel is also a consultant to UNDP, where he has participated in making some environmental laws. He is currently a visiting lecturer, researcher, and scholar to Kampala International University (KIU), Uganda. Oke-Samuel is a visiting scholar to City University, Mogadishu, Somalia.

Ngozi Chinwa Ole, PhD, is the Research Team Leader, Environment and Water Regulation Unit, African Centre of Excellence for Water and Environmental Research (ACEWATER), Nigeria. She is a Senior Lecturer and Acting Head – Department of Public Law and Director of Clinical Legal Education, Federal University, Oye-Ekiti, Nigeria. Ngozi is a Consultant-managing associate at Alliance Law Firm, a top-tier energy law firm in Africa. She holds a PhD in Energy and Environmental Law, following three years of doctoral study, as an Elphinstone Scholar at the University of Aberdeen, Scotland. She has a master's in law from the same University. le has a first-class degree in law from Ebonyi State University, Nigeria, where she garnered several academic awards, including the Gani Fawehnimi Endowment Award for Best Graduating Law Students in Nigeria. She also emerged as the Best Graduating Student in the 2012 Bar 2 Exams, Nigerian Law School, Augustine Nnamani Campus. Ngozi is a barrister and solicitor of the Supreme Court of Nigeria. She is currently the National Publicity Secretary of the Association of Environmental Lawyers of Nigeria. She is also a recipient of several international awards and grants, including the AIPN 2018 Conference Award, Recognition of Excellence in Service as the President of the University of Aberdeen AIPN Club, Elphinstone Scholarship Award, International Ambassador of Peace Award, NULAI/ Open Society Initiative Grant on Clinical Legal Education, etc. She has been part of the teaching team for several courses, including renewable energy law at the University of Aberdeen, Scotland; Redeemers University, Nigeria, etc. Her publications cut across various themes in energy law, including decommissioning of petroleum installations, health and safety of persons in the petroleum industry, local content, off-grid renewable energy, environmental protection in the petroleum industry, etc. The latter is published in several international journals and government gazettes.

Adesola Omotola is a barrister and solicitor of the Supreme Court of Nigeria. She is a law teacher and a researcher at Federal University, Oye-Ekiti, Nigeria. She holds a degree in law and an LLM from Ekiti State University, Nigeria.

Opeyemi Omotuyi is a law teacher and researcher, with a specialty in energy law and sustainable development. She holds an LLB from Obafemi Awolowo University, Nigeria. She holds an LLM from the University of Cape Town, and a PhD from the University of the Witwatersrand, all in South Africa. She currently teaches Law at the Federal University Oye-Ekiti, Nigeria.

Dickson E. Omukoro is a Lecturer in the Faculty of Law, University of Port Harcourt. He holds a PhD (2017) in Energy and Environmental Law from the University of Aberdeen, United Kingdom, and an LLM (2011) in Oil and Gas Law (with distinction) from the same university. He obtained his LLB (2007) from the Niger Delta University, Bayelsa State, and was called to the Nigerian Bar before proceeding to the University of Aberdeen for his postgraduate studies. His area of interest includes energy and natural resources law, environmental law, sustainable development, public and private partnerships in infrastructural development, international commercial law, decommissioning, and renewable energy law. He has presented papers at several local and international conferences, and is the recipient of several local and international research grants, academic awards, and scholarships. Omukoro has also published in several local and international peer-reviewed outlets. He consults for organisations and continues to engage in legal practice with special focus on energy and environmental justice issues.

Peter Kayode Oniemola, PhD, is a Senior Lecturer in the Faculty of Law at the University of Ibadan, Nigeria, and the Energy Law Programme Coordinator at the University of Ibadan's Centre for Petroleum, Energy Economics and Law. He holds a Master of Laws (LLM) from the University of Ibadan and a PhD in Renewable Energy Law from the University of Aberdeen, United Kingdom. Prior to the latter, he obtained a Bachelor of Laws (LLB) from the University of Ilorin, Nigeria, and is a graduate of the Nigerian Law School. Oniemola was a MacArthur Foundation Scholar at the University of Ibadan from 2008 to 2009. He is a barrister and solicitor of the Supreme Court of Nigeria. He has accumulated decades of experience in advising various government bodies on energy-related issues. He is widely published and referenced. His publication features in several international journals, including the ones published by Oxford University Press.

Kingsley Osinachi Onu is a Lecturer at the Faculty of Law, Adeleke University, Nigeria. He is a barrister and solicitor of the Supreme Court of Nigeria. Kingsley is the Managing Editor of *Adeleke University Law Journal*, a journal publication of the Faculty of Law, Adeleke University. He is a Chartered Manager and a Professional Negotiator and Meditator. He holds a Certificate in International Water Law from the University of Geneva, Switzerland. He is a rapacious researcher with a particular interest in environmental, energy, and space laws. He has published in several learned peer-reviewed national and international journals.

Eduardo G. Pereira is a worldwide recognised scholar specialising in Natural Resources and Energy Law. He is a founding partner at the International Energy Law Training and Research Company as well as at the International Energy Law Advisory Group. He has been active in the natural resources and energy industry for more than 15 years and is an international expert on oil and gas contracts and regulations. His experience in this area – both academic and practical – is extensive. He has practical experience in over 50 jurisdictions covering America, Europe, Africa, and Asia. He possesses professorial and related scholarly positions as a full-time, part-time, honorary, adjunct, and/or visiting scholar in several leading academic institutions worldwide. He is also a managing editor for *Global South Energy and Natural Resources Law Journal* (*GSENRLJ*) and an associate editor of *Oil, Gas & Energy Law* (*OGEL*). He is also the author and editor of several leading oil and gas textbooks. Further information about his profile can be found at www.eduardogpereira.com.

Tajudeen Sanni was a Lecturer in Kampala International University, Uganda. He holds a PhD in law from the Nelson Mandela University, Port Elizabeth, South Africa. Prior to this, he obtained a master degree in law from the Liverpool John Moors University, United Kingdom. Sanni is a One Ocean Hub (University of Strathclyde) affiliated Senior Research Fellow at Nelson Mandela University. He specialises in Law of the Sea, International Economic Law, Environmental Law, and International Energy Law. He is also interested in Comparative and Islamic Law. He has taught and supervised theses in these disciplines at undergraduate and postgraduate levels. He co-founded and is heading the Environmental and Energy Law Research Forum, Uganda. Previously, Sanni taught Environmental Law at the Nigerian Study Center of Houdegbe North America University. He has many academic publications to his credit and has presented academic papers in Africa, Europe, America, and Asia. Sanni is a member of a number of professional associations such as International Law Association, African Environmental Law Lecturers Association, Society of International Economic Law and Society of Legal Scholars, UK, among others.

María Serna is a partner in Telenza Energy Consulting, Mexico. She holds a masters degree in Energy, Environment, and Natural Resources Law, from the University of Houston; and a Bachelor of Laws from Escuela Libre de Derecho. María Serna has a broad experience in transactional and regulatory energy matters from the private and public sectors. She has advised international energy companies in compliance and advocacy before energy regulators. María served as director of the National Hydrocarbons Commission's bidding department for the award of exploration and extraction contracts during the first three calls of Round 1, and served as a legal counsel for the Legal Affairs of ASEA, participating in the drafting of regulation. Her experience in the private sector includes being an associate in top-tier energy and finance law firms. Abroad, María advised an international energy company for their Africa and Latin America policy and public affairs.

1 General Introduction

Conceptualising Off-Grid Renewable Electricity in Global Energy Transition

Ngozi Chinwa Ole

1.1 Introduction

This book analyses the extent to which laws and policies support the development of off-grid renewable electricity (OGRE) in the context of global energy transition. Consequent upon which it discusses how laws and policies can be strengthened to optimise the development of OGRE. It uses analysis of international instruments and selected case studies in Africa, Asia, and the Americas to provide a full gaze of how law interacts with OGRE development. It seeks to ascertain the extent to which such interaction enables the development of OGRE. The regulation of any aspect of RE, including cuts across energy security, trade, and global environmental protection,[1] renders it an area of transnational law.[2] In the words of Kelly, 'renewable energy regulation transcends domestic legal systems thereby rendering them areas of transnational law'.[3] Thus, the analyses of selected case studies from diverse continents buttress the global picture of how law interacts with OGRE in several respects.

Renewable energy encapsulates energy derived from sources that are replenishable by nature including the wind, tidal waves, water, sun, and bio materials. OGRE generally

> refers to electricity produced from renewable energy sources from smaller electricity plants at or near the point of sale. The distribution is directly to end-users without connecting to a national grid. It encompasses electricity from stand-alone or mini-grid renewable energy technologies.[4]

While the latter is its typical definition, jurisdictions like China[5] stretch it to incorporate mini-grid renewable electricity which feeds some of its load to a national grid. Besides the point mentioned, OGRE is distinguishable from on-grid by scale. While on-grid electricity is produced from large scale technologies and projects developed by major sector actors, OGRE is produced from smaller technologies mostly developed by small actors including individual consumers.

OGRE is growing in popularity in developed[6] and developing countries[7] as a cost competitive option of electrification given its peculiar features and benefits. Its small scale in comparison to on-grid and its abundant replenishable source

DOI: 10.4324/9781003178088-1

makes it a better fit as a backup in the context of unreliable electricity from the national grid. It is a viable option for the electrification of isolated cities, local communities, and islands because of its independent feature.[8] For such areas, it is impracticable to extend the national grid given the characteristics of sparse population and/or physical features.[9] Addressing climate change and attaining sustainable development entails a paradigm change from fossil fuel electricity to renewable energy including OGRE.[10] It is for this reason that the Director of IRENA opines that 'off-grid renewables are a fundamental component of the energy transition and a pillar of sustainable development'.[11] Among other benefits, the imperatives for OGRE continue to be critically exacerbated by COVID-19.[12] Recovery from the pandemic entails access to crucial energy infrastructures including OGRE.[13]

Unsurprisingly, the development of OGRE has grown significantly over the years.[14] Renewable electricity including OGRE continues to expand at double the rate of every other source.[15] The United Nations Sustainable Development Goals aim is to achieve 100% access to affordable, reliable, and modern energy including electricity for all by 2030.[16] While the world is gravitating towards attaining these goals, over 13% do not have access to electricity.[17] The majority of the population without access to electricity are domiciled in remote rural areas of Asia and Africa where OGRE is a viable option for electrification.[18] The Intergovernmental Panel on Climate Change (IPCC) has cautioned that avoiding the worst of climate change implies limiting global warming to 1.5°C by 2050.[19] 'A consensus has formed that an energy transition grounded in renewable sources and technologies that increases efficiency and conservation is the only way to give us a fighting chance of limiting global warming to 1.5°C by 2050'.[20] Thus, there is a strong impetus to develop OGRE beyond its current capacity in the context of universal access to electricity and addressing climate change.

OGRE is also very beneficial in the creation of jobs. In the words of the International Labour Organisation (ILO), 'Renewable energy has a demonstrated job creation effect … energy created through solar photovoltaic cells, landfill gas, or biomass plants have a higher number of jobs created per unit of energy produced than energy produced through conventional sources'.[21] The latter is because renewable energy, including OGRE, is more labour intensive in comparison to competing sources.[22] OGRE is already creating ten times more jobs than it has displaced in the sector.[23] There is also room for increased job creation especially in unelectrified areas.[24] Thus, the optimal development of OGRE is crucial to job creation across the globe.

Importantly, the optimal development of OGRE cannot be attainable without effective policies and laws in place at every sphere of governance. In the words of IRENA, 'policies for … off-grid renewable energy are [*sic*] seen as central to a just and inclusive energy transition that supports the achievement of both socioeconomic and global climate ambitions'.[25] The effectiveness of the laws will be measured against its ability to enable the optimal development of OGRE at every level.[26] Notably, its effectiveness implies that law is able to garner private and public investments in the sector by facilitating the removal of barriers to their participation.[27] It should also have an implied benefit of accelerating the employment benefits of OGRE. Law must be sturdy enough to be responsive to emerging challenges (including) that attempt to hinder the development of the sector.

An effective legal framework that will underpin the development of OGRE should be founded on a clear understanding of its imperatives in a given jurisdiction. These frameworks should consist of a Licensing and Environmental Impact Assessment (EIA) that enables rather than impedes the growth of the sector. It should include tariff structure and contractual vehicles that is sensitive to the peculiar needs of the sector. Among other things, such frameworks must embolden investors to develop the sectors by supporting them in areas where they are particularly constrained. It should factor in local circumstances such as communities' perception of renewable electricity and make it better. Prior literatures have analysed the barriers and enablers to the development of OGRE in a given jurisdiction. However, still lacking is a comprehensive, extensive, and book-depth analysis of the barriers, legal enablers, and supportive framework for OGRE that cuts across various jurisdictions. It is the mentioned lacuna that this book addresses.

As reiterated, this book analyses the extent to which laws and policies support or enable the development of off-grid renewable electricity in the context of global energy transition. It contextualises the place of OGRE in achieving access to sustainable electricity and the extent of international institutions and instruments in enabling its development. Some of the institutions and instruments include the International Renewable Energy Agency, United Nations Sustainable Development Goals, the Paris Climate Change Agreement 2015, etc. Following, it analyses the governance, contractual, fiscal and support framework including licensing, EIA, social impact assessment, tariff structure, and power purchase agreement in selected jurisdictions in Asia, Africa, and North and South America. Notably, the mentioned continents are where the imperatives for OGRE are strongest. Given the identified enablers and impediments to the development of OGRE as espoused in the case studies, recommendations are made on how law can be strengthened to optimally support its development in the context of global energy transition.

This chapter is divided into five sections. The first section is the introduction; Section 1.2 discusses the trend in the global development of OGRE, providing a picture of its status against where it ordinarily should be given the need for it. Section 1.3 provides an overview of the identified barriers to the development of OGRE. Section 1.4 contains an overview of the entire book, an expatiated justification for the book, and its unique selling points. Section 1.5 contains the conclusion.

1.2 Trends in OGRE Development

The point has been made that the imperatives for OGRE are strongest in Africa, Asia, and North and South America. These continents currently have populations without access to electricity, with Africa having the most.[28] Asia, North and South

America are closing in on their electricity deficit.[29] However, Africa is still lagging behind in their electrification rate with Nigeria, the Democratic Republic of Congo, and Ethiopia having the largest electricity deficit.[30] The optimal development of OGRE has, and will continue to play, the crucial role in the attainment of universal access to electricity in the abovementioned regions. Hydro and solar OGRE has, since 2010, maintained the lead in the sector globally (see Pie Chart).[31] According to the IRENA Off-grid Renewable Energy Statistics 2021, the development of hydro-based OGRE stood at 753 MW/per hour while solar was 404 MW in 2019.[32] Other OGRE like bio-electricity are also recording progress in this regard.[33] Asia is at the forefront of the manufacturing and development of hydro and solar followed by Africa.[34] North and South America are also recognised as having improved in the development of OGRE.[35] However, the limited development of OGRE in the latter may be accounted for by the high rate of electrification in the region in comparison to Africa and Asia. Countries like China are renowned not just for the development of OGRE but in its global manufacturing.[36] Kenya, the Philippines, Indonesia, South Africa, Nigeria, Uganda, Congo, Brazil, Argentina, Mexico, etc. were reported as progressing in the development of the sector.[37]

OGRE has also been developed sparingly in other regions like Europe and the Middle East. In Europe, the use of OGRE has increased in comparison to what it was in 2009[38] particularly for the electrification of islands[39] and communities.[40] The European Union initiated a policy called the 'Clean Energy Package', which embeds OGRE as an option for communities.[41] In Germany, some residential households use OGRE solar photovoltaic for self-electrification.[42] Some communities in Scotland use off-grid renewable electricity, for example, the Island of Eigg[43] and Aberdeen River Don community.[44] In the Middle East, countries like the United Arab Emirates, Palestine, Oceania, and Fiji are listed as countries that have experienced the development of OGRE.

Irrespective of what is mentioned above, the development of OGRE is not as it should, and could, be in the context of the need for it. As stated, there are still some populations in remote rural areas of the world without access to electricity, especially in Africa and Asia. Fossil fuel electricity still continues to maintain the

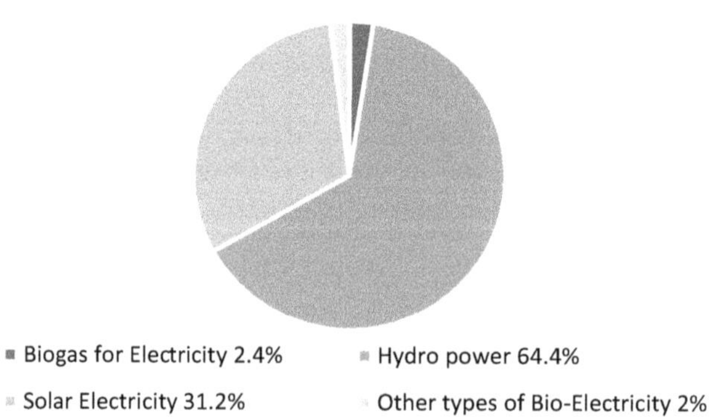

Figure 1.1 Global Off-grid Renewable Electricity Development Statistics

lead even in the off-grid electricity sector. Addressing the environmental ramifications of such fossil fuel use entails a shift to OGRE. The inability of OGRE to develop its full potential has been attributed to a host of social, economic, and legal barriers which will be summarised in the following section and detailed in subsequent chapters in this book.

1.3 Barriers to OGRE Development

Some factors commonly inhibit the global development of OGRE. These include investors' inability to afford the high initial capital costs of OGRE projects, unfavourable licensing requirements, and existing subsidies cum market for conventional electricity.[45] Renewable electricity, including OGRE, has become cost competitive with fossil fuel at per unit cost in most jurisdictions.[46] However, its initial capital cost is remarkably higher than competing fossil fuel electricity options even though the running costs are lower than the latter.[47] Investors and prosumers cannot easily afford the initial capital costs of OGRE without recourse to assistance from government and/or financial institutions.[48] Unfortunately, such assistance is not available to the extent needed from most governments in Africa, Asia, and North and South America.[49] Financial institutions are reluctant to lend to OGRE investors because of its relatively small-scale nature, low opportunity cost profile, and high risks in comparison to fossil fuel electricity and on-grid renewable electricity options.[50] Fossil fuel electricity has accumulated a prolonged period of experience with usage, established market, and subsidies which translates to lower cost and increased acceptability in comparison to OGRE.[51] It is for this reason that fossil fuel electricity is still growing and leading the energy industry.[52]

In some instances, the nature of the barrier is country-specific ranging from lack of political will, multiple permit systems, capacity barriers, lack of awareness of the benefits of OGRE, etc.[53] The Mexican OGRE sector is heavily constrained by the absence of political will by the government towards its development.[54] Such lack of will is motivated by the goal of energy sovereignty and the desire to support state-owned companies.[55] While China is leading in the development of renewable electricity, some of its rural dwellers are still in the dark as to the benefits of OGRE.[56] Thus, its wide spread is limited in those areas. In Kenya, community aversion to OGRE development halted the construction of some projects including Kinangop's wind power project.[57] The federalist structure in Nigeria was argued to have militated against the increased participation by states in the development of the sector.[58] In jurisdictions like Trinidad and Tobago, the sector grapples with the absence of technological capacity to manufacture and maintain the underpinning technologies.[59]

1.4 OGRE Regulatory Support

All 193 United Nations member states are parties to the 17 Sustainable Development Goals (SDGs), to create the anticipated future for the world.[60] Achieving a universal access to sustainable electricity entails the development of renewable electricity, increasing its share in the global electricity mix, and developing relevant infrastructures. In the same vein, the Paris Climate Change Agreement 2015 aims to limit global warming to well below 2°C, with renewable energy including OGRE being proposed by most states as their low greenhouse gas emission strategies.[61] It is trite that international laws and soft initiatives remain the catalyst for adopting domestic legal measures on a given issue.[62] International instruments can also underpin support, including financial and technological, for the development of a given sector. Thus, law at all spheres is rudimentary to the development of renewable electricity. However, its usefulness is highly contingent on its effectiveness. Extrapolating the best from the OGRE sector in the context of climate change, job creation, and energy security entails having efficient policy and legal enablers at all levels. Strengthening the role of law to support OGRE must be predicated on a thorough analysis of existing laws to identify its extent, strengths, and limitations against which commendations and recommendations will be made on how law can be strengthened to achieve the optimal development of OGRE. It is against the above background that this book investigates the role of laws and policies at all spheres in supporting the development of OGRE with a view to strengthening it to optimise its benefits.

1.4.1 Scope

As reiterated, still lacking is a book-depth analysis of the role of law at all spheres in supporting the global development of OGRE. This book uses doctrinal, non-doctrinal, methodological holism and methodological institutionalism to provide a theoretical and empirical analysis of the extent to which law-including policies support the development of the sector. It starts by contextualising the place of OGRE in achieving global energy transition including achieving sustainable electricity and addressing climate change. Relevant international instruments like the Sustainable Development Goals and the Paris Climate Change Agreement 2015 were analysed in line with the central theme of this book.

The latter is followed by country case studies analyses. It is trite that case studies analysis enjoys superiority over an ordinary exposition of a given subject because it reflects the lived reality and recommendations based on it are more realistically applicable. The role of law in enabling or supporting OGRE in Kenya, Nigeria, Uganda, China, the Philippines, Indonesia, Brazil, Mexico, and Trinidad and Tobago are analysed. Governance frameworks such as licensing, environmental impact assessment, social impact assessment, and tariff regulation are analysed to determine the extent of their support for OGRE. Contractual mechanisms and frameworks were examined. Each case study contains an analysis of the role of support schemes in enabling the optimal development of OGRE in each national context. Such analysis is at all instances preceded by a discussion of the need for OGRE in the given jurisdiction, the status and barriers to its development. The lessons extrapolated from the country case studies are discussed.

Drawing from the case studies analysis, recommendations are made on how law can be strengthened to play an efficient role in supporting OGRE development.

The recommendations will discuss how law can best provide for the need for OGRE and align targets that will drive its development in context. It will establish best practices on how law at all levels can effectively support the development of OGRE including by addressing its barriers. Electricity sector transformation plans are recommended which will be tailored towards eliminating the legal barriers to the development of the sector. Best contractual mechanisms that will foster the optimal development of the sector are also advocated.

1.4.2 Significance

The unique selling point of this book is its original contribution to the discourse on the role of law in supporting the development of OGRE. This makes it useful for contributing to developing or reforming any legal framework for the sector in any state of the world. As reiterated, no literature has considered, in a book-depth manner, the role of law in effectively supporting the development of OGRE.[63] Also lacking is a literature that considers the interaction of law with OGRE on a global scale. Thus, this book is a novel in this regard. The point that the regulation of OGRE is transnational has already been made.[64] Thus, save for national circumstances, the content and design of the legal framework for the support of a functional aspect of RE are similar across the globe.[65] Consequently, the analysis of country case studies provided in this book will give valuable insights and lessons on how another country can develop or reform its legal framework for similar functions.

What is more, this book will contribute to the discourse on the role of the international climate change regime in driving the national development of OGRE. The relationship between climate change and renewable energy has already been emphasised. There is a dearth of academic literature on the role of the international climate change regime in driving the national development of RE. Ferrey has analysed the role of the Kyoto Protocol 1997 in driving the national development of RE in developing countries.[66] Woolley analysed the role of the Paris Agreement 2015 in driving and supporting the global development of RE.[67] Neither author, however, considers the relationship between the international climate change regime and the development of OGRE in developing countries. On the other hand, the analysis that will be provided in this book will certainly contribute to, if not provide the first literature on, the relationship between the international climate change regime and the national development of OGRE.

In all, this book will be a useful resource to a wide range of audiences. For those in the academics, it represents a strong starting and finishing pointing on the role of law in driving the development of OGRE in the light of energy security and combating climate change. It will also be resourceful to researchers investigating any topics bordering on climate change, energy law, renewable energy, planning, development, and related themes. This book will appeal to legal counsels whether in-house or external seeking to know about the legal environment for investing in OGRE in the country case studies analysed. This text will be a valuable resource to government authorities seeking to fine-tune their laws to best support the development of OGRE. It will benefit international bodies and NGOs who are interested in the subject matter of renewable energy, climate change, and energy financing.

1.4.3 Structure

Besides the introduction, the book is divided into five parts. It starts with a consideration of the role of international law in the development of OGRE and graduates to the national legal frameworks. The first part is 'Sustainable Electricity and OGRE: Critical Issues and Challenges'. The second, third, and fourth parts contain the country case studies analysis in Africa, Asia, and North and South America, respectively. The final part reflects on the lessons learnt from the case studies analyses and recommends on how best law can be positioned to support the development of OGRE. The first part contextualises OGRE as a sustainable form of electricity, analysing the role of international institutions and frameworks in enabling its development. Chapter 2 analyses the international institutions including IRENA that promote sustainable electricity and the extent to which their activities affect the development of OGRE. Chapter 3 discusses sustainable development goals 2030 as a driver for OGRE development. Chapter 4 analyses the Paris Climate Change Agreement 2015 as a catalyst for the development of OGRE. It looks at the extent to which the Financial and Technological Mechanisms of the Paris Climate Change Agreement facilitates the removal of the financial and capacity barriers to the OGRE sector.

Chapters 5 through 13 cover the country case studies analysis including Kenya, Nigeria, Uganda, China, the Philippines, Indonesia, Brazil, Mexico, and Trinidad and Tobago. A common framework for the analysis was adopted which includes to the extent applicable imperatives, status, and barriers for OGRE development. The latter is typically followed by an analysis of the regulatory context for OGRE development including licensing, environmental impact assessment, social impact assessment, and tariff framework. Relevant contractual regulatory frameworks for OGRE are analysed including power purchase agreements. The support frameworks for OGRE are analysed to establish the extent to which they enable its development including by removing the identified barriers to it. Chapter 14 contains recommendations informed by lessons extrapolated from the country case studies.

1.5 Conclusion

Indomitably, transition to renewable electricity, including OGRE, 'are found to be the central cog in efforts to mitigate runaway climate change and ensuring ever-growing energy needs'.[68] The lack of access to reliable and sustainable electricity especially in remote rural areas of developing countries present the perfect opportunities for the development of OGRE. States of the world have recognised and pledged to develop renewable electricity including OGRE. Thus, it is not surprising that the growth of renewable electricity is the most sporadic in the energy market. Irrespective, there is still a substantial gap between where the development of renewable electricity including OGRE ought to be and where it is now. The gap, if not addressed, will continue to widen with the exponential growth in population.

Law is crucial to the optimal development of OGRE to attain universal access to sustainable electricity. The burning question remains: how will the international community and states position laws and policies to support the optimal development of OGRE? This book provides an in-depth analysis of the role of law in supporting the development of OGRE at a multi-dimensional level using country case studies in some respects. Thus, the thorough analysis provided will illuminate the path to positioning the law to optimally support the development of the sector.

Notes

1 Jessup Philip, *Transnational Law* (Yale University Press 1950) 1. See Peer Zumbesan, 'Transnational Law' in J Smith (ed), *Encyclopedia of Comparative Law* (Oxford Press 2006) 740; Thijs Etty, 'Energy Transition in a Transnational World' (2021) 10(2) Transnational Environmental Law 197. See also Fazil Jamal, 'Legal Aspects of Transnational Energy Pipelines: A Critical Appraisal' (2015) Eur. Networks L. & Reg. Q. 103.
2 Bengt Johansson, 'Security Aspects of Future Renewable Energy Systems: A Short Overview' (2013) 61 Energy 598. Paolo Farah and Elena Cima, 'Energy Trade and the WTO: Implications for Renewable Energy and the OPEC Cartel' (2013) 16(3) JIEL 707.
3 Camoroun Kelly, 'Comparative Law as an Instrument in Transnational Law: The Example of Large-Scale Renewable Energy Regulation' (2016) 25(1) EEELR 25.
4 Ngozi Chinwa Ole, 'The Role of Renewable Energy Law 2005 in Supporting the Development of Off-Grid Renewable Electricity in China' (2019) 5 International Energy Law Review 132.
5 Gopal K Sarangi and others, 'Working Paper Decentralised Renewable Energy Systems in China, India and Thailand: Assessing the Role of Policies and Incentive Structures' (2017) 18 <www.prospernet.com> accessed 6th June 2022.
6 Christopher S McCallum and others, 'Renewable Electricity Generation for Off Grid Remote Communities; Life Cycle Assessment Study in Alaska, USA' (2021) 299 Applied Energy 117325. See also Mohamed M Elkadragy and others, 'Off-Grid and Decentralized Hybrid Renewable Electricity Systems Data Analysis Platform (OSDAP): A Building Block of a Comprehensive Techno-Economic Approach Based on Contrastive Case Studies in Sub-Saharan Africa and Canada' (2021) 34 Journal of Energy Storage 101965.
7 Majbaul Alam and Subhes Bhattacharyya, 'Are the Off-Grid Customers Ready to Pay for Electricity from the Decentralized Renewable Hybrid Mini-Grids? A Study of Willingness to Pay in Rural Bangladesh' (2017) 139(15) Energy 433–446.
8 In the United States, roof top solar photovoltaic have been used by a great number of residence for electrification in Colorado, New Jersey, Arizona, Nevada etc. In Brazil and Canada, OGRE has experienced marginal growth particularly for the electrification of local communities. See IEA, 'IEA-PVPs' (2013) 18–20 <http://www.iea-pvps.org/fileadmin/dam/public/report/statistics/FINAL_TRENDS_v1.02.pdf> accessed 1st June 2020.
9 Juan Pablo Viteri and others, 'Optimizing the Insertion of Renewable Energy in the Off-Grid Regions of Colombia' (2019) 235 Journal of Cleaner Production 535.
10 Steven Ferrey, 'The Failure of International Global Warming Regulation to Promote Needed Renewable Energy' (2010) 37(1) BC Env ALR 68.

11 IRENA, 'Energy Community Meets to Advance Off-Grid Renewables in Pursuit of SDGs and Climate Goals' (2021) <https://www.irena.org/newsroom/pressreleases/2021/Nov/Energy-Community-Meets-to-Advance-Off-Grid-Renewables-in-Pursuit-of-Development-and-Climate-Goals> accessed 14th June 2022.
12 Abdulrasheed Isah and Gylych Jelilov, 'The Impact of COVID-19 on the Off-Grid Renewable Energy Sector in Nigeria' (2020) <https://www.iaee.org/en/publications/newsletterdl.aspx?id=887> accessed 14th June 2022.
13 Ibid.
14 IRENA, *Off-Grid Renewable Energy Statistics 2021* (IRENA 2021).
15 IRENA, *World Energy Transition Outlook* (IRENA 2021) 17.
16 United Nations, 'Transforming Our World: The 2030 Agenda for Sustainable Development A/RES/70/1' <https://sustainabledevelopment.un.org/content/documents/21252030%20Agenda%20for%20Sustainable%20Development%20web.pdf> accessed 15th June 2022.
17 IEA, 'Access to Electricity' (2021) <https://www.iea.org/reports/sdg7-data-and-projections/access-to-electricity> accessed 15th June 2022.
18 Ibid.
19 IPCC, 'Global Warming of 1.5° C' (2019) <https://www.ipcc.ch/site/assets/uploads/sites/2/2019/06/SR15_Full_Report_High_Res.pdf> accessed 15th June 2022.
20 IRENA, *World Energy Transition Outlook* (n 15).
21 ILO, 'Green Jobs and Renewable Energy: Low Carbon, High Employment' (2021) <https://www.ilo.org/wcmsp5/groups/public/---ed_emp/---emp_ent/documents/publication/wcms_250690.pdf> accessed 23rd June 2022.
22 Ibid.
23 Evan Mills, 'Job Creation and Energy Savings through a Transition to Modern Off-Grid Lighting' (2016) 33 Energy for Sustainable Development 155–164.
24 Ibid.
25 IRENA, 'Energy Community Meets to Advance Off-Grid Renewables in Pursuit of SDGs and Climate Goals' (n 11).
26 Ibid.
27 Ibid.
28 Moussa P Blimpo and Malcolm Cosgrove-Davies, *Electricity Access in Sub-Saharan Africa* (World Bank 2019) 11–39.
29 World Bank, 'Report: Universal Access to Sustainable Energy Will Remain Elusive without Addressing Inequalities' (2021) <https://www.worldbank.org/en/news/press-release/2021/06/07/report-universal-access-to-sustainable-energy-will-remain-elusive-without-addressing-inequalities> accessed 20th June 2022.
30 Ibid.
31 IRENA, *Off-Grid Renewable Energy Statistics 2021* (IRENA 2021) 3–32.
32 Ibid.
33 Ibid.
34 Ibid.
35 Ibid.
36 IRENA, *Off-Grid Renewable Energy: Global and Regional Status and Trends* (IRENA 2018) 3.
37 IRENA, *Off-Grid Renewable Energy Statistics 2021* (n 14).
38 The European Union Directive on Renewable Energy 2009 created the impetus for the accelerated development of OGRE. See Directive 2009/28/EC of the European Parliament and, European Council.
39 26 European Union Islands have formerly initiated a transition process to OGRE under the Clean Energy for EU Islands. See EU, '26 European Union Islands launch the Clean Energy Transition' (2019) <https://ec.europa.eu/info/news/26-european-islands-launch-clean-energy-transition-2019-feb-18_en> accessed 1st December 2019. Currently, some

islands in Greece, United Kingdom (Scottish Island and Orkney), France, Ireland, Italy and Spain are using OGRE system of electrification. See IEA, 'IEA-PVPs' (Ibid) 14.

40 Council of European Energy Regulators, 'Regulatory Aspects of Self Consumption and Energy Communities' (2019) CER Report: C18-CRM9_DS7-05-03, 11, 39–49.

41 The European Commission, 'Clean Energy for All Europeans' (EU 2019) 13.

42 IRENA, *Off-Grid Renewable Energy: Global and Regional Status and Trends* (IRENA 2018) 3.

43 Zbigniew Chmiel and Subhes C Bhattacharyya, 'Analysis of Off-Grid Electricity System at Isle of Eigg (Scotland): Lessons for Developing Countries' (2015) 81 Renewable Energy 578.

44 Ibid, See A Yadoo and others, 'Low-Carbon Off-Grid Electrification for Rural Areas in the United Kingdom: Lessons from the Developing World' (2011) 39(10) Energy Policy 6400e7.

45 Ibid.

46 Dmitrii Bogdanov and others, 'Low-Cost Renewable Electricity as the Key Driver of the Global Energy Transition towards Sustainability' (2021) 227 Energy 120467.

47 Hwayoung Jeon, 'Renewable versus Non-Renewable: The Role of Electricity Generation to Economic Growth' (2022) 35(6) The Electricity Journal 107140.

48 Hooman Peimani, 'Financial Barriers to Development of Renewable and Green Energy Projects in Asia' (2018) <https://www.adb.org/sites/default/files/publication/445156/adbi-wp862.pdf> accessed 22nd June 2022.

49 Ibid, See Ouedraogo Nadia, 'Opportunities, Barriers and Issues with Renewable Energy Development in Africa: A Comprehensible Review' (2019) 6 Current Sustainable/Renewable Energy Reports 52–60.

50 Ibid.

51 Ibid.

52 Scott Foster and David Elzinga, 'The Role of Fossil Fuels in a Sustainable Energy System' (2015) <https://www.un.org/en/chronicle/article/role-fossil-fuels-sustainable-energy-system> accessed 22nd June 2022.

53 A Johnson and S Jacobsson, 'Inducement and Blocking Mechanisms in the Development of a New Industry: The Case of Renewable Energy Technology in Sweden' in R Coombs and others (eds), *Technology and the Market Demand, Users and Innovation* (Edwar Elgar Publishing Ltd 2000) 89–111.

54 Juan C Percino-Picazo and others, 'Analysis of Restructuring the Mexican Electricity Sector to Operate in a Wholesale Energy Market' (2021) 14 Energies 3331, 3346.

55 Ibid.

56 Chapter 8.

57 Chapter 5.

58 Chapter 6.

59 Chapter 11.

60 UN, 'The 17 Goals' <https://sdgs.un.org/goals> accessed 23rd June 2022.

61 IRENA, *NDCs and Renewable Energy Targets in 2021* (IRENA 2021).

62 Ngozi Chinwa Ole, 'Examining the Legal Framework on Financial Securities for Decommissioning in Nigeria' (2017) 15(1) Oil, Gas and Energy Law 12.

63 The closest literature is Subhes Bhattacharyya, 'To Regulate or Not to Regulate Off-Grid Electricity Access in Developing Countries' (2013) 63 Energy Policy 493. It discusses the best licensing models for OGRE in developing countries.

64 Centre for International Governance Innovation, *Emerging Issues to International and Transnational Law Related to Climate Change* (CIGI 2015) 3, 6–9. Fazil Jamal, 'Legal Aspects of Transnational Energy Pipelines: A Critical Appraisal' (2015) Eur. Networks L. & Reg. Q. 103. See Camoroun Kelly, 'Comparative Law as an Instrument in Transnational Law: The Example of Large-Scale Renewable Energy Regulation' (2016) 25(1) EEELR 25.

65 For instance, the regulation of the financial support of on-grid renewable energy through the design of a Feed-in-Tariff is a global practice. See United Nations Environmental Programme, *Feed in Tariff as a Policy Instrument for Promoting Renewable Energies and Green Economies in Developing Countries* (UNEP 2010) 17–24.
66 Steven Ferrey, 'The Failure of International Global Warming Regulation to Promote Needed Renewable Energy' (2010) 37(1) BC Env ALR 68.
67 Olivia Woolley, 'The Paris Climate Change Agreement: A New Stimulus for International Efforts to Promote Renewable Energy Development?' (2016) 28(5) Environmental Law and Management 185.
68 Manish Ram and others, 'Global Energy Transition to 100% Renewables by 2050: Not Fiction, but Much Needed Impetus for Developing Economies to Leapfrog into a Sustainable Future' (2022) 246 Energy 123419.

References

Alam M and Bhattacharyya S, 'Are the Off-grid Customers Ready to Pay for Electricity from the Decentralized Renewable Hybrid Mini-grids? A Study of Willingness to Pay in Rural Bangladesh' (2017) 139(15) Energy 433–446.

Blimpo M P and Cosgrove-Davies M, *Electricity Access in Sub-Saharan Africa* (World Bank 2019) 11–39.

Bogdanov D and others, 'Low-cost Renewable Electricity as the Key Driver of the Global Energy Transition Towards Sustainability' (2021) 227 Energy 120467.

Council of European Energy Regulators, 'Regulatory Aspects of Self Consumption and Energy Communities' (2019) 11 CER Report: C18-CRM9_DS7-05-03, 39–49.

Etty T, 'Energy Transition in a Transnational World' (2021) 10(2) Transnational Environmental Law 197.

Elkadragy M M and others, 'Off-grid and Decentralized Hybrid Renewable Electricity Systems Data Analysis Platform (OSDAP): A Building Block of a Comprehensive Techno-Economic Approach Based on Contrastive Case Studies in Sub-Saharan Africa and Canada' (2021) 34 Journal of Energy Storage 101965.

Farah P and Cima E, 'Energy Trade and the WTO: Implications for Renewable Energy and the OPEC Cartel' (2013) 16(3) JIEL 707.

Ferrey S, 'The Failure of International Global Warming Regulation to Promote Needed Renewable Energy' (2010) 37(1) BC Env ALR 68.

Foster S and Elzinga D, 'The Role of Fossil Fuels in a Sustainable Energy System' (2015) <https://www.un.org/en/chronicle/article/role-fossil-fuels-sustainable-energy-system> accessed 22nd June 2022.

IEA, 'Access to Electricity' (2021) <https://www.iea.org/reports/sdg7-data-and-projections/access-to-electricity> accessed 15th June 2022.

IEA, 'IEA-PVPs' (2013) 18–20 <http://www.iea-pvps.org/fileadmin/dam/public/report/statistics/FINAL_TRENDS_v1.02.pdf> accessed 1st June 2020.

ILO, 'Green Jobs and Renewable Energy: Low Carbon, High Employment' (2021) <https://www.ilo.org/wcmsp5/groups/public/---ed_emp/---emp_ent/documents/publication/wcms_250690.pdf> accessed 23rd June 2022.

IPCC, 'Global Warming of 1.5^0 C' (2019) <https://www.ipcc.ch/site/assets/uploads/sites/2/2019/06/SR15_Full_Report_High_Res.pdf> accessed 15th June 2022.

IRENA, 'Energy Community Meets to Advance Off-Grid Renewables in Pursuit of SDGs and Climate Goals' (2021) <https://www.irena.org/newsroom/pressreleases/2021/Nov/Energy-Community-Meets-to-Advance-Off-Grid-Renewables-in-Pursuit-of-Development-and-Climate-Goals> accessed 14th June 2022.

IRENA, *NDCs and Renewable Energy Targets in 2021* (IRENA 2021).
IRENA, *Off-grid Renewable Energy: Global and Regional Status and Trends* (IRENA 2018).
IRENA, *Off-grid Renewable Energy Statistics 2021* (IRENA 2021).
IRENA, *World Energy Transition Outlook* (IRENA 2021) 17.
Isah A and Jelilov G, 'The Impact of COVID-19 on the Off-grid Renewable Energy Sector in Nigeria' (2020) <https://www.iaee.org/en/publications/newsletterdl.aspx?id=887> accessed 14th June 2022.
Jamal F, 'Legal Aspects of Transnational Energy Pipelines: A Critical Appraisal' (2015) European Networks Law and Regulation 103.
Jeon H, 'Renewable versus Non-Renewable: The Role of Electricity Generation to Economic Growth' (2022) 35(6) The Electricity Journal 107140.
Johansson B, 'Security Aspects of Future Renewable Energy Systems: A Short Overview' (2013) 61 Energy 598.
Johnson A and Jacobsson S, 'Inducement and Blocking Mechanisms in the Development of a New Industry: The Case of Renewable Energy Technology in Sweden' in R Coombs and others (eds), *Technology and the Market Demand, Users and Innovation* (Edwar Elgar Publishing Ltd 2000) 89–111.
Kelly C, 'Comparative Law as an Instrument in Transnational Law: The Example of Large-Scale Renewable Energy Regulation' (2016) 25(1) EEELR 25.
McCallum C S and others, 'Renewable Electricity Generation for Off-grid Remote Communities; Life Cycle Assessment Study in Alaska, USA' (2021) 299 Applied Energy 117325.
Mills E, 'Job Creation and Energy Savings Through a Transition to Modern Off-grid Lighting' (2016) 33 Energy for Sustainable Development 155–164.
Nadia O, 'Opportunities, Barriers and Issues with Renewable Energy Development in Africa: A Comprehensible Review' (2019) 6 Current Sustainable/Renewable Energy Reports 52–60.
Ole N C, 'Examining the Legal Framework on Financial Securities for Decommissioning in Nigeria' (2017) 15(1) Oil, Gas and Energy Law 12.
Ole N C, 'The Role of Renewable Energy Law 2005 in Supporting the Development of Off-grid Renewable Electricity in China' (2019) 5 International Energy Law Review 132.
Peimani H, 'Financial Barriers to Development of Renewable and Green Energy Projects in Asia' (2018) <https://www.adb.org/sites/default/files/publication/445156/adbi-wp862.pdf> accessed 22nd June 2022.
Percino-Picazo J C and others, 'Analysis of Restructuring the Mexican Electricity Sector to Operate in a Wholesale Energy Market' (2021) 14 Energies 3331, 3346.
Philip J, *Transnational Law* (Yale University Press 1950).
Sarangi G K and others, 'Working Paper Decentralised Renewable Energy Systems in China, India and Thailand: Assessing the Role of Policies and Incentive Structures' (2017) 18 <www.prospernet.com> accessed 6th June 2022.
United Nations, 'Transforming Our World: The 2030 Agenda for Sustainable Development A/RES/70/1' <https://sustainabledevelopment.un.org/content/documents/21252030%20Agenda%20for%20Sustainable%20Development%20web.pdf> accessed 15th June 2022.
Viteri P J and others, 'Optimizing the Insertion of Renewable Energy in the Off-grid Regions of Colombia' (2019) 235 Journal of Cleaner Production 535.
World Bank, 'Report: Universal Access to Sustainable Energy will Remain Elusive Without Addressing Inequalities' (2021) <https://www.worldbank.org/en/news/press

-release/2021/06/07/report-universal-access-to-sustainable-energy-will-remain-elusive-without-addressing-inequalities> accessed 20th June 2022.

Yadoo A and others, 'Low-Carbon Off-grid Electrification for Rural Areas in the United Kingdom: Lessons from the Developing World' (2011) 39(10) Energy Policy 6400e7.

Zbigniew C and Bhattacharyya S C, 'Analysis of Off-grid Electricity System at Isle of Eigg (Scotland): Lessons for Developing Countries' (2015) 81 Renewable Energy 578.

Zumbesan P, 'Transnational Law' in J Smith (ed), *Encyclopedia of Comparative Law* (Oxford Press 2006) 898–925.

Part 1

Sustainable Electricity and Off-grid Renewable Energy

Critical Issues and Challenges

2 International Legal Framework for Sustainable Energy

Peter Kayode Oniemola and Adejumoke J Akinbusoye

2.1 Introduction

Unsustainable energy use is an important trigger for the challenges of climate change.[1] Fears were expressed that the international community had not addressed the need for a safe and sustainable energy future with the expected level of cooperation and urgency.[2] Challenges such as funding, international cooperation, and technical assistance, especially for developing countries, are potential barriers to developing an international framework for sustainable energy.[3] An international framework is akin to a global architecture that brings together public and private institutions under a coordinated platform. It offers an interaction which is relevant in global politics.[4] Such a structure or framework for energy exists at the international level, but only in a fragmented manner.[5] The need for concerted action on sustainable energy is made clear by major international global energy challenges such as energy security and climate change. Sustainable development patterns can be easily understood by illustrating patterns of unsustainable development.

Unsustainable development involves development decisions or models that put pressure on natural resources, thus resulting in negative ecological changes and negative human impacts in the long run.[6] It also connotes the use of natural resources in a manner that neglects the effects on nature or future generations. Thus, sustainable development seeks to achieve a higher sense of harmony between humans and their environment. This ideal development model is an attempt to leave a legacy of continuity for generations to come. One of the means of achieving this is through sustainable energy utilisation. The concept of just energy transition has also become much more operationalised.[7] With the trend in just energy transition, there is the implication that none should be left behind in the attainment of sustainable energy.[8]

The contentions of this chapter in the international framework for sustainable energy law cannot be said to have clearly emerged. The promotion of sustainable development in international law has received much attention. International energy governance is an evolving branch of international law seeking to address transboundary energy problems, including energy security, energy financing, economic development, international security, environmental sustainability, and other energy challenges. There is still no effective coordination of the governance

DOI: 10.4324/9781003178088-3

of energy at a global level.[9] Several stakeholders and institutions are involved in international energy governance, including governments, regional alliances, the private sector, multinational and national oil companies, and civil societies at different levels.[10] Sustainable energy sources, as energy sources that endure through the entirety of time, are relevant to the human race and therefore contribute to the sustainability of the ecosystem.[11] It covers the production and consumption of energy in ways that support holistic human development and ecological balance for the long term.[12]

Although international support for sustainable energy is growing, the international community is not ready for more binding or formalised structures for sustainable energy governance.[13] This chapter explores the international framework for sustainable energy as a component of the global energy regime. It examines whether there is an international framework for sustainable energy and considers what the framework for sustainable energy looks like by analysing major global energy institutions at the forefront of international energy policy discussions. The chapter further considers some barriers to developing a comprehensive legal framework for sustainable energy and concludes with the position that there is a need for concrete international commitment on substantiable energy with a national framework for sustainable energy governance well-coordinated at domestic levels.

2.2 Sustainable Energy in International Law

Overall, sustainable energy involves looking at energy and energy sources from the lens of the sustainable development principle. It involves a transition to cleaner, low-carbon energy sources. Sustainable use of energy implies environmentally friendly energy options that take into consideration the rights of people everywhere to have access to energy and the rights of future generations to energy sources. Sustainable development can be viewed as a principle in international law. Explicitly put, the principle of sustainable development receives much patronage from international environmental law. From the populist view, it is expected that the principle of sustainable development should be one that ensures that the needs of the current generation are met without jeopardising those of future generations. By the same token, sustainable development is one that strives to meet economic, environmental, and social objectives. It is generally perceived that international law has not crafted a clear mechanism for the systematic development of the framework for sustainable energy. The principles of sustainable development have support in other relevant principles such as the precautionary, polluter pays, common but differentiated responsibility, among others.[14] Precautionary principles support the notion that promotion of sustainable energy is a way of taking precautionary measures to prevent the consequences of climate change or other unsustainable energy utilisation. Polluter pay may be viewed from the perspective that polluters should take measures to address the consequences of their actions, and such measures could be through the financial deployment of support for sustainable energy. The principle of common but differentiated responsibility is to the effect that those countries well positioned should offer support to countries that are not in the position to deploy sustainable energy. The idea is that there is a common goal of striving to enhance sustainable energy, but the countries with the most capacity should give technical and financial support to the developing ones, who should also strive to achieve sustainability.

The United Nations Charter calls for the cooperation of the members of the United Nations.[15] Such cooperation, it is expected, will have far-reaching implications for the promotion of sustainable energy for the common good of all through the UN systems. There is no specific treaty that has been set out for the actualisation of sustainable development. Nevertheless, it is seen that sustainable development is usually articulated in a series of treaties, particularly environmental-related treaties and current investment or trade-related treaties.

Soft law instruments have much relevance in the shaping of the international framework for energy.[16] Sustainable development has shown much international prominence since the 1972 Stockholm Declaration.[17] The World Commission of Environment and Development further set the agenda through popularising sustainable development and setting in motion the acceptance and integration of international and domestic affairs.[18] Agenda 21,[19] as well as the Rio Declaration on Environment and Development,[20] articulate relevant provisions projecting sustainable energy. The provisions of the Johannesburg Plan of action well appreciates and give further impetus to sustainable energy.[21] With the coming into effect of the United Nations Framework Convention on Climate Change (UNFCCC),[22] a new approach or view may be said to have been clearly determined for the projection of sustainable energy. The ultimate objective of the UNFCCC is: 'stabilization of greenhouse gas concentrations in the atmosphere at a level that would prevent dangerous anthropogenic interference with the climate system ... enable economic development to proceed in a sustainable manner'.[23] The UNFCCC requires: 'the Parties have a right to, and should, promote sustainable development'.[24] There should also be cooperation among the parties 'to promote a supportive and open international economic system that would lead to sustainable economic growth and development'.[25] Members also commit to sustainable management of greenhouse gases.[26] It follows, therefore, that there are a series of provisions that offer support for sustainable energy through the promotion of sustainable development, the call for the development of new and renewable energy sources, as well as the promotion of energy efficiency and environmentally sound technologies.[27] The Kyoto Protocol has mechanisms that also encourage clean energy projects. For example, the clean development mechanism framework under the Kyoto Protocol enabled the implementation of projects by developed countries in developing countries, with such projects reducing greenhouse gases and promoting sustainable development.[28] Examples of projects that are implemented form part of sustainable energy projects.[29] The Paris Agreement is also replete with provisions that support sustainable energy promotion. Its overall objective is to reduce global atmospheric temperature to 'below 2 °C above pre-industrial levels and to pursue efforts to limit the temperature increase to 1.5 °C above pre-industrial levels'.[30] Every state party is obligated to implement what are known as its nationally determined contributions (NDCs) towards mitigating the effects of climate change in a manner that is consistent with the fundamental goals of the Paris Agreement. The NDCs that can be achieved through implementation of the mechanisms for project development in Article 6, may be in the form of the voluntary cooperation of nations towards the accomplishment of their own NDCs through projects.[31] It is stipulated that parties would be able to satisfy their NDCs on a voluntary basis by making use of internationally transferable mitigation outcomes from another country.[32] The experience under the UNFCCC regime shows the disposition of countries to multilateral attempts to address energy-related externalities in the interest of sustainable development.[33]

Initially called the European Energy Treaty, the Energy Charter Treaty (ECT) is an international investment treaty that is now open to all countries. The ECT came into force in 1998 to promote energy security through the promotion of competitive and transparent energy markets based on the principles of sustainable development.[34] The ECT has 53 signatories and can best be described as an economic or trade organisation that seeks to protect its member states' interests in the energy market. It focuses on promoting equality among contracting parties, dispute resolution among its participating states, promoting energy efficiency, and minimising negative environmental impacts. Its provisions are legally binding. Though the ECT contains several environmentally friendly provisions, these provisions have been described as ambiguous and ineffective. The ECT lacks broad coverage and has not enjoyed much patronage unlike the World Trade Organisation.[35] The ECT process does not sufficiently recognise climate change concerns.[36] Further criticism from civil society has been levelled against the ECT for its provisions which protect foreign investments but indirectly hinder countries that choose to protect their citizens from climate change by stopping fossil fuel projects.[37]

Within the framework of the United Nations, the UN Millennium Development Goals (MDGs) were implemented with the view that these goals would lead to development. There has been the general assertion that the attainment of the MDGs would only be realistic by bringing energy to the fore.[38] With the expiration of the goal, the UN brought about the Sustainable Development Goals for the actualisation of rapid and equitable development in the world by 2030.[39] The Sustainable Development Goals contained a set of 17 goals brought forward to project world development. Specifically, Goal 7 of the SDG is on clean and affordable energy. Invariably, the goal supports the deployment of sustainable energy.[40] The UN General Assembly Resolution 65/151 on the International Year of Sustainable Energy for All is relevant; and furthermore, the 2012 declaration of years 2014–2024 as the Decade of Sustainable Energy for All[41] are an appetite wetter for global facilitation of sustainable energy.[42]

In connection with global climate action, these goals give greater consideration to sustainable energy in that the way of actualising climate change mitigation will be for the energy sector to be such that it is sustainable through the deployment of clean energy sources such as renewable energy. These will have far-reaching implications for a just and equitable transition.[43]

There appears no doubt that the idea of sustainable development, which has now been fully integrated into sustainable energy promotion, has crystallised into customary international law. While the debate has been ongoing on the state of international law, it is clear that a series of binding international law instruments have expressions of sustainable development. Likewise, domestic legislation relevant to the energy sector has expressed sustainability as one of its integral considerations.[44] The international aspect of sustainable energy has an impact on domestic law, thereby giving an avenue for the gradual change of rules and a shift to sustainability.

The International Energy Charter (IEC) of 2015 is a political instrument that is designed to create a political platform for the projection of energy governance.[45] It is expressed as a document that is pivotal to promotion of sustainability, and the UNFCCC and Paris Agreement. The IEC is an offshoot of the ECT and part of the Energy Charter process. Unlike the ECT, the IEC is a non-binding political declaration that has been designed as an upgrade to the provisions of the ECT, extending the global reach of the Energy Charter.[46] It is expected that the dialogues opened by participating in the process should translate into domestic legal measures that promote sustainable energy.

2.3 The Law of International Institutions and Sustainable Energy

International institutions are rule makers in international law.[47] International economic law, for example, has been largely a product of the interplay of international financial institutions, which operate as part of its subset.[48] Thus, the activities of the World Bank, International Monetary Fund, and its conditionalities have given much direction to domestic systems that have embarked on public sector reform which has led to the process of such reform leading to changes in law to enhance or give validity to the reform process.[49] An extrapolation from the above would give an indication that international institutions relevant to sustainable energy may contribute to the shaping of the energy space in domestic legal systems.[50] The UN and its specialised institutions and agencies have spearheaded dialogue and initiatives that have created platforms for international momentum towards a transition to sustainable development. The United Nations Environment Programme (UNEP), United Nations Department of Economic and Social Affairs (UNDESA), and United Nations Development Program (UNDP) feature prominently in these initiatives. UNEP coordinates global environmental issues, which by implication includes sustainable energy, and oversees the sustainable use of natural resources. The UNDESA is the leading platform overseeing sustainable development and is facilitating the SDGs. Additionally, the UNDP is the lead agency on international development and its approach to sustainable energy has been in the context of development. It has played a vital role in assisting countries in formulating policies, leadership and governance skills, and strengthening institutional capacity towards sustainable development. Another central platform for advancing the sustainable energy agenda is the United Nations Commission on Sustainable Development. It has the mandate to review the progress made in implementing Agenda 21 and the Rio Declaration. It also offers policy guidance at local, national, regional, and international levels on enhancing the application of the Johannesburg Plan of Implementation.

Besides, there are emerging non-governmental international institutions in the process. These institutions have drawn inspiration among themselves, leading to the replication of ideas.[51] Institutions are coming up with rules that have aftermath impact in the way countries would put in place measures at the domestic level.[52] The World Bank now champions sustainable energy and would readily support energy generation projects from sustainable energy sources. The implication of this is that countries may find it more promising to implement commendable legal changes in the energy sector to make them more attractive to funding coming from the World Bank or other multilateral financial institutions of similar standing or position to support sustainable energy. The IEA is an Organisation for Economic Cooperation and Development (OECD) organisation and one of the leading and influential global energy agencies. It provides policy guidance to governments and valuable data to other stakeholders in the energy sector. Established in 1974 with the mandate to promote energy security and cooperation in the event of a sudden interruption in global oil supply, the IEA is concerned with securing sustainable energy for all.[53] Through its annual publication, *World Energy Outlook*, it gives insights into trends and developments in the energy market and provides direction on the implication of such trends for sustainable development. The IEA has expanded its mandate to cover contemporary energy challenges, including energy efficiency, climate change, international partnerships, technology collaborations, etc.

IRENA is the first major international organisation that aims to ensure sustainable energy use in renewable energy. IRENA's main objective is to promote the development of all sources of renewable energy among its member countries. Established in 2009, IRENA works with its 164 member countries by offering policy guidance and support for countries in all aspects of renewable energy development, including capacity building and technology transfer.[54] It also provides opportunities for partnerships and collaborative research. The emergence of IRENA is considered to be a significant step towards developing a more robust framework for sustainable energy.[55]

2.4 Judicial Pronouncements Supporting Sustainable Energy in International Law

Sustainable development has received pronouncements from international tribunals. Notable of these is the International Court of Justice in *Gabčikovo-Nagymaros Project Case (Hungary v. Slovakia)*,[56] where sustainable development was articulated in the context of energy projects. The Permanent Court of Arbitration Tribunal in *Iron Rhine (Belgium v. Netherlands)*[57] acknowledged the importance of sustainable development. Judicial decisions have also played a role in affirming general principles relating to sustainable energy generation, production, and consumption. They have contributed to the jurisprudence on international sustainable energy governance and set the stage for developing a framework for sustainable energy. Increasingly, national courts are being persuaded to affirm, in local regulations, that unsustainable energy practices resulting in excessive GHG emissions constitute violations of first-generation rights, such as the right to life and right to respect for family life.[58] The courts have been found to be playing a lead role in the championing of sustainable energy. The role of the courts has been to utilise and apply the direct provisions of international treaties or principles of international law relevant to sustainable energy in a domestic setting. These authorities have recognised international law provisions, rules, and principles as being relevant in projecting sustainable energy. In *Taralga Landscape Guardians Inc. v. Minister for Planning and RES Southern Cross Property Limited*,[59] where the court found that on the bases of precautionary principle, inter-generational equity, and commitment to the international climate change regime, the promotion of renewable energy should be prioritised. In *Hindustan Zinc Limited v. Rajasthan Electricity Regulatory Commission*,[60] the Supreme Court of India also made a pronouncement justifying the promotion of renewable energy on grounds of keeping with the international commitment of India under the Kyoto Protocol.

In the *State of the Netherlands v. Urgenda Foundation*,[61] the Dutch Supreme Court upheld the position that the Dutch government has a legal duty, in line with the provisions of the European Convention on Human Rights, to protect the right to private and family life of its citizens by taking appropriate measures to mitigate carbon emissions.[62] In *Vereniging Milieudefensie and Others v. Royal Dutch Shell (RDS) Plc*,[63] the Hague District Court ruled that RDS must reduce its CO_2 emissions by 45% by 2030 compared with 2019 emissions. The decision is particularly significant as it means that similar cases may be brought against multinational companies operating in other countries, compelling them to embrace the transition to low-carbon energy at a faster rate.

The cases identified above have direct implications as they validate the sustainable development principle and contribute to the advancement of international law on sustainable energy. They provide pointers to the future direction where the protection of fundamental human rights could be a persuasive ground for further advances towards a more concrete international agreement on sustainable energy. The point therefore is that the court has continued to make pronouncements that give more justification to the deployment of sustainable energy. Consequently, the decisions could lead to changes in the law to promote sustainable energy and may be taken into account by the executives or the parliament in shaping domestic law on sustainable energy. Likewise, it will offer much inspiration public interest or strategic litigation for sustainable energy which would invariably influence the decision of the government.

2.5 Sustainable Energy Trends at the Centre of the Law

Internationally, sustainable energy has been at the centre of the changes to the rules of the game on the perception of some fields of the law. In international trade law, particularly within the jurisprudence of the WTO, domestic mechanisms have been challenged where they fall short of the WTO provisions of trade liberalisation.[64] The WTO, however, has sustainable development as one of its main objectives.[65] Support will usually come from the government in the form of regulations and incentives.[66] Support mechanisms for renewable energy in particular have been subjected to judicial analysis through the WTO dispute settlement understanding, with rules emerging on the ideal ways that legal regimes should craft their domestic laws so that they do not fall foul of commitments under the WTO.[67] Sustainable energy promotion, insofar as it is done for legitimate pursuits, is most likely to receive tacit approval.[68] Similarly, international investment law jurisprudence has seen tribunals making rules on how domestic law should treat support for sustainable energy. This has largely been affected by the changes in law or support cuts for renewable energy, which were interpreted to have fallen short of the standards of protection such as fair and equitable treatment and unlawful expropriation.[69] Aside from these dictates, the emerging notion of business and human rights has also been seen as a plus for the deployment of renewable energy or other sustainable energy sources.[70] Attaining human rights has placed importance on the role of sustainable energy, which is clean and affordable. More so, energy is viewed as central to human development.[71] By the same token, sustainable energy may be promoted as a legal obligation to promote human rights or in the form of corporate social responsibility to promote human rights through clean energy access.

The promotion of sustainable energy in international law has not been straightforward. It is a fact that the peculiarities of countries vary, as do their respective priorities. The process of the emergence of sustainable energy as a key consideration in international law is largely a question of the nations' choice. This has been further complicated by the fact that all states have their own sovereign right to determine governance.[72]

Thus, states may decide to employ mechanisms that may not directly project sustainable energy. The international climate change regime confirms that countries may, on their own volition, not be bound by obligations. Likewise, countries may decide not to follow the respective provisions to which they have committed themselves. For example, the United States withdrew from the Kyoto Protocol.[73] The Donald Trump administration withdrew the US from the Paris Agreement.[74] However, the current US administration has now committed the US to the Paris Agreement once again.[75]

There is a proliferation of rules and mechanisms that have been put in place. The climate change debate has triggered the consideration, and one can argue that projecting sustainable energy would need to be systematic and not divorced from the international climate change regime which has developed over time.[76] Sustainable energy for all is an important initiative that will need to be further concretised within the framework of the United Nations to put in place a more binding structure and criteria for countries to key into and structure the domestic framework to accept or adopt the integration of renewable energy.

2.6 The Challenge for Sustainable Energy Governance and Directions for Global Sustainable Energy Law

There is a need for a global energy revolution to embrace sustainable energy pathways.[77] There is no clear-cut global commitment to sustainable energy.[78] Development and its related aspects can be a delicate issue at the international level. International cooperation implies mutual interest, information exchange, support for weaker countries, information exchange, technology transfer, etc. States are often not interested in being parties to global agreements that do not provide clear economic advantages. International politics is coming to the fore as countries continue to resist legally binding provisions of international laws that seek to compel specific actions or which place a heavier burden of responsibility on some groups of countries. States cannot be forced to follow a particular pattern of development. However, they can be persuaded to do so. The sovereign rights of states over their natural resources further establish the right to regulate energy within their territories. As seen in recent examples like the withdrawal of the US from the Paris Agreement, state parties are free to withdraw or decline from international cooperation due to national interests. Energy geopolitics plays a huge role in stalling consensus on global sustainable energy governance.[79] For example, oil-producing companies with stakes in downstream operations are generally not willing to embrace a rapid shift away from oil and, even where other countries may be open to such transition, it must be a gradual transition as several businesses are tied to the oil and gas industries.

Typically, the regulation of energy naturally falls within national jurisdiction based on the principle of permanent sovereignty over natural resources. Thus, countries have the prerogative to determine their energy future and can dictate their choices in line with their national aspirations or priorities. The exercise of such a sovereign right should be done with restraint.[80] However, they are constrained by international principles which prohibit transboundary harm. Existing multilateral treaties have, at best, persuasive authority over states in the use of their energy resources.[81] In essence, national energy governance dominates regional or global environmental governance.[82] The discussions around energy, and the regulation of energy at the international level, are highly political as the whole idea involves energy sector reform at state level as well as the transition to cleaner fuels.[83] States have often resisted external control over their use of natural resources, particularly energy resources, based on the principle of permanent sovereignty over natural resources. Thus, the international community is yet to achieve a united front over enforceable guidelines on energy usage.

Due to globalisation and energy-related externalities, states are gradually losing control over their energy policies, but they resist the call to act jointly.[84] The transition to sustainable energy sources involves a high level of politics. Additionally, the transition to cleaner and more sustainable energy sources and practices involves an overhaul of entrenched systems in countries with carbon lock-ins and countries with rich oil and gas reserves that are heavily dependent on oil. The move to sustainable energy has further indicated a change that sees much support from non-fossil fuel sources. This is therefore not coming without a challenge, as fossil fuel companies have resorted to investor state arbitration to protect their investments and get compensation for regulator support that denies them benefits based on their investments and projections on fossil fuel. Such countries are poised to maintain their advantages as oil-producing and oil-exporting countries and are quick to guard their sovereign prerogatives in relation to energy policy.

There are already many international forums for dialogue on energy issues. Some treaties, agencies, or framework agreements address sustainable energy directly, while some do so indirectly. Where cross-cutting areas of policy are involved, and multiple organisations are handling several policy aspects, no single forum may be the right one to develop a new law due to tilts towards certain areas of interest.[85] International energy organisations which attempt to promote sustainable energy are not perceived as productive or legitimate.[86] These two characteristics must be present for countries to fully buy into an international regime for sustainable energy. For example, it has been observed that even though the IEA is efficient in meeting its objectives, it has not yet attained legitimacy in the eyes of certain countries like China and India.[87]

Most agencies within the UN system may strategically exercise their mandates in furtherance of sustainable energy use. While the UNGA is said to bring a level of consistency to the policies and activities of diverse institutions, there is concern that some UN agencies have overlapping functions and can be inefficient. The overlapping nature of energy concerns could pose a challenge to developing a more robust international governance framework due to the complexity of energy concerns and their interrelatedness with other issues where international laws already exist. Also, it is said that the UNGA does not have the power to direct the policies of organisations such as the WTO, IEA, and the World Bank and that the best it can do is to make recommendations to these bodies.[88] There is a concern that the diversity of preferences, heavy bureaucracy in the development of policies, and general distrust at the UN level are likely to hinder a treaty on sustainable energy use.

On the other hand, supranational entities such as the European Union (EU) have been at the forefront of committing to the deployment of sustainable energy. The EU energy law is developing, with renewable energy being a major integral lead in the development of sustainable energy to meet the energy security and environmental objectives of the EU.[89] The projections made within the EU have been quite promising. The EU has committed that 32% of energy consumption should come from renewable energy sources under its Renewable Energy Directive.[90] The directives are required to be transposed and implemented by member states at the domestic level. At African levels, there are no indications that there are efforts channelled towards concretising the Economic Community for West Africa (ECOWAS) Protocol and the Southern African Development Community (SADC) Protocol to project regional development of sustainable energy. The commitment of the EU, with its clear structural mechanism for enforcement, would be worthy of emulation, adjusting it to regional peculiarities.

The international framework for energy governance is evident in the fragmentation of the global energy governance structure.[91] A look at the history of the global energy governance regime shows a complex regime with myriad rules and institutions that further complicate the international energy governance framework.[92] Thus, the fragmented structure of global energy architecture as a challenge is largely evident. Thus, among the institutions championing energy is the question of the conflicting priorities of economic and environmental energy institutions. Numerous international institutions are addressing several aspects of energy governance; as such, progress in one institution can imply regression in another.[93] The international framework is expedient and should be complemented by implementation through domestic laws. This is because the barriers that have limited the deployment of sustainable energy can be addressed through international collaboration and innovative partnerships.[94] Effort should be made to ensure that energy transition should also take into account the realities on the ground such as the need for stability and support from the private sector.[95]

2.7 Conclusion

This chapter has identified sustainable energy production, distribution, and consumption as key components of sustainable development. We also note that as challenges associated with a regime complex have plagued the energy governance regime, similar challenges hinder the growth of a comprehensive international framework for sustainable energy. While progress has been made in the last two decades, it appears that the global community is not yet ready for binding commitments on energy resource management. Changing energy systems is a significant long-term process involving political will at high levels of energy governance, institutional cooperation, and system overhauls. At the forefront of global discussions on sustainable energy is the UN. The UN remains the most important platform for global action towards sustainable energy, primarily through its SDGs. Notwithstanding these efforts, there is no clear binding global agreement or commitment towards the advancement of sustainable energy at the international level. It is pertinent too that a stronger consensus towards a concrete international framework for sustainable energy can emerge under the platform of the UN or other prominent international agencies associated with sustainable energy should be developed.

The agencies and institutions considered above can inspire far-reaching action towards a concrete legal agreement on sustainable energy. However, based on experiences in the international climate governance regime, it is clear that for momentum to increase, political will must be present across major economies to arrive at a consensus on future energy trends. The development of concrete international regimes will positively impact off-grid renewable energy promotion across countries. The commitment for support becomes useful as there will be more impetus to advance an international duty through domestic legal measures focusing on sustainable energy such as off-grid energy.

Notes

1 IPCC, Climate Change 2022: Mitigation of Climate Change: Summary for Policymakers-Contribution of Working Group III to the Sixth Assessment Report of the Intergovernmental Panel on Climate Change (IPCC 2022) 44.
2 WCED, *Our Common Future* (Oxford University 1987) par. 62.
3 Ibid.
4 Frank Biermann et al., 'The Fragmentation of Global Governance Architectures: A Framework for Analysis' (2009) 9 Global Environmental Politics 14, 15.
5 Alexandra Wawryk, 'International Energy Law: An Emerging Academic Discipline' in Paul Babie and Paul Leadbeter (eds), *Law as Change: Engaging with the Life and Scholarship of Adrian Bradbrook* (University of Adelaide Press, 2014) 223.
6 Herman E Daly and Joshua Farley, *Ecological Economics Principles and Applications* (Island Press 2004); Patricia Birnie, Alan Boyle and Catherine Redgwell, *Birnie, Boyle, and Redgwell's International Law and the Environment* (4th edn, Oxford University Press 2021) 41.
7 Romain Mauger, 'Making Sense of Changing Concepts for the Energy Transition: An Energy Transition Concepts Nexus for the Development of Policy and Law', in Ruven Fleming, Kaisa Huhta and Leonie Reins (eds), *Sustainable Energy Democracy and the Law* (Brill/Nijhoff 2021) 38–43.
8 Dorothée Cambou, 'Uncovering Injustices in the Green Transition: Sámi Rights in the Development of Wind Energy in Sweden' (2020) 11 Arctic Review on Law and Politics 310–333, 312–313.
9 Neil Gunningham, 'Confronting the Challenge of Energy Governance' (2012) 1(1) Transnational Environmental Law 119, 130.
10 A Cherp, J Jewell and A Goldthau, 'Governing Global Energy: Systems, Transitions, Complexity' (2011) 2(1) Global Policy 75, 59.
11 Henrik Lund, *Renewable Energy Systems: A Smart Energy Systems Approach to the Choice and Modeling of 100% Renewable Solutions* (2nd edn, Academic Press 2014) 11.
12 UNDP, *World Energy Assessment: Energy and the Challenge of Sustainability* (UNDP 2015) 3.
13 Catherine Redgwell, 'International Soft Law and Globalization' in Barry Barton et al (eds), *Regulating Energy and Natural Resources* (Oxford University Press 2006) 89, 98–9.
14 Raphael J Heffron, Anita Rønne, Joseph P Tomain, Adrian Bradbrook and Kim Talus, 'A Treatise for Energy Law' (2018) 11(1) Journal of World Energy Law and Business 34–48, 35–36.
15 United Nations Charter, Art. 1.
16 Duncan French, *International Law and Policy of Sustainable Development* (Juris and Manchester University Press 2005) 37.
17 Stockholm Declaration on the Human Environment, Report of the United Nations Conference on the Human Environment, UN Doc.A/CONF.48/14 (1972).
18 WCED (n 2).
19 UN Doc A/Conf.151/26 (vol. I), Agenda 21: Programme of Action for Sustainable Development (adopted 14 June 1992, UN GAOR, 46th Sess., Agenda Item 21).
20 Rio Declaration on Environment and Development, Report of the United Nations Conference on Environment and Development, UN Doc. A/CONF.151/26 (1992).
21 Plan of Implementation of the World Summit on Sustainable Development (adopted 4 September 2002, A/CONF.199/20).
22 See United Nations Framework Convention on Climate Change, May 9, 1992, Art. 3, S. Treaty Doc. No. 102–38 (1992), 1771 UNTS 107.
23 UNFCCC, art. 2.
24 Ibid., art. 3.4.
25 Ibid., art. 3.5.

26 Ibid., art. 4.1(d).
27 See Alessandro Monti and Beatriz Martinez Romera, 'Fifty Shades of Binding: Appraising the Enforcement Toolkit for the EU's 2030 Renewable Energy Targets' (2020) 29 Review of European, Comparative & International Environmental Law 222–223.
28 See art. 12.
29 See UNFCCC, 'List of Sectoral Scopes', available at: http://cdm.unfccc.int/DOE/scopelst.pdf, accessed 28 June 2022.
30 See Paris Agreement, art. 2 (Dec. 13, 2015), UNFCCC COP Report No. 21, UN Doc. FCCC/CP/2015/10/Add.1 (January 29, 2016).
31 Ibid, art. 6.1.
32 Ibid, arts. 6.2 and 6.3.
33 Harold Hongju Koh, *The Trump Administration and International Law* (Oxford University Press 2019) 221.
34 The Energy Charter Treaty, 2080 UNTS 95.
35 Andrei Belyi, 'The Energy Charter Process in the Face of Uncertainties' (2021) 14 Journal of World Energy Law and Business 363, 363–364.
36 Morakinyo Adedayo Ayoade, 'Bridging the Gap between Climate Change and Energy Policy Options: What Next for Nigeria?' in Patricia Kameri-Mbote, Alexander Paterson, Oliver C Ruppel, Bibobra Bello Orubebe and Emmanuel D Kam Yogo (eds), *Law, Environment and Africa* (Nomos Verlagsgesellschaft 2019) 83.
37 The Transnational Institute, 'ECT's Dirty Secrets' (2020), <https://energy-charter-dirty-secrets.org/> accessed 5, 22 June 2022.
38 Adrian J Bradbrook and Judith G Gardam, 'Placing Access to Energy Services within a Human Rights Framework' (2006) 28 Human Rights Quarterly 389, 399.
39 United Nations, *Transforming Our World: The 2030 Agenda for Sustainable Development*, https://sustainabledevelopment.un.org/content/documents/21252030%20Agenda%20for%20Sustainable%20Development%20web.pdf, accessed 4 July 2022.
40 Paola Villavicencio Calzadilla and Romain Mauger, 'The UN's New Sustainable Development Agenda and Renewable Energy: The Challenge to Reach SDG7 while Achieving Energy Justice' (2017) 36 Journal of Energy & Natural Resources Law 233–254, 233–254.
41 UN GA Res. 67/215 of 21 December 2012.
42 Peter Kayode Oniemola, 'International Law on Renewable Energy: The Need for a Worldwide Treaty' (2013) 56 German Yearbook of International Law 239, 250–251; Tedd Moya Mose, 'Toward a Harmonized Framework for International Regulation of Renewable Energy' (2018) 23(2) Uniform Law Review 373, 384.
43 Godswill A Agbaitoro and Kester I Oyibo, 'Realizing the United Nations Sustainable Development Goals 7 and 13 in Sub-Saharan Africa by 2030: Synergizing Energy and Climate Justice Perspectives' (2022) Journal of World Energy Law and Business, https://academic.oup.com/jwelb/advance-article/doi/10.1093/jwelb/jwac009/6564681, accessed 30 June 2022.
44 See for example Petroleum Act 2021 (Nigeria).
45 Eghosa Osa Ekhator and Godswill Agbaitoro, 'Energy Law and Policy in Nigeria with Reflection on the International Energy Charter and Domestication of the African Charter' in Romola Adeola and Ademola Oluborode Jegede (eds), *Governance in Nigeria Post-1999: Revisiting the Democratic 'New Dawn' of the Fourth Republic* (Pretoria University Law Press 2019) 118–119.
46 'International Energy Charter Annual Report' (2020) <https://www.energycharter.org/fileadmin/ImagesMedia/Publications/IEC_Annual_Report_2020_WEB_1.pdf> accessed 6 August 2021.
47 José E Alvarez, *International Organizations as Law-Makers* (Oxford University Press 2005); José E Alvarez, *The Impact of International Organizations on International Law* (Brill/Nijhoff 2017) 346–347.

48 It has been contended that international economic law has nexus with energy activities which also impact public international law. See K Talus, 'Internationalization of Energy Law' in K Talus (ed), *Research Handbook on International Energy Law* (Edward Elgar Publishing Limited 2014) 5–8.
49 See Jakob Skovgaard, *The Economisation of Climate Change How the G20, the OECD and the IMF Address Fossil Fuel Subsidies and Climate Finance* (Cambridge University Press 2021).
50 Likewise international law has been much impactful in influencing domestic regulation. See Catherine Redgwell, 'International Regulation of Energy Activities' in Martha Roggenkamp et al (eds), *Energy Law in Europe: National, EU and International Law and Institutions* (Oxford University Press, 2nd ed, 2007) 13.
51 Raphael J Heffron and Kim Talus, 'The Evolution of Energy Law and Energy Jurisprudence: Insights for Energy Analysts and Researchers' (2016) 19 Energy Research & Social Science 1.
52 The rules or guide from these institutions offers guide law and policy formulation at domestic levels. See Benjamin K Sovacool and Michael H Dworkin, *Global Energy Justice: Problems, Principles, and Practices* (Cambridge University Press 2014) 25–26.
53 IEA, 'Mission: The IEA Works with Governments and Industry to Shape a Secure and Sustainable Energy Future for All' (last updated 20 November 2020) <https://www.iea.org/about/mission> accessed 5 August 2021.
54 Statute of the International Renewable Energy Agency (adopted 36 January 2009), art. IV(A).
55 Thijs Van de Graaf and Jeff Colgan, 'Global Energy Governance: A Review and Research Agenda' (2016) 2 Palgrave Communications 1, 2, 7.
56 Case Concerning the Gabcikovo-Nagymaros Project (Hungaryislovakia) Judgment (25 September 1997) Separate Opinion of Vice-President Weeramantry [1997] ICJ Reports 7, 94.
57 Permanent Court of Arbitration (2005) 1, 59–57.
58 Peter Kayode Oniemola, 'A Proposal for Transnational Litigation against Climate Change Violations in Africa' (2021) 38(2) Wisconsin International Law Journal 301, 301–330.
59 (2007) 161 LGERA 1.
60 Civil Appeal No. 4417. The judgment of the Supreme Court is available at http://www.iexindia.com/Uploads/NewsUpdate/18_05_2015Final%20Order_Hindustan%20Zinc%20v%20RERC%20(2).pdf, accessed 30 June 2020.
61 ECLI:NL:HR:2019:2007, Judgment (Sup. Ct. Neth. Dec. 20, 2019) (Neth.) see M Meguro, 'State of the Netherlands v. Urgenda Foundation' (2020) 114 American Journal of International Law 729, 729–735.
62 *The State of the Netherlands (Ministry of Economic Affairs and Climate Change) v Stichting Urgenda* Supreme Court of Netherlands (20 December 2019) <https://www.urgenda.nl/wp-content/uploads/ENG-Dutch-Supreme-Court-Urgenda-v-Netherlands-20-12-2019.pdf> accessed 6 August 2021.
63 [2021] Hague District Court C/09/571932 / HA ZA 19-379.
64 William J Davey, *Non-discrimination in the World Trade Organization: The Rules and Exceptions* (Martinus Nijhoff Publishers 2012) 31.
65 Rafael Leal-Arcas and Andrew Filis, 'Renewable Energy Disputes in the World Trade Organization' (2015) 13(3) OGEL, https://www.ogel.org/article.asp?key=3535, accessed 4 July 2022.
66 Kim Talus, 'Introduction - Renewable Energy Disputes in the Europe and Beyond: An Overview of Current Cases' (2015) 13(3) OGEL, https://www.ogel.org/article.asp?key=3534, accessed 4 July 2022.
67 Rick A Waltman, 'Renewable Energy Development for WTO Member Nations' (2016) 14 Santa Clara Journal of International Law 543, 560.

68 Jan-Christoph Kuntze and Tom Moerenthout, 'Are Feed-in Tariff Schemes with Local Content Requirements Consistent with WTO Law?' in Freya Baetens and José Caiado (eds), *Frontiers of International Economic Law: Legal Tools to Confront Interdisciplinary Challenges* (Brill Nijhoff 2014) 151; WTO, *Canada–Certain Measures Affecting the Renewable Energy Generation Sector/Canada–Measures Relating to the Feed-In Tariff Program* (Appellate Body Reports) WT/DS412/AB/R and WT/DS426/AB/R.

69 A Boute, 'The Potential Contribution of International Investment Protection Law to Combat Climate Change' (2009) 27(3) Journal of Energy and Natural Resources Law 333, 335.

70 See generally Peter Kayode Oniemola, 'Business Enterprises in Renewable Energy Projects in Africa and the Human Rights Questions Arising from the Duty to Protect' in Damilola S Olawuyi and Oyeniyi Abe (eds), *Business and Human Rights Law and Practice in Africa* (Edward Elgar 2022).

71 Jenny Sin-hang Ngai, 'Energy as a Human Right in Armed Conflict: A Question of Universal Need, Survival, and Human Dignity' (2012) 37 Brooklyn Journal of International 579, 611.

72 James Crawford, 'Sovereignty as a Legal Value' in James Crawford and Martti Koskenniemi (eds), *The Cambridge Companion to International Law* (Cambridge University Press 2012) 117.

73 Aarthi S Anand, 'The Importance of Being Factual: The U.S., China, and The Future of the Kyoto Protocol' (2013) 24(1) Duke Environmental Law & Policy Forum 1, 8.

74 Johannes Urpelainen and Thijs Van de Graaf, 'United States Non-Cooperation and the Paris Agreement' (2018) 18(7) Climate Policy 839.

75 Antony J Blinken, *Secretary of State, Press Statement: The United States Officially Rejoins the Paris Agreement* (19 February 2021), https://www.state.gov/the-united-states-officially-rejoins-the-paris-agreement/, accesses 4 July 2022.

76 See Peter Kayode Oniemola, 'International Law on Renewable Energy: The Need for a Worldwide Treaty' (2013) 56 German Yearbook of International Law 239, 252–255.

77 International Energy Agency, *World Energy Outlook 2008* (IEA 2008) 37.

78 David Hodas, 'International Law and Sustainable Energy: A Portrait of Failure' in Jamie Benidickson, Ben Boer, Antonio Herman Benjamin and Karen Morrow (eds), *Environmental Law and Sustainability after Rio* (Edward Elgar 2011) 257, 271.

79 Andreas Goldthau and Jan-Martin Witte, 'The Role of Rules and Institutions in Global Energy: An Introduction' in Andreas Goldthau and Jan-Martin Witte (eds), *Global Energy Governance. The New Rules of the Game* (Brookings Press 2010) 2.

80 Nico Schrijver, *Sovereignty over Natural Resources: Balancing Rights and Duties* (Cambridge University Press 1997) 395.

81 See UNCLOS, arts. 56, 73, 77 and 193; Energy Charter Treaty, art. 18.

82 Van de Graaf and Colgan (n 55) 4.

83 Michaël Aklin and Johannes Urpelainen, 'Political Competition, Path Dependence, and the Strategy of Sustainable Energy Transitions' (2013) 57 American Journal of Political Science 643, 643.

84 Van de Graaf and Colgan (n 55) 4

85 Birnie, Boyle and Redgwell (n 6).

86 Van de Graaf and Colgan (n 55) 8.

87 Ibid.

88 Birnie, Boyle and Redgwell (n 6) 62.

89 See Kaisa Huhta, 'The Scope of State Sovereignty under Article 194(2) TFEU and the Evolution of EU Competences in the Energy Sector' (2021) 70 International and Comparative Law Quarterly 991, 1006–1008.

90 Directive (EU) 2018/2001 of the European Parliament and of the Council of 11 December 2018 on the promotion of the use of energy from renewable sources [2001] OJ L328/82.

91 K Abbott and D Snidal, 'Strengthening International Regulation through Transnational New Governance: Overcoming the Orchestration Deficit' (2009) 42(2) Vanderbilt Journal of Transnational Law 501.
92 Jeff D Colgan, Robert O Keohane and Thijs Van de Graaf, 'Punctuated Equilibrium in the Energy Regime Complex' (2012) 7 The Review of International Organizations 117,118.
93 Ibid, 5.
94 Anthony Polack, *Enabling Frameworks for Sustainable Energy Transition* (Commonwealth Secretariat 2020) 12.
95 Kaisa Huhta, 'Anchoring the Energy Transition with Legal Certainty in EU Law' 2020 27(4) Maastricht Journal of European and Comparative Law 425, 427.

Bibliography

Cases

Gabčikovo-Nagymaros Project, Hungary v Slovakia ICJ GL No 92, [1997] ICJ Rep 7, [1997]

Hindustan Zinc Limited v. Rajasthan Electricity Regulatory Commission (Civil Appeal No. 4417 of 2015)

Iron Rhine Arbitration, Belgium v Netherlands, Award ICGJ 373 (PCA 2005), 24th May 2005, Permanent Court of Arbitration [PCA]

Vereniging Milieudefensie and Others v Royal Dutch Shell (RDS) Plc ECLI:NL:RBDHA:2021:5339

Taralga Landscape Guardians Inc. v Minister for Planning and RES Southern Cross Property Limited (2007) 161 LGERA 1

International Legislation

Convention on the Law of the Sea, Dec. 10, 1982, 1833 U.N.T.S. 397

Energy Charter Treaty, 2080 UNTS 100

Paris Agreement to the United Nations Framework Convention on Climate Change, Dec. 12, 2015, T.I.A.S. No. 16-1104. UNFCCC COP Report No. 21, UN Doc. FCCC/CP/2015/10/Add.1

United Nations Framework Convention on Climate Change. New York: United Nations, General Assembly, 1994, A/RES/48/189

Books

Anthony Polack, *Enabling Frameworks for Sustainable Energy Transition* (Commonwealth Secretariat 2020)

Benjamin K Sovacool and Michael H Dworkin, *Global Energy Justice: Problems, Principles, and Practices* (Cambridge University Press 2014)

Duncan French, *International Law and Policy of Sustainable Development* (Juris and Manchester University Press, 2005)

Harold Hongju Koh, *The Trump Administration and International Law* (Oxford University Press 2019)

Henrik Lund, *Renewable Energy Systems: A Smart Energy Systems Approach to the Choice and Modelling of 100% Renewable Solutions* (2nd edn, Academic Press 2014)

Herman E Daly and Joshua Farley, *Ecological Economics Principles and Applications* (Island Press 2004)
Jakob Skovgaard, *The Economisation of Climate Change How the G20, the OECD and the IMF Address Fossil Fuel Subsidies and Climate Finance* (Cambridge University Press 2021)
José E Alvarez, *International Organizations as Law-Makers* (Oxford University Press 2005)
José E Alvarez, *The Impact of International Organizations on International Law* (Brill/Nijhoff 2017)
K Talus, 'Internationalization of Energy Law' in K Talus (ed.), *Research Handbook on International Energy Law* (Edward Elgar Publishing Limited 2014) 5–8
Nico Schrijver, *Sovereignty over Natural Resources: Balancing Rights and Duties* (Cambridge University Press 1997)
Patricia Birnie, Alan Boyle and Catherine Redgwell, *International Law and the Environment* (4th edn, Oxford University Press 2021)
WCED, *Our Common Future* (Oxford University 1987)
William J Davey, *Non-discrimination in the World Trade Organization: The Rules and Exceptions* (Martinus Nijhoff Publishers 2012)

Contributions to Edited Books

Alexandra Wawryk, 'International Energy Law: An Emerging Academic Discipline' in Paul Babie and Paul Leadbeter (eds), *Law as Change: Engaging with the Life and Scholarship of Adrian Bradbrook* (University of Adelaide Press 2014)
Andreas Goldthau and Jan-Martin Witte, 'The Role of Rules and Institutions in Global Energy: An Introduction' in Andreas Goldthau and Jan-Martin Witte (eds), *Global Energy Governance. The New Rules of the Game* (Brookings Press 2010)
Catherine Redgwell, 'International Regulation of Energy Activities' in Martha Roggenkamp et al (eds), *Energy Law in Europe: National, EU and International Law and Institutions* (2nd edn, Oxford University Press 2007)
Catherine Redgwell, 'International Soft Law and Globalization' in Barry Barton et al (eds), *Regulating Energy and Natural Resources* (Oxford University Press 2006)
David Hodas, 'International Law and Sustainable Energy: A Portrait of Failure' in Jamie Benidickson, Ben Boer, Antonio Herman Benjamin and Karen Morrow(eds), *Environmental Law and Sustainability After Rio* (Edward Elgar 2011)
Eghosa Osa Ekhator and Godswill Agbaitoro, 'Energy Law and Policy in Nigeria with Reflection on the International Energy Charter and Domestication of the African Charter' in Romola Adeola and Ademola Oluborode Jegede (eds), *Governance in Nigeria Post-1999: Revisiting the Democratic 'New Dawn' of the Fourth Republic* (Pretoria University Law Press 2019)
James Crawford, 'Sovereignty as a Legal Value' in James Crawford and Martti Koskenniemi (eds), *The Cambridge Companion to International Law* (Cambridge University Press 2012)
Jan-Christoph Kuntze and Tom Moerenthout, 'Are Feed-in Tariff Schemes with Local Content Requirements Consistent with WTO Law?' in Freya Baetens and José Caiado (eds), *Frontiers of International Economic Law: Legal Tools to Confront Interdisciplinary Challenges* (Brill Nijhoff 2014)
K Talus, 'Internationalization of Energy Law' in K Talus (ed.), *Research Handbook on International Energy Law* (Edward Elgar Publishing Limited 2014)

Romain Mauger, 'Making Sense of Changing Concepts for the Energy Transition: An Energy Transition Concepts Nexus for the Development of Policy and Law', in Ruven Fleming, Kaisa Huhta and Leonie Reins (eds), *Sustainable Energy Democracy and the Law* (Brill/Nijhoff 2021)

Morakinyo Adedayo Ayoade, 'Bridging the Gap Between Climate Change and Energy Policy Options: What Next for Nigeria?' in Patricia Kameri-Mbote, Alexander Paterson, Oliver C Ruppel, Bibobra Bello Orubebe and Emmanuel D Kam Yogo (eds), *Law, Environment and Africa* (Nomos Verlagsgesellschaft 2019)

Peter Kayode Oniemola, 'Business Enterprises in Renewable Energy Projects in Africa and The Human Rights Questions Arising from the Duty to Protect' in Damilola S Olawuyi and Oyeniyi Abe (eds), *Business and Human Rights Law and Practice in Africa* (Edward Elgar 2022)

Journal Articles

A Boute, 'The Potential Contribution of International Investment Protection Law to Combat Climate Change' (2009) 27(3) *Journal of Energy and Natural Resources Law* 333–376

A Cherp, J Jewell and A Goldthau, 'Governing Global Energy: Systems, Transitions, Complexity' (2011) 2(1) Global Policy 75-87

Adrian J Bradbrook and Judith G Gardam, 'Placing Access to Energy Services Within a Human Rights Framework' (2006) 28 Human Rights Quarterly 389–415

Andrei Belyi, 'The Energy Charter Process in the Face of Uncertainties' (2021) 14 Journal of World Energy Law and Business 363–375

Dorothée Cambou, 'Uncovering Injustices in the Green Transition: Sámi Rights in the Development of Wind Energy in Sweden' (2020) 11 Arctic Review on Law and Politics 310–333

Frank Biermann et al, 'The Fragmentation of Global Governance Architectures: A Framework for Analysis' (2009) 9 Global Environmental Politics 14–40

K Abbott and D Snidal, 'Strengthening International Regulation through Transnational New Governance: Overcoming the Orchestration Deficit' (2009) 42(2) Vanderbilt Journal of Transnational Law 501–578

Jeff D Colgan, Robert O Keohane, and Thijs Van de Graaf, 'Punctuated Equilibrium in the Energy Regime Complex' (2012) 7 The Review of International Organizations 117–143

Jenny Sin-hang Ngai, 'Energy as a Human Right in Armed Conflict: A Question of Universal Need, Survival, and Human Dignity' (2012) 37 Brooklyn Journal of International 579–622

Johannes Urpelainen and Thijs Van de Graaf, 'United States Non-cooperation and the Paris Agreement' (2018) 18(7) Climate Policy 839–851

Kaisa Huhta, 'Anchoring the Energy Transition with Legal Certainty in EU Law' 2020 27(4) Maastricht Journal of European and Comparative Law 425–444

Kaisa Huhta, 'The Scope of State Sovereignty Under Article 194 (2) TFEU and the Evolution of EU Competences in the Energy Sector' (2021) 70 International and Comparative Law Quarterly 991–1010

Kim Talus, 'Introduction - Renewable Energy Disputes in the Europe and Beyond: An Overview of Current Cases' (2015) 13(3) OGEL, https://www.ogel.org/article.asp?key =3534, accessed 4 July 2022

Michaël Aklin and Johannes Urpelainen, 'Political Competition, Path Dependence, and the Strategy of Sustainable Energy Transitions' (2013) 57 American Journal of Political Science 643–658

Neil Gunningham, 'Confronting the Challenge of Energy Governance' (2012) 1(1) Transnational Environmental Law 119–135

Paola Villavicencio Calzadilla and Romain Mauger, 'The UN's New Sustainable Development Agenda and Renewable Energy: The Challenge to Reach SDG7 While Achieving Energy Justice' (2017) 36 Journal of Energy & Natural Resources Law 233–254

Peter Kayode Oniemola, 'A Proposal for Transnational Litigation Against Climate Change Violations in Africa' (2021) 38(2) Wisconsin International Law Journal 301–329

Peter Kayode Oniemola, 'International Law on Renewable Energy: The Need for a Worldwide Treaty' (2013) 56 German Yearbook of International Law 281–314

Rafael Leal-Arcas and Andrew Filis, 'Renewable Energy Disputes in the World Trade Organization' (2015) 13(3) OGEL, https://www.ogel.org/article.asp?key=3535, accessed 4 July 2022.

Raphael J Heffron, Anita Rønne, Joseph P Tomain, Adrian Bradbrook and Kim Talus, 'A Treatise for Energy Law' (2018) 11(1) Journal of World Energy Law and Business 34–48

Raphael J Heffron and Kim Talus, 'The Evolution of Energy Law and Energy Jurisprudence: Insights for Energy Analysts and Researchers' (2016) 19 Energy Research & Social Science 1–10

Rick A Waltman, 'Renewable Energy Development for WTO Member Nations' (2016) 14 Santa Clara Journal of International Law 543–566

Tedd Moya Mose, 'Toward a Harmonized Framework for International Regulation of Renewable Energy' (2018) 23(2) Uniform Law Review 373–396

Thijs Van de Graaf and Jeff Colgan, 'Global Energy Governance: A Review and Research Agenda' (2016) 2 Palgrave Communications 1–12

Internet Sources

Antony J Blinken, 'Secretary of State, Press Statement: The United States Officially Rejoins the Paris Agreement' (19 February 2021), https://www.state.gov/the-united-states-officially-rejoins-the-paris-agreement/, accessed 4 July 2022.

Godswill A Agbaitoro and Kester I Oyibo, 'Realizing the United Nations Sustainable Development Goals 7 and 13 in Sub-Saharan Africa by 2030: Synergizing Energy and Climate Justice Perspectives' (2022) Journal of World Energy Law and Business, https://academic.oup.com/jwelb/advance-article/doi/10.1093/jwelb/jwac009/6564681, accessed 30 June 2022

IEA, 'Mission: The IEA Works with Governments and Industry to Shape a Secure and Sustainable Energy Future for All' (last updated 20 November 2020), https://www.iea.org/about/mission, accessed 5 August 2021

International Energy Charter Annual Report (2020), https://www.energycharter.org/fileadmin/ImagesMedia/Publications/IEC_Annual_Report_2020_WEB_1.pdf, accessed 6 August 2021

The State of the Netherlands (Ministry of Economic Affairs and Climate Change) v Stichting Urgenda Supreme Court of Netherlands (20 December 2019), https://www.urgenda.nl/wp-content/uploads/ENG-Dutch-Supreme-Court-Urgenda-v-Netherlands-20-12-2019.pdf, accessed 6 August 2021

The Transnational Institute, 'ECT's Dirty Secrets' (2020), https://energy-charter-dirty-secrets.org/, accessed 22 June 2022

United Nations, *Transforming Our World: The 2030 Agenda for Sustainable Development*, https://sustainabledevelopment.un.org/content/documents/21252030%20Agenda%20for%20Sustainable%20Development%20web.pdf, accessed 4 July 2022

3 Sustainable Development and Off-Grid Renewable Electricity

Current Status and Challenges

Chitzi C. Ogbumgbada, Rasaki Stephen Dauda, and Eduardo G. Pereira

3.1 Introduction

By contributing to the expansion of access to electricity and by strengthening the response to climate change, off-grid renewable electricity (OGRE) has an important role to play in achieving sustainable development. OGRE refers to electricity generated from renewable energy sources such as solar, wind, geothermal, bioenergy, hydropower, and ocean through the utilisation of stand-alone and mini-grid infrastructure.[1] Electrification – which is a prime property of access to energy – has the instrumental role of powering whole households and communities,[2] in addition to facilitating other valuable productive activities which have profound economic implications.[3] There is also evidence to suggest that improved electricity access can positively influence business profits and revenues.[4] In effect, at the household level, electricity as a form of energy can enable people to lead flourishing lives through valued energy-facilitated capabilities such as education, health and safety, communication, and entertainment, among others.[5] All these are significant for sustainable development, which effectively aims to balance competing notions from environmental protection, economic development, and social development.[6]

Despite the above, access to electricity continues to pose a problem for large swathes of the world, with hundreds of millions of people lacking electricity access. In sub-Saharan Africa (SSA) alone, an estimated 597 million people were without access to electricity in 2021.[7] The COVID-19 pandemic has further worsened the situation resulting in a rise in the share of sub-Saharan Africa's global population without access to electricity.[8]

Table 3.1 below provides information on electricity access by region from 2000 to 2020. As evident in the table, the SSA is the most affected region, with more than half of the population lacking access to electricity since 2000. This diverges substantially from the figures in North America and Europe & Central Asia whose population boasts of 100% electricity access as well as East Asia & Pacific, Latin America & Caribbean, Middle East & North Africa, and South Asia where close to 99% of the population enjoy access to electricity.

Energy deficits can deprive people from leading productive and fully functioning lives, and retard the sustainability of nations' development. The central

DOI: 10.4324/9781003178088-4

Table 3.1 Summary of Electricity Access by Region, 2000-2020

Region/Year	*2000 (%)*	*2005 (%)*	*2010 (%)*	*2015 (%)*	*2020 (%)*
Eastern and Southern Africa	19.81	23.49	27.56	33.87	45.80
East Asia & Pacific	92.22	93.84	95.54	97.10	98.01
Europe & Central Asia	99.95	99.95	99.97	99.26	100.00
Latin America & Caribbean	91.74	93.57	95.86	97.28	98.52
Middle East & North Africa	92.52	94.04	95.74	96.52	97.36
North America	100.00	100.00	100.00	100.00	100.00
South Asia	56.24	64.81	73.19	84.61	95.78
Sub-Saharan Africa	25.57	29.29	33.30	39.07	48.35
Western and Central Africa	34.15	37.85	41.80	46.77	52.11
World	78.22	80.65	83.22	86.62	90.52

Source: Generated by Authors from the World Bank (2022)[9]

role of energy in advancing human development was brought to the fore when it was included as Goal 7 of the United Nations Sustainable Development Goals (SDGs). Goal 7 aims for universal access to energy services[10] suggestive of the fact that energy is at the heart of sustainable development. OGRE is central to realising the objectives of Goal 7 within ecological limits.

Within the last decade, OGRE has become a mainstream energy solution, developing exponentially in many parts of the world.[11] The success of OGRE is aided by many factors including rapidly declining costs in technologies such as solar, and increased awareness of the benefits of OGRE to the subject of access to modern and reliable energy.[12] Despite this growth, however, challenges remain that hamper the full realisation of the benefits of OGRE. As countries have varied experiences in the deployment of OGRE, these challenges would necessarily vary from country to country.

Ensuing from the above, therefore, this chapter considers OGRE within the context of sustainable development, in particular Goal 7 of the SDGs. Following this introduction, the chapter examines the relationship between OGRE and sustainable development, with a view to mapping out the significance of OGRE to the achievement of sustainable development. It then examines the status of OGRE deployment around the world. To help in this discussion, the chapter presents and interrogates statistics from international organisations and bodies which have as part of their core mandates and work programmes the monitoring of trends and the compilation of statistics relating to OGRE. Some of these organisations include the International Renewable Energy Agency (IRENA), and the International Energy Agency (IEA). In addition, the chapter assesses a number of barriers which OGRE developers face in the market. A concluding section follows thereafter.

3.2 Nexus between OGRE and Sustainable Development

This section examines the relationship between OGRE and sustainable development, and it is therefore helpful to begin with the argument that OGRE is significant to the achievement of sustainable development, since it mainly relates to the expansion of access to modern energy, which itself is a primary concern of sustainable development.[13] The significance of energy – whether fossil fuel or clean energy – to the advancement of society can hardly be overstated.[14] Modern energy, for example, played a crucial role in ushering in the Industrial Revolution – a period characterised by greater productivity and economic prosperity, mainly powered by coal-propelled engines and machines.[15] As has been observed about the usefulness of having sufficient energy supplies:

> The gulf between a highly developed and a subsistence economy does not exist because the researcher examining samples of moon rock is an inherently superior human being to the hill shepherd. It depends almost entirely on the relative availability of energy within their two societies. Like some great hovercraft, industrial society is lifted and maintained above concern with the elemental necessities of life by a prodigious expenditure of energy. Without this energy supply, the sophisticated skills of the industrial world are merely a burden in the struggle for survival.[16]

Notwithstanding the utility of all forms of energy resources to societal progress and development, there is the important issue of sustainability.[17] This mainly accounts for the shift from high-polluting energy sources based on coal, gas, and crude oil to energy systems and units built on clean and renewable sources of energy.[18] The latter is more aptly conceived as the global energy transition.[19] It is recognised that renewable energy resources such as wind, solar, geothermal, modern biomass, and tidal could help to mitigate the effects of climate change,[20] promote access to energy services such as electricity and clean cooking,[21] and enhance economic development through the production of green employment.[22]

The centrality of energy usage and consumption is recognised in fora such as the SDGs. The SDGs are at the heart of the 2030 Agenda for Sustainable Development adopted by the UN General Assembly during the UN Sustainable Development Summit in 2015. The SDGs supplanted the Millennium Development Goals. Goal 7 of the SDGs has the target of 'ensuring access to affordable, reliable, sustainable and modern energy for all by 2030'.[23] One of its targets is to increase the use of renewable energy around the globe. Importantly, the goal seeks to facilitate international finance to developing countries to support renewable energy research, development, and production in those countries.[24] Goal 7 establishes the importance of energy, not just as an intrinsic resource in its own right, but as an instrumental medium for achieving other goals under the SDS.

Mention may be made of some criticisms of the goal. Thus, Goal 7 has been criticised as having procedural issues, including for being framed in terms that do not recognise the genesis of energy poverty among developing countries.[25] These criticisms do not however downgrade the importance of the goal in achieving the SDGs.

The point may be restated that access to energy has a profound effect on society and modern living. As has been argued, 'access to reliable and affordable electricity [...] can have an immediate and transformative impact on quality of life, access to basic services (e.g., health, education) and livelihoods'.[26] Conversely, a lack of access to energy services may be likened to life in primordial times: solitary, poor, nasty, brutish, and short.[27] As has been pointed out:

> This lack of access to modern energy limits income generation, blunts efforts to escape poverty, affects the health of women and children, and contributes to global deforestation and climate change.[28]

This arguably explains why access to energy occupies a central position within the SDGs, manifesting as a campaign to 'ensure access to affordable, reliable, sustainable, and modern energy for all'.[29]

It is recognised that OGRE has the potential to contribute to the rapid expansion of access to electricity in a cost- and time-efficient manner.[30] This places OGRE at the heart of sustainable development. Extending grid connection to remote areas is an expensive process, coupled with the unrelated but relevant fact that centralised electricity systems that run on fossil fuels have environmental sustainability issues not least of which is the fact that they produce greenhouse gas emissions which contribute to global warming.[31] As already argued in this chapter, energy usage should always address the question of whether such usage is sustainable or not.

Grid-connected electricity also has other associated issues. Oftentimes, for example, users who dwell in rural areas are mostly not connected to the central grid while contemporaneously constituting the group with the most need for access to electricity.[32] It happens that companies that provide electric services oftentimes demonstrate a reluctance to connect rural areas to the central electric system, not just because of difficulties associated with terrain and costs of connection, but more importantly, because of the conception that such connection may prove to be commercially unviable and therefore unprofitable.[33] OGRE can thus provide electricity access either through stand-alone systems like solar home systems or mini-grid technologies.[34] In this sense, OGRE is contributing to the larger objective of access to energy services which is an important pathway to achieving the overarching goal of sustainable development.[35] Furthermore, OGRE is highly critical for economic growth and development. This is reinforced by empirical findings, which report positive and significant impact of renewable energy on growth and development in both advanced and developing economies.[36] This in effect represents a strong case for its continued deployment across the globe.

The availability of OGRE has the capacity to supply energy not only at the household level but also to industries and infrastructure, which help to boost the level of productivity, raise income, and enhance growth and development. It is 'a solution to meet growing energy demand' for industrial production 'while sharply reducing carbon emissions'; it can also be regarded 'as a potential engine for economic growth and diversification'.[37] Energy is considered an important input into production processes, and since OGRE has the potential to reduce the cost of energy generation, it follows that output can expand because of deploying more OGRE in the economy.[38] In analysing the linkages that exist between energy system and the global economy 'within a single quantitative framework', it was reported that if the share of 'renewables in the global energy mix' is doubled through increased investment in its deployment, the world's Gross Domestic Product will increase by about 1.1% in 2030, which translates into around US$1.3 trillion and which is capable of engendering sustainable development.[39]

A panel study of 25 developing countries found that renewable electricity generation contributes positively and significantly to economic growth in both the short and long run.[31] In fact, the analysis discovered that a bidirectional relationship exists between growth and renewable electricity generation; this implies that both have the capacity to affect each other positively. In other words, renewable electricity generation enhances growth while growth in turn boosts renewable electricity generation. The findings of this study are also consistent with the outcome of another work on 38 developed and developing countries.[40]

Furthermore, OGRE fosters green economic growth and development. This implies that as countries implement growth and development policies, which employ OGRE, pollution emissions reduce drastically while output rises. However, this may be hampered in the absence of appropriate technological advancement,[41] particularly in developing economies that are technologically backward.

Similarly, OGRE can contribute to employment generation both directly and indirectly. In many developing countries, the level of unemployment is very high; so, OGRE is considered an important policy option to address this problem, if the opportunities it provides are properly harnessed. In this regard, the International Labour Organization has argued that 'energy created through solar photovoltaic cells, landfill gas, or biomass plants have a higher number of jobs created per unit of energy produced than energy produced through conventional sources'.[42] Equally, employment generation through OGRE occurs along the entire value chain, across manufacturing and distribution of equipment; production of inputs like chemicals; in the services sector like 'project management, installation, operation, and maintenance'; in the agricultural sector and so on, from entry-level to highly skilled positions of employment.[43]

Moreover, most of the skills required 'to install, operate and maintain off-grid systems can be developed locally, providing access to training and employment opportunities, especially for youths and women'.[44] Furthermore, it is projected that given 'the distributed and labour-intensive nature of renewable energy', it will be able to generate about 24.4 million direct and indirect jobs by 2030.[31] A study on job creation capacity of the off-grid solar industry, covering East, West, and Central Africa as well as South Asia, reported that the sector as at 2019 supported about 370,000 full time equivalent (FTE) jobs, and by 2022, the figure would have risen to approximately 1.3 million FTE jobs across the four regions.[45] The U.S. Bureau of Labor Statistics predicted around 61% and 51% growth in the number of wind turbine service technicians and solar photovoltaic installers that would be employed between 2019 and 2029.[46]

Apart from the contribution of OGRE to economic growth and development, as well as employment generation, it also has the capacity to provide and boost household income in the society. Most of the persons employed directly and indirectly because of activities generated through introduction of OGRE will earn income. For example, one report reveals that in the US 'rural landowners who host wind farms on their property' earn about US$222 million per annum while farmers who grow crops to be used as biofuels make a lot of income.[47] The reduction in the cost of energy will also contribute to boosting the purchasing power of the households. As noted by Folk (2019), 'switching to renewable energy is an excellent way for residential, commercial and industrial energy customers to save money on their bills'.[48]

3.3 Barriers to OGRE Development

OGRE faces challenges and barriers that may hamper its development. These barriers are mainly legal and economic in nature and are both discussed below.

3.3.1 Legal and Regulatory Barriers

It is acknowledged that a robust legal, regulatory, and policy environment could facilitate both public and private investments in the development and consequent deployment and expansion of OGRE in areas where this form of electricity is required to boost electricity access for households and industries to enable them to lead fully functioning lives and engage in productive activities.[49] An effective legal ecosystem could provide certainty and stability, among other things, that are necessary for OGRE investments to take hold and flourish. Despite the recognition of the importance of an effective legal climate for OGRE, there have been noticeable problems of a legal nature detected in parts of the world which concern OGRE.

An example of one such problem is the absence of legally binding renewable energy generation targets and incentives, especially within the sub-Saharan African region.[50] This means that there is the absence of effective legislative measures that mandate electricity to be sourced from renewable sources such as solar, wind, geothermal, and so on. Implicitly, these could hinder the adoption of effective measures that would remove barriers to private investment in OGRE.[51]

Another problem relates to the absence of bespoke or dedicated legislations on renewable energy development,[52] observed mostly in countries in sub-Saharan

Africa.[53] A bespoke legislation helps to streamline expectations, as well as influence renewable energy policy towards a desired direction. It helps to communicate the intention of the political class, demonstrating their seriousness in promoting not just OGRE, but also the use of renewable energy more generally. Such legislation could ultimately help to incentivise investments in the development of OGRE. A bespoke renewable energy legislation or law could also help to enshrine renewable energy electricity generation targets, which could help to expand the use of OGRE.

Finally, a further common challenge concerns the institutional structure or architecture on OGRE. Effective and well-functioning institutions are necessary for OGRE to flourish and become mainstream.[54] Such institutions could be set up to formulate policies on OGRE, in addition to enforcing and executing those policies. In contrast to this, there are situations where there seems to be either a proliferation of institutions on renewable energy or where institutions with ill-defined objectives exist.[55] Concerning the first problem, a proliferation of institutional structures on OGRE could lead to an overlap of functions. These could potentially produce uncertainty in respect to issues such as costs of licences and permits to embark on construction of OGRE technologies.[56]

3.3.2 Economic Barriers

Economic challenges faced by OGRE, which are more pronounced in low-income countries, cover issues relating to investment in infrastructure, human capital development, market constraint, and economic policy environment.

OGRE requires high levels of investment in infrastructure to flourish and grow. This places investments at the heart of OGRE discussions. A good number of the infrastructure/facilities needed for OGRE projects must be procured, installed and run, and most of these require initial capital, and subsequent finance for operation. For example, in the case of solar energy, there is need to invest in solar photovoltaic facilities, which many households may not be able to afford in low-income countries. In most African countries, the facilities are imported among dwindling incomes and degenerating exchange rates, which places a heavy burden on investors in the sector, which in most cases are private individuals.

It has been shown that capital cost for building and installing OGRE facilities is the main challenge, and for instance, the average cost of installing solar systems as at 2017 hovers around US$2,000 per kilowatt for large-scale systems and US$3,700 for residential systems.[57] Currently, residential systems will cost between US$15,000 and US$25,000 on average 'prior to tax credits or incentives'.[58] Most households in developing countries cannot afford this cost while some investors in the project may find it difficult to access funds from financial institutions due to high rates of interest in low-income countries and lack of collateral security. Moreover, 'higher construction costs might make financial institutions more likely to perceive renewables as risky, lending money at higher rates and making it harder for utilities or developers to justify the investment'.[59]

Another economic barrier is the issue of market constraint. This occurs in the area of entry due to high competition with operators in the existing energy industry. Reporting on the US economy as in most other countries, the electricity industry is dominated by some major wealthy players, who have invested heavily in the technologies that use fossil fuels (coal, oil, and natural gas) and nuclear energy; and who also benefit from existing infrastructure, expertise, and government policies; as well as wield enormous market power.[60] A 2016 Energy Information Administration (EIA) report on international energy outlook as cited by Seetharaman and others shows that in 2016, investment in fossil fuels still accounted for 55% of global energy investment compared with just 16% for renewable energy, and that by 2040, fossil fuels will still supply around 78% of global energy use.[61] In fact, a current report indicates that as at 2020, fossil fuels made up about 80% of global energy supply while renewables accounted for just 12%.[62] All these still serve as great barriers for renewable energy.

The issue of human capital development for the technical skills required for installation and maintenance of OGRE facilities is another critical economic factor. In several low-income counties, particularly in sub-Saharan Africa, people are lacking in such skills, which is a challenge that must be surmounted. Moreover, users of the products require some level of education to understand the usage. A study conducted in Nigeria, reported 'lack of skilled personnel to meet a code of standard procedure, to install and maintain solar PV in the country'.[63] This is in addition to the risk of potential customers failing to embrace the product due to 'inadequate experience and lack of technical know-how'.[64]

Economic policy environment also constrains the deployment and usage of OGRE in developing countries. For example, government's provision of grants and subsidies to conventional energy producers to generate electricity from fossil fuel sources operate against renewable energy by preventing investors in the sector from competing favourably with non-renewable energy investors. In addition, OGRE faces the challenge of tariff charge, especially in rural areas, which are mainly inhabited by very low-income earners, who cannot afford the tariff which providers may charge. For off-grid system developers to cover investment and operational costs, in comparison with cheaper grid-based electricity, they must charge considerably high tariffs, which most low-income earners may hardly be able to afford.[65]

3.4 The Way Forward

OGRE deployment has increased exponentially across many parts of the world. This growth has been mainly precipitated by rapidly declining technology costs, innovative models of deployment, creative financing frameworks, as well as the involvement of a diverse range of stakeholders and necessary actors composed of local entrepreneurs, the international private sector, and financial institutions.[66] The statistics and numbers seem to buttress this point. Thus, according to the IRENA, which regularly publishes statistics and numbers on the global deployment of OGRE, it is estimated that 'some 133 million people accessed lighting and other electricity services using off-grid renewable energy solutions in 2016'.[67] This represented a six-fold global expansion since 2011,[68] indicating the desirability and the utility of off-grid renewable energy solutions in general, and OGRE in particular, in servicing the imperatives of access to modern and sustainable energy. In terms of OGRE capacity, it may be noted that 'off-grid renewable energy capacity has witnessed a spectacular three-fold increase from under 2 gigawatts (GW) in 2008 to over 6.5 GW in 2017'.[69]

Furthermore, despite the outbreak of COVID-19, which hampered progress in several sectors across the globe, renewable energy in 2020 still witnessed some form of growth. The sector was reported to have 'set a record in new power capacity in 2020 and was the only source of electricity generation to register a net increase in total capacity', alongside rising investment for the third consecutive year as more nations shifted towards renewables for the electrification of heat.[70] However, it may be pointed out that more needs to be done to ensure OGRE technologies are available in countries and regions that are struggling with energy access problems.

Given the strategic importance of OGRE for growth and sustainable development of nations, it is important that the constraints to its progress be addressed. This has to take multifaceted approaches, ranging from legal to economic. From a legal perspective, there is the need for a consistent and clear framework to incentivise the private sector to invest in the OGRE activities to develop the required infrastructure, manpower, and the market itself which would be in line with the current climate targets and environmental, social, governance practices.

With respect to the economic dimension, it is essential that an enabling environment be provided to ease investment in OGRE infrastructure development. Financial institutions should be encouraged to provide funding at affordable interest rates to investors in the sector. Moreover, regular technical training should be provided for human resources that will engage in the installation and operation of OGRE facilities. In addition to the latter, users at the household level should be given regular education on how to deploy the facilities for domestic and community usage. Furthermore, the issue of policy environment covering tariff charge should be given priority. Governments should consider giving grants and subsidies to investors in the system to prevent them from charging high tariffs in order to make the services affordable for low-income earners, particularly in the developing countries of Africa. Finally, the issue of market constraints cannot be over emphasised. Firms in the OGRE sector should be given certain protection by governments to enable them to compete with the wealthy firms in the non-renewable energy sector.

3.5 Conclusion

In summary, this chapter discussed the relationship between OGRE and sustainable development. It was shown that OGRE could help to meet the aims and objectives of sustainable development through the provision of energy access to communities, households, and other entities requiring the same to lead functioning lives and engage in productive activities. The chapter also outlined statistics showing that OGRE has expanded greatly in recent times. Nevertheless, it was shown that more deployment of OGRE is required, especially to remediate energy deficits in many parts of the world. Increasing the deployment of OGRE technologies would necessitate overcoming barriers to such deployment. Finally, the legal and economic challenges which developers face when considering OGRE investments should be properly addressed. Strategies are required to overcome these challenges and to ensure that OGRE continues to play a main role in the provision of energy access to millions requiring such access otherwise SDG 7 will not be met for millions of people in the developing world.

Notes

1 IRENA, *Off-Grid Renewable Energy Systems: Status and Methodological Issues* (IRENA Abu Dhabi 2015) 5.
2 Kristine Bos, Duncan Chaplin and Arif Mamun, 'Benefits and Challenges of Expanding Grid Electricity in Africa: A Review of Rigorous Evidence on Household Impacts in Developing Countries' (2018) 44 Energy for Sustainable Development 64–77.
3 Joseph M Ngowi, Lennart Bångens and Erik O Ahlgren, 'Benefits and Challenges to Productive Use of Off-Grid Rural Electrification: The Case of Mini-Hydropower in Bulongwa-Tanzania' (2019) 53 Energy for Sustainable Development 97–103.
4 See Charles Kirubi and others, 'Community-Based Electric Micro-Grids Can Contribute to Rural Development: Evidence from Kenya' (2009) 37(7) World Development 1208–1221.
5 Sushil Rajagopalan, 'Who Benefits and How? A Capabilities Perspective on Solar Micro-Grids in India' (2021) Journal of Human Development and Capabilities 1 at 7–13.
6 Marie-Claire Cordonier Segger and Ashfaq Khalfan, *Sustainable Development Law: Principles, Practices, & Prospects* (OUP 2004) 1–2.
7 IEA, 'Global Population without Access to Electricity by Region, 2000 – 2021' (IEA 2021) <https://www.iea.org/data-and-statistics/charts/global-population-without-access-to-electricity-by-region-2000-2021-2> accessed 11 July 2022. See also IEA, 'SDG7: Data and Projections' (2022) <https://www.iea.org/data-and-statistics/data-product/sdg7-database)> accessed 11 July 2022.
8 IEA, 'The Covid-19 Crisis is Reversing Progress on Energy Access in Africa' (2020) <The Covid-19 crisis is reversing progress on energy access in Africa – Analysis - IEA> accessed 11 July 2022.
9 World Bank, 'World development indicators' (2022) <https://data.worldbank.org/indicator> accessed 11 July 2022.
10 United Nations Department of Economic and Social Affairs, 'Goal 7: Ensure Access to Affordable, Reliable, Sustainable and Modern Energy for All' (Department of Economic and Social Affairs) <https://www.un.org/en/chronicle/article/goal-7-ensure-access-affordable-reliable-sustainable-and-modern-energy-all#:~:text=SDG%207%E2%80%94to%20%E2%80%9Censure%20access,statement%20of%20the%20goal%20itself> accessed 11 July 2022.
11 IRENA, *Off-Grid Renewable Energy Statistics 2019*, <irena.org> accessed 11 July 2022.
12 IRENA, *Off-Grid Renewable Energy Solutions: Global and Regional Status and Trends* (IRENA Abu Dhabi 2018) 1.

13 Binayak Bhandari and Sung-Hoon Ahn, 'Off-Grid Hybrid Renewable Energy Systems and Their Contribution to Sustainable Development Goals' in Asmae Berrada and Rachid El Mrabet (eds), *Hybrid Energy System Models* (Academic Press 2021) 75–89.
14 See Ripudaman Malhotra (ed), *Fossil Energy* (2nd edn, Springer 2020).
15 Yinka Omorogbe, 'Legal Dimensions of Access to Modern Energy Services in Africa: Lessons from Nigeria, Ghana, and Rwanda' in Íñigo del Guayo et al (eds), *Energy Justice and Energy Law* (OUP 2020) 331.
16 Gerald Foley, *The Energy Question* (4th edn, Penguin Books Ltd. 1992) 3 quoted in Yinka Omorogbe, 'Promoting Sustainable Development through the Use of Renewable Energy: The Role of Law' in Don Zillman and others (eds), *Beyond the Carbon Economy: Energy Law in Transition* (OUP 2008) 41.
17 See David L Goldblatt, *Sustainable Energy Consumption and Society: Personal, Technological, or Social Change* (Springer 2005).
18 IRENA, 'Energy Transition' <irena.org> accessed 20 February 2022.
19 See Peter D Cameron, Xiaoyi Mu and Volker Roeben (eds), *The Global Energy Transition: Law, Policy and Economics for Energy in the 21st Century* (Hart Publishing 2020).
20 IPCC, 'Global Warming of 1.5°, an IPCC Special Report on the Impacts of Global Warming of 1.5° Above Pre-Industrial Levels and Related Global Greenhouse Gas Emission Pathways, in the Context of Strengthening the Global Response to the Threat of Climate Change, Sustainable Development, and Efforts to Eradicate Poverty' (2018).
21 'Sustainable Energy for All: A Vision Statement by Ban Ki-Moon, Secretary-General of the United Nations' (United Nations, November 2011).
22 IRENA, *Renewable Energy Benefits: Measuring the Economics* (IRENA Abu Dhabi 2016).
23 United Nations Department of Economic and Social Affairs, 'Goal 7: Ensure Access to Affordable, Reliable, Sustainable and Modern Energy for All' (Department of Economic and Social Affairs) <https://www.un.org/en/chronicle/article/goal-7-ensure-access-affordable-reliable-sustainable-and-modern-energy-all#:~:text=SDG%207%E2%80%94to%20%E2%80%9Censure%20access,statement%20of%20the%20goal%20itself> accessed 11 July 2022.
24 Ibid.
25 See Paul Munro, Greg van der Horst and Stephen Healy, 'Energy Justice for All? Rethinking Sustainable Development Goal 7 through Struggles over Traditional Energy Practices in Sierra Leone' (2017) 105 Energy Policy 635–641.
26 IRENA, '*Off-Grid Renewable Energy Solutions: Global and Regional Status and Trends* (IRENA Abu Dhabi 2018) 1.
27 With apologies to Thomas Hobbes. See Thomas Hobbes, *Leviathan* (Richard Tuck ed, Cambridge University Press 1991).
28 Benjamin K Sovacool, 'Deploying Off-Grid Technology to Eradicate Energy Poverty' (2012) 338 Science 47 at 47.
29 United Nations Department of Economic and Social Affairs, 'Goal 7: Ensure Access to Affordable, Reliable, Sustainable and Modern Energy for All' (Department of Economic and Social Affairs) <https://www.un.org/en/chronicle/article/goal-7-ensure-access-affordable-reliable-sustainable-and-modern-energy-all#:~:text=SDG%207%E2%80%94to%20%E2%80%9Censure%20access,statement%20of%20the%20goal%20itself> accessed 11 July 2022.
30 United Nations Department of Economic and Social Affairs, 'Accelerating SDG7 Achievement, Policy Brief 24, Energy Sector Transformation: Decentralized Renewable Energy for Universal Energy Access' (2018) 3 <17589PB24.pdf (un.org)> accessed 11 July 2022.

31 Carlo Vezzoli and others, *Designing Sustainable Energy for All: Sustainable Product-Service System Design Applied to Distributed Renewable Energy* (Springer International Publishing AG 2018) 23.

32 United Nations Department of Economic and Social Affairs, 'Accelerating SDG7 Achievement, Policy Brief 24, Energy Sector Transformation: Decentralized Renewable Energy for Universal Energy Access' (2018) 3 <17589PB24.pdf (un.org)> accessed 11 July 2022.

33 Morgan Brazilian and others, 'More Heat and Light' (2010) 38(10) Energy Policy 5409 at 5411.

34 IRENA, *Off-Grid Renewable Energy Solutions: Global and Regional Status and Trends* (IRENA Abu Dhabi 2018) 1.

35 Ibid.

36 Aslan Alper and Ocal Oguz, 'The Role of Renewable Energy Consumption in Economic Growth: Evidence from Asymmetric Causality' (2016) 60 Renewable and Sustainable Energy Reviews 953–959.

37 IRENA, *Renewable Energy Benefits: Measuring the Economics* (IRENA Abu Dhabi 2016).

38 OECD, 'OECD Green Growth Studies: Energy' (2011) <ACKNOWLEDGEMENTS (oecd.org)> accessed 11 July 2022. See also Steve Keen, Robert U Ayres and Russell Standish, 'A Note on the Role of Energy in Production' (2019) 157 Ecological Economics 40–46; Zahid Asghar, 'Energy–GDP Relationship: A Causal Analysis for the Five Countries of South Asia' (2008) 8–1 Applied Econometrics and International Development 167–180.

39 IRENA, *Renewable Energy Benefits: Measuring the Economics* (IRENA Abu Dhabi 2016).

40 Muhammad Shahbaz and others, 'The Effect of Renewable Energy Consumption on Economic Growth: Evidence from the Renewable Energy Country Attractive Index' (2020) 207 Energy 118162.

41 Fangming Xie and others, 'How to Coordinate the Relationship between Renewable Energy Consumption and Green Economic Development: From the Perspective of Technological Advancement' (2020) 32 (71) Environmental Sciences Europe 1–15

42 ILO, 'Green Jobs and Renewable Energy: Low Carbon, High Employment' <https://www.ilo.org/wcmsp5/ groups/public/---ed_emp/---emp_ent/documents/publication/wcms_250690.pdf> accessed 11 July 2022.

43 Ibid.

44 IRENA, 'Off-Grid Renewable Energy Solutions to Expand Electricity Access: An Opportunity not to Be Missed' <https://www.irena.org/-/media/Files/IRENA/Agency/Publication/2019/Jan/IRENA_Off-grid_RE_Access_2019.pdf> accessed 11 July 2022.

45 Vivid Economics, 'Off-Grid Solar: A Growth Engine for Jobs' <https://www.gogla.org/sites/default/files/resource_docs/gogla_off_grid_solar_a_growth_engine_for_jobs_web_opt.pdf> accessed 11 July 2022.

46 U.S. Bureau of Labor Statistics, 'Fastest Growing Occupations' <https://www.bls.gov/ooh/fastest-growing.htm> accessed 11 July 2022.

47 Emily Folk, 'The Many Economic Benefits of Renewable Energy' <https://www.renewableenergy magazine.com/emily-folk/the-many-economic-benefits-of-renewable-energy-20190312> accessed 11 July 2022.

48 Ibid.

49 Peter Kayode Oniemola, 'Powering Nigeria through Renewable Electricity Investments: Legal Framework for Progressive Realization' (2015) 6(1) Journal of Sustainable Development Law and Policy 83–108.

50 See Ngozi Chinwa Ole, 'The Nigerian Electricity Regulatory Framework: Hotspots and Challenges for Off-Grid Renewable Electricity Development' (2020) 38(4) Journal of Energy & Natural Resources Law 367.

51 Ngozi Chinwa Ole, 'The Nigerian Electricity Regulatory Framework: Hotspots and Challenges for Off-grid Renewable Electricity Development' (2020) 38(4) Journal of Energy & Natural Resources Law 367 at 369.
52 See generally Chitzi C. Ogbumgbada, 'Developing an Effective Legal Framework for Renewable Energy Utilization in Nigeria' (2018) 8(3) Renewable Energy Law and Policy Review 45 at 51.
53 See, e.g., Ngozi Chinwa Ole, 'The Nigerian Electricity Regulatory Framework: Hotspots and Challenges for Off-Grid Renewable Electricity Development' (2020) 38(4) Journal of Energy & Natural Resources Law 367.
54 Michaël Aklin, 'The Off-Grid Catch-22: Effective Institutions as a Prerequisite for the Global Deployment of Distributed Renewable Power' (2020) Energy Research and Social Science, *Forthcoming* <https://papers.ssrn.com/sol3/papers.cfm?abstract_id=3717818> accessed 11 July 2022.
55 Peter Kayode Oniemola, 'Powering Nigeria through Renewable Electricity Investments: Legal Framework for Progressive Realization' (2015) 6(1) Journal of Sustainable Development Law and Policy 83–108.
56 Ngozi Chinwa Ole, 'The Nigerian Electricity Regulatory Framework: Hotspots and Challenges for Off-Grid Renewable Electricity Development' (2020) 38(4) Journal of Energy & Natural Resources Law 367.
57 Union of Concerned Scientists, 'Barriers to Renewable Energy Technologies' (2017) <https://www.ucsusa.org/resources/barriers-renewable-energy-technologies> accessed 11 July 2022.
58 Centre for Sustainable Energy, 'How Much Does a Typical Residential Solar Electric System Cost?' (2022) <https://sites.energycenter.org/solar/homeowners/cost> accessed 11 July 2022.
59 Union of Concerned Scientists, 'Barriers to Renewable Energy Technologies' (2017) <https://www.ucsusa.org/resources/barriers-renewable-energy-technologies> accessed 11 July 2022.
60 Ibid.
61 Seetharaman and others, 'Breaking Barriers in Deployment of Renewable Energy' (2019) 5(1) Heliyon e01166.
62 Noah Browning, 'World Must Triple Clean Energy Investment by 2030 to Curb Climate Change – IEA' <Investment in renewable energy needs to triple, say IEA | World Economic Forum (weforum.org)> accessed 11 July 2022.
63 Dahiru Abdullahi and others, 'Barriers for Implementing Solar Energy Initiatives in Nigeria: An Empirical Study' (2021) Smart and Sustainable Built Environment (*advance publication*).
64 Ibid.
65 Inken Hoeck and others, 'Challenges for Off-Grid Electrification in Rural Areas. Assessment of the Situation in Namibia Using the Examples of Gam and Tsumkwe' (2021) Energy, Ecology and Environment.
66 IRENA, *Off-Grid Renewable Energy Solutions: Global and Regional Status and Trends* (IRENA Abu Dhabi 2018) 1.
67 Ibid.
68 Ibid.
69 Ibid, 2.
70 REN21, 'Renewables 2021: Global Status Report' (2021) (Paris: REN21 Secretariat) <GSR2021_Full_Report.pdf (ren21.net)> accessed 11 July 2022.

Bibliography

Abdullahi D and others, 'Barriers for Implementing Solar Energy Initiatives in Nigeria: An Empirical Study' (2021) Smart and Sustainable Built Environment, *ahead of print.*

Aklin M, 'The Off-Grid Catch-22: Effective Institutions as a Prerequisite for the Global Deployment of Distributed Renewable Power' (2020) Energy Research and Social Science, *Forthcoming* <https://papers.ssrn.com/sol3/papers.cfm?abstract_id=3717818> accessed 11 July 2022.

Alper A and Oguz O, 'The Role of Renewable Energy Consumption in Economic Growth: Evidence from Asymmetric Causality' (2016) 60 Renewable and Sustainable Energy Reviews 953–959.

Asghar Z, 'Energy–GDP Relationship: A Causal Analysis for the Five Countries of South Asia' (2008) 8–1 Applied Econometrics and International Development 167–180.

Bhandari B and Ahn S-H, 'Off-grid Hybrid Renewable Energy Systems and Their Contribution to Sustainable Development Goals' in Asmae Berrada and Rachid El Mrabet (eds), *Hybrid Energy System Models* (Academic Press 2021).

Bos K, Chaplin D, and Mamun A, 'Benefits and Challenges of Expanding Grid Electricity in Africa: A Review of Rigorous Evidence on Household Impacts in Developing Countries' (2018) 44 Energy for Sustainable Development 64–77.

Brazilian M and others, 'More Heat and Light' (2010) 38(10) Energy Policy 5409–5412.

Browning N, 'World Must Triple Clean Energy Investment by 2030 to Curb Climate Change – IEA' <Investment in renewable energy needs to triple, say IEA | World Economic Forum (weforum.org)> accessed 11 July 2022.

Cameron P D, Mu X and Roeben V (eds), *The Global Energy Transition: Law, Policy and Economics for Energy in the 21st Century* (Hart Publishing 2020).

Centre for Sustainable Energy, 'How Much Does a Typical Residential Solar Electric System Cost?' (2022) <https://sites.energycenter.org/solar/homeowners/cost> accessed 11 July 2022.

Foley G, *The Energy Question* (4th edn, Penguin Books Ltd, 1992) 3 quoted in Omorogbe Y, 'Promoting Sustainable Development Through the Use of Renewable Energy: The Role of Law' in Zillman D and others (eds), *Beyond the Carbon Economy: Energy Law in Transition* (OUP 2008).

Folk E, 'The Many Economic Benefits of Renewable Energy' <https://www.renewableenergymagazine.com/emily-folk/the-many-economic-benefits-of-renewable-energy-20190312> accessed 11 July 2022.

Goldblatt D L, *Sustainable Energy Consumption and Society: Personal, Technological, or Social Change* (Springer 2005).

Hobbes T, *Leviathan* (Richard T ed, Cambridge University Press 1991).

Hoeck I and others, 'Challenges for Off-grid Electrification in Rural Areas. Assessment of the Situation in Namibia Using the Examples of Gam and Tsumkwe' (2021) Energy, Ecology and Environment 508–522.

IEA, 'Global Population without Access to Electricity by Region, 2000–2021' (IEA 2021) <https://www.iea.org/data-and-statistics/charts/global-population-without-access-to-electricity-by-region-2000-2021-2> accessed 11 July 2022.

IEA, 'SDG7: Data and Projections' <SDG7Database-Dataproduct-IEA> accessed 11 July 2022.

ILO, 'Green Jobs and Renewable Energy: Low Carbon, High Employment' <https://www.ilo.org/wcmsp5/groups/public/---ed_emp/---emp_ent/documents/publication/wcms_250690.pdf> accessed 11 July 2022.

IPCC, 'Global Warming of 1.5°, an IPCC Special Report on the Impacts of Global Warming of 1.5° Above Pre-industrial Levels and Related Global Greenhouse Gas Emission Pathways, in the Context of Strengthening the Global Response to the Threat of Climate Change, Sustainable Development, and Efforts to Eradicate Poverty' (2018).

IRENA, 'Energy Transition' <EnergyTransition(irena.org)> accessed 11 July 2022.

IRENA, *Off-grid Renewable Energy Solutions: Global and Regional Status and Trends* (IRENA Abu Dhabi 2018).

IRENA, 'Off-grid Renewable Energy Statistics 2019' <Off-gridrenewableenergystatistics2019(irena.org)> accessed 11 July 2022

IRENA, *Off-grid Renewable Energy Systems: Status and Methodological Issues* (IRENA, Abu Dhabi 2015).

IRENA, *Renewable Energy Benefits: Measuring the Economics* (IRENA, Abu Dhabi, 2016).

Keen S, Ayres R U and Standish R, 'A Note on the Role of Energy in Production' (2019) 157 Ecological Economics 40–46.

Kirubi C and others, 'Community-Based Electric Micro-Grids Can Contribute to Rural Development: Evidence from Kenya' (2009) 37(7) World Development 1208–1221.

Malhotra R (ed), *Fossil Energy* (2nd edn Springer, 2020).

Munro P, Horst van der G, and Healy S, 'Energy Justice for All? Rethinking Sustainable Development Goal 7 Through Struggles Over Traditional Energy Practices in Sierra Leone' (2017) 105 Energy Policy 635–641.

Ngowi J M, Bångens L and Ahlgren E O, 'Benefits and Challenges to Productive Use of Off-grid Rural Electrification: The Case of Mini-hydropower in Bulongwa-Tanzania' (2019) 53 Energy for Sustainable Development 97–103.

OECD, 'OECD Green Growth Studies: Energy' (2011) <ACKNOWLEDGEMENTS(oecd.org)> accessed 11 July 2022.

Ogbumgbada C C, 'Developing an Effective Legal Framework for Renewable Energy Utilization in Nigeria' (2018) 8(3) Renewable Energy Law and Policy Review 45–52.

Ole N C, 'The Nigerian Electricity Regulatory Framework: Hotspots and Challenges for Off-grid Renewable Electricity Development' (2020) 38(4) Journal of Energy & Natural Resources Law 367–390.

Omorogbe Y, 'Legal Dimensions of Access to Modern Energy Services in Africa: Lessons from Nigeria, Ghana, and Rwanda' in Íñigo del Guayo et al (eds), *Energy Justice and Energy Law* (OUP 2020).

Oniemola P K, 'Powering Nigeria Through Renewable Electricity Investments: Legal Framework for Progressive Realization' (2015) 6(1) Journal of Sustainable Development Law and Policy 83–108.

Rajagopalan S, 'Who Benefits and How? A Capabilities Perspective on Solar Micro-grids in India' (2021) Journal of Human Development and Capabilities 316–335.

REN21, 'Renewables 2021: Global Status Report' (2021) (Paris: REN21 Secretariat) <GSR2021_Full_Report.pdf (ren21.net)> accessed 11 July 2022.

Seetharaman and others, 'Breaking Barriers in Deployment of Renewable Energy' 5(1) Heliyon e01166 (2019) 1–23.

Segger M-C and Khalfan A, *Sustainable Development Law: Principles, Practices, & Prospects* (OUP 2004).

Shahbaz M and others, 'The Effect of Renewable Energy Consumption on Economic Growth: Evidence from the Renewable Energy Country Attractive Index' (2020) 207 Energy 1–14.

Sovacool B K, 'Deploying Off-Grid Technology to Eradicate Energy Poverty' (2012) 338 Science 47–48.

United Nations Department of Economic and Social Affairs, 'Accelerating SDG7 Achievement, Policy Brief 24, Energy Sector Transformation: Decentralized Renewable Energy for Universal Energy Access' (2018) 3 <17589PB24.pdf(un.org)> accessed 11 July 2022.

United Nations Department of Economic and Social Affairs, 'Goal 7: Ensure Access to Affordable, Reliable, Sustainable and Modern Energy for All' <Goal 7 | Department of Economic and Social Affairs (un.org)> accessed 11 July 2022.

Union of Concerned Scientists, 'Barriers to Renewable Energy Technologies' (2017) <https://www.ucsusa.org/resources/barriers-renewable-energy-technologies> accessed 11 July 2022.

U.S. Bureau of Labor Statistics, 'Fastest Growing Occupations' <https://www.bls.gov/ooh/fastest-growing.htm> accessed 11 July 2022.

Vezzoli C and others, *Designing Sustainable Energy for All: Sustainable Product-Service System Design Applied to Distributed Renewable Energy* (Springer International Publishing AG 2018).

Vivid Economics, 'Off-Grid Solar: A Growth Engine for Jobs' <https://www.gogla.org/sites/default/files/resource_docs/gogla_off_grid_solar_a_growth_engine_for_jobs_web_opt.pdf> accessed 11 July 2022.

Xie F and others, 'How to Coordinate the Relationship Between Renewable Energy Consumption and Green Economic Development: From the Perspective of Technological Advancement' (2020) 32(71) Environmental Sciences Europe 1–15.

4 The International Climate Change Regime and Off-Grid Renewable Electricity

Ngozi Chinwa Ole, Empire Hechime Nyekwere, Cosmos Nike Nwedu, Kingsley Osinachi Onu, and Adesola Omotola

4.1 Introduction

There is now a universal consensus and a broad scientific agreement that climate change is a global problem that needs urgent international attention and response. What once was called 'climate change' is now indeed a 'climate crisis', and what was once called 'global warming' has more accurately become 'global heating'.[1] The statement by the US Global Research Programme and the American Geophysical Union shows the general scientific consensus of the reality of climate change and the human role in its current pattern:

> Evidence for climate change abounds, from the top of the atmosphere to the depth of the oceans. Scientists and engineers from around the world have meticulously collected this evidence, using satellites and networks of weather balloons, observing and measuring changes in location and behaviours of species and the functioning of ecosystems. Taken together, this evidence tells an unambiguous story: the planet is warming, and over the half-century, this warming has been driven primarily by human activity. Humanity is the significant influence on the global climate change observed over the past 50 years. Rapid societal responses can significantly lessen adverse outcomes.[2]

The electricity sector contributes the most to the global problem of climate change.[3] It is axiomatic that the use of fossil fuel electricity technologies has resulted in, and exacerbated the global problem of, climate change.[4] Thus, addressing climate change entails a paradigm shift 'of the technological base for power-generation from fossil fuels to renewable energy resources'.[5] In the context of remote rural areas and islands, off-grid renewable electricity remains the way forward to address climate change and provide a low carbon future regarding environmental friendliness, cost, and availability.[6]

While renewable energy development, including OGRE, has grown significantly, fossil fuel is still dominating the global energy mix, accounting for 84%.[7] The latter is irrespective of the recognised role of renewable energy in

DOI: 10.4324/9781003178088-5

mitigating climate change and its superior benefits in attaining energy security.[8] The limited development of renewable energy is mostly in developed countries, save for China and India.[9] Several barriers impede the development of renewable electricity, including OGRE limiting its development to less than what it should and can be.[10] The primary barrier to its development is that its initial capital cost is remarkably higher than competing fossil fuel electricity technologies.[11] As such, fossil fuel electricity technologies still remain more attractive to investors.[12] The problem of the initial capital cost is further complicated by the absence of a level playing field between fossil fuel and renewable electricity.[13] In the words of Hart, 'fossil fuel and other established sources of energy retain the accumulated advantages of decades of public support in the form of subsidies, loans, guarantees and various favourable laws and regulations'[14] In addition, they have enjoyed long-term experience with usage, which all translates to cheaper costs.[15] Another significant barrier to developing renewable electricity, including OGRE, is the absence of the needed technological capacity to manufacture and maintain it.[16]

The international climate change regime is a conglomerate of the United Nations Framework Convention on Climate Change (UNFCCC) 1992, the Kyoto Protocol 1997, and the Paris Climate Change Agreement 2015.[17] It aims to address the global problem of climate change to attain the 'stabilisation of greenhouse gas concentrations in the atmosphere at a level that would prevent dangerous anthropogenic interference with the climate system'.[18] Given that climate change is a strong impetus for developing renewable energy, the regime is expected to be a strong driver for the development of OGRE.[19] Against this background, this chapter analyses the role of international climate change in driving renewable energy development, including OGRE. It argues that while the international climate change regime has contributed positively in driving the development of renewable energy, including OGRE, several factors limit its effect. These factors include, among other things, the lack of a definition of environmentally sound technologies (ESTs), the adoption of the concessionary form of climate finance, lack of a concrete obligation to transfer ESTs. It was further argued that while subsequent meetings of the Conference of the Parties (COP) including the Glasgow 26th Meetings of the COP, have acknowledged some of the identified lapses: they are yet to be addressed. It concludes that the Paris Climate Change Agreement 2015 remains a progressive document. Thus, there are possibilities that it will be fine-tuned to appropriately drive the development of renewable energy, including OGRE. The chapter is divided into four sections. Section 4.2 contains an analysis of the extent to which the provisions of the international climate change regime is a driver of renewable energy including OGRE in member states. Section 4.3 discusses the extent to which the support mechanisms have promoted the development of OGRE by addressing the capacity and financial barriers to it in developing countries. Section 4.4 concludes the chapter with key recommendations on how the international climate change regime will be strengthened to sustain OGRE.

4.2 International Climate Change: Drive to Develop OGRE

4.2.1 Introduction

The United Nations Framework Convention on Climate Change (UNFCCC) 1992 was the first international instrument to address the global problem of climate change.[20] In line with its overall objective, it encourages member states to cooperate in developing environmentally sound technologies.[21] While the UNFCCC did not define environmentally sound technologies, in the third meeting of the Conference of Parties (COP), renewable energy was mentioned as the best option in the context.[22] Regrettably, the UNFCCC did not record remarkable progress in driving the development of renewable energy. The reasons for the latter include the non-mandatory nature of provisions relevant to the development of OGRE, the absence of the definition of environmentally sound technologies, and the dependency of developing countries on the developed ones.[23] The Kyoto Protocol was adopted under the auspices of the UNFCCC 1992. Like the UNFCCC, it failed to drive the development of renewable electricity, including OGRE, because of several reasons, including an express recognition of renewable energy as a priority in the context of addressing climate change. Another flaw was the exclusion of developing countries from their commitments.[24] For this reason, Ferrey remarks 'the first problem is that Kyoto is not shifting the world's energy base to renewable power in lieu of fossil-fuel–fired power resources'.[25] What is more, the commitment period under the Kyoto Protocol ends in 2020. In addition to the failure of the UNFCCC and Kyoto Protocol to drive the development of renewable energy, 'they were widely regarded as inadequately implemented'.[26] Against this background, the Paris Climate Change Agreement was adopted in 2015.[27] Its provisions will be analysed in line with the central research question.

4.2.2 Obligation to Develop OGRE

The Paris Climate Change Agreement 2015 aims 'to strengthen the global response to the threat of climate change in the context of sustainable development'.[28] It sets a target of holding the temperature under 2°C and pursuing measures to limit the temperature.[29] In the context of the mentioned target, it provides that 'each party shall prepare, communicate and maintain successive nationally determined contributions that it intends to achieve. Parties shall pursue domestic mitigation measures, with the aim of achieving the objectives of such contributions'.[30] Article 2 contains the overall objectives of the Agreement to control the global temperature to well below 2°C above pre-industrial levels and pursue efforts to limit the temperature increase to 1.5°C above pre-industrial levels, recognising that this would significantly reduce the risks and impacts of climate change.[31] To achieve the latter, member states are obligated to adopt low greenhouse gas emission development strategies and communicate them in their National Determined Contributions (NDCs).[32] The NDCs set out targets, policies, and targets that the government aims to implement in response to climate change and contribute to global climate action. Key to these contributions is the concept of national determination, meaning that the plans are based on national contexts, needs, and preferences.[33]

It is apt to mention that there is no definition of low greenhouse gas emission strategies. Regardless, academic literature, policies, and stakeholders' comments in the international climate change regime lend credence to the conclusion that renewable energy, including OGRE, is at the very front of it.[34] In the just-concluded 26th Meeting of the COP, in Glasgow COP 2021, the UN Climate Change Executive Secretary acknowledged that the accelerated development of renewable energy is at the centre of low greenhouse gas emission strategies.[35] It is not surprising that most of the NDCs commit to developing renewable energy in one form or the other. In the words of the International Renewable Energy Agency (IRENA), 'Renewable energy features prominently in most of these NDCs, confirming that the transition to a renewable energy future has come to be recognised globally as central to addressing climate change'.[36] True to these words, out of the 188 Parties that had submitted NDCs as of early December 2020, 170 (or 90% of the total) mentioned renewables, while 134 (or 71%) included quantified renewable energy targets (OGRE in some context) for electricity generation in their NDCs.[37] Thus, it is right to say that the mentioned provisions of the Paris Agreement 2015 has pushed the development of OGRE in the context of where it is needed. Implementing the relevant provisions of the NDC would go a long way in engineering a paradigm shift away from these GHGs emission pathways in developing countries.[38]

However, the absence of a definition of low greenhouse development strategies to explicitly be renewable electricity reduces the effectiveness of this provision in driving the development of OGRE. While member states have elected and are developing renewable electricity, including OGRE as their low greenhouse gas emission strategies, they have also pledged to develop more fossil fuel electricity options.[39] While the fossil fuel strategies are termed more 'efficient and cleaner' compared to older generations of the same, they are still more damaging to the environment than renewable electricity, including OGRE.[40] Notably, every fossil fuel electricity that will be developed implies less room for developing renewable options, including OGRE. The latter is regardless of the debilitating effects of the continuous use of fossil fuel electricity options.[41] In the just-concluded COP Meeting 2021, it was acknowledged that the development of renewable energy (including OGRE) is still not as it should be in the light of climate change mitigation.[42]

To crown it all, while the obligation to develop low greenhouse gas emission strategies has necessitated a pledge to develop renewable electricity in various NDCs, we argue that it is whittled down by the absence of a definition of low greenhouse development strategies.

4.3 Support Mechanisms Relevant to OGRE

4.3.1 Introduction

There are three support mechanisms established under the Paris Climate Change Agreement: financial, technological, and capacity building. This section discusses the support these mechanisms provide to environmentally sound projects, how this support positively affects the development of OGRE, and the limitations of this mechanism in supporting the development of OGRE.

4.3.2 The Financial Mechanism

The Paris Climate Agreement contains provisions on financial support for developing climate-friendly technologies such as OGRE in developing member states. The Agreement provides that 'developed country parties shall provide financial resources to assist developing country parties with respect to both mitigation and adaptation in continuation of their existing obligations under the UNFCCC'.[43]Such support for mitigation includes implementing low greenhouse gas emission strategies adopted by developing member states to mitigate climate change. There is no definition of low greenhouse gas emission under the Agreement. However, the Paris Decision (the Decision of the COP adopting the Paris Agreement) contains an indication that renewable energy, including OGRE, has a role to play in these low greenhouse gas strategies. The Decision provides that the COP acknowledges 'the need to promote universal access to sustainable energy in developing member states, particularly Africa, through *the enhanced deployment of renewable energy*'.[44] This means that it is expected that such financial support would contribute to the development of renewable energy, including OGRE, in developing countries. The Global Environment Facility (GEF)[45] and the Green Climate Fund (GCF)[46] are the financial mechanisms designated under the Paris Agreement to disburse these financial resources.[47] Since the GEF and GCF adoption as these financial mechanisms, they have been disbursing counterpart funds to sponsor low carbon projects, including OGRE.[48] Some of the projects are geared towards providing resilience to climate change.[49] As such, one can safely say that the provisions on disbursal of financial assistance have positively affected the development of OGRE.

In addition to this financial support, the Agreement mandates developed countries to mobilise climate funding from various sources. It provides that developed member countries should continue to take the lead in mobilising climate finance from a wide variety of sources, instruments, and channels, noting the significant role of public funds, through various actions, taking into account the needs and priorities of developing country Parties.[50]

It further enjoins developing countries who can do so to join in voluntarily in the mobilisation of climate funds.[51] The annual target of such mobilisation is a floor of USD 100 billion by 2020.[52] This target is expected to be achieved by voluntary pledges from member states.[53] Commenting on this, Zahar writes that 'as with mitigation effort itself, the plan for raising climate finance is voluntary pledging, plus peer pressure'.[54] The Agreement requires developed member states to report their contributions to such mobilisation of climate finance.[55] They can justify such contributions to climate finance by showing that it is just and fair based on their national circumstances.[56]

This obligation on developed member states to mobilise climate finance increases the finances available for the OGRE project in developing countries.

The nature of this mobilisation is not the traditional climate finance aid; it is expected to be in various forms, including private and public foreign investment packages.[57] There is also an expectation that such flow of finance would consider the peculiarity of the national circumstances of developing member states.[58] This entails a consideration of the peculiar impediments to the development of climate-friendly technologies such as OGRE in developing member countries. Some bodies including the World Bank, the International Renewable Energy Agency, and the African Development Bank, are making their flow of finances align with this goal of galvanising support for ESTs, including renewable electricity.[59] There is evidence that the latter is connected to the provisions of the Paris Agreement. For instance, the World Bank alone has mobilised and disbursed USD 30 billion to low carbon projects in furtherance to the provisions of the Paris Agreement.[60] It is anticipated that private and public actors would make more finances available to benefit renewable electricity, including OGRE, in developing countries.

Though the Paris Agreement impacts positively on the development of renewable electricity, some factors impede its contribution to addressing the above-mentioned financial barriers to the development of OGRE. The Paris Decision adopts concessionary funding as the definition of financial resources under the Agreement.[61] The Paris Agreement provides that 'the financial mechanism of the convention … shall serve as the financial mechanism of this Agreement'.[62] The implication of concessionary funding is that an investor seeking to benefit from the financial mechanism must show evidence of possessing the minimum capital required for the cheapest energy project in such a category.[63] The mechanism will only give the funding needed to translate such a project into a clean energy one.[64] Thus, for the investors in developing countries struggling with the affordability and accessibility of the capital needed to develop OGRE, the financial assistance disbursed by the financial mechanism is of no help.[65] The mentioned provision of the Agreement is constrained to the extent to which it can help remove the affordability and accessibility of the initial capital cost of OGRE projects in developing countries. In the 26th COP Meeting 2021, some developing countries mentioned the inability of investors to access the concessionary form of climate finance as a problem.[66] As a result, the COP called on the financial mechanism to streamline the eligibility criteria to incorporate the differing needs of developing countries.[67]

Furthermore, the obligation on developed member states to take the lead in mobilising climate finance is recommendatory and, therefore, non-binding. The Agreement provides that 'developed member countries *should* continue to take the lead in mobilising climate finance from a wide variety of sources … taking into account the needs and priorities of developing country Parties'. The word *should* in the above provision means that there is no legal obligation created on developed member states to take the lead in mobilising climate finance (a minimum of USD 100 billion a year). Commenting on this, Rajamani writes that the use of the word 'should' in any provision of the Paris Agreement means that a party is merely encouraged rather than mandated to fulfil the obligation created in the provision.[68] Bodansky also writes that the particular character of a provision is usually determined by the choice of verb: for example, 'shall' generally denotes that a provision in a treaty creates a legal obligation, 'should' (and to a lesser degree, 'encourage') that the provision is a recommendation, 'may' that it creates a licence or permission.[69]

In the light of the recommendatory nature of this obligation, the extent to which they can affect the development of OGRE is constrained.[70] The non-binding nature of this provision implies that it is entirely up to individual developed states to voluntarily join in the mobilisation of climate finance. This renders the possible outcome of this provision uncertain, given that it is not possible to envisage the direction of such voluntary behaviour of developed sovereign states and private actors. In the COP Meeting 2021, it was noted 'with deep regret that the goal of developed country parties to mobilise jointly USD 100 billion per year by 2020 in the context of meaningful mitigation actions and transparency on implementation has not yet been met'.[71] The latter may be averted if there was some element of bindingness in the relevant provision.

4.3.3 The Technology Mechanism[72]

The Paris Agreement 2015 contains provisions on technology transfer. The Agreement recognises the role of ESTs in mitigating climate change. As such, it provides that 'parties, noting the importance of technology for the implementation of mitigation and adaptation actions ... and recognising existing technology deployment and dissemination efforts, shall strengthen cooperative action on technology development and transfer'.[73] It provided the basis for the adoption of a framework for technology transfer.[74]

However, the provision of the Paris Agreement 2015 on technology transfer is whittled down considerably by the absence of an express obligation on developed member states to transfer such ESTs, including OGRE, to developing member countries. The Agreement provides that 'Parties ... shall strengthen cooperative action on technology development and transfer'.[75] The obligation of developed member states, just like other member parties, is limited to cooperating with other member states to achieve technology development, transfer, and support for developing member states to achieve this cooperation. One dimension of such obligation on member states to cooperate in developing and transferring ESTs is that it introduces competition.

Cedric noted that under such cooperative measures, countries 'are also interested in helping their own companies to take the lead in the international economic competition'. Hence, such transfer of ESTs goes typically to developing countries with good investment climates where such companies would retain their competitive edge while making adequate returns for their investments.[76] Parties from developing countries that have a poor investment climate[77] cannot compete favourably with other developing countries with a good investment climate to attract such technology transfer. The absence of such express obligation on developed countries to transfer ESTs to developing member countries affects its effectiveness in this context.

Furthermore, the Paris Agreement 2015 provides that 'support, including financial support, shall be provided to developing country parties to implement this Article, including for strengthening cooperative action on technology development and transfer at different stages of the technology cycle'.[78] This provision has been interpreted to mean that the obligation is opened to developed and developing countries to provide support for developing countries voluntarily. The rationale for leaving this obligation open to developing countries is that patenting and development of ESTs occur in some developing countries like China, India, the Philippines, and Brazil. Thus, they are able to provide support to other developing countries on technology transfer and development.[79] This wider ambit of those under the obligation on technology transfer of OGRE makes for greater possibilities that the same will be transferred to several other developing countries. While there has been evidence of the flow of transfer of renewable electricity technology to developing countries, it is not quite clear whether they are driven by market forces or by the provisions of the Paris Agreement or both.[80] The global stocktake is scheduled to commence this year – 2021 through to 2023.[81] It will provide a collective assessment of progress made in implementing the Agreement, including the Technology Mechanism provisions.[82] Presumably, the overall effect of the mentioned provisions on the influx of renewable electricity technology to developing countries will be known.

4.3.4 The Capacity-Building Mechanism[83]

The Paris Agreement 2015 contains provisions on capacity building. The Agreement provides that 'developed country parties should enhance support for capacity-building actions in developing country Parties'.[84] The support is defined to be to the extent that should enhance the capacity and ability of developing country Parties … to take effective climate change action, including, inter alia, to implement adaptation and mitigation actions, and should facilitate technology development, dissemination and deployment, access to climate finance, relevant aspects of education, training and public awareness, and the transparent, timely and accurate communication of information.[85]

It is firmly expressed that such capacity building should be responsive to the national needs of individual developing member countries.[86] The Agreement provides for the establishment of an institution that will enhance capacity-building activities.[87] Pursuant to the latter provision, the Paris Agreement Committee on Capacity Building (PCCB) was adopted in the 22nd Meeting of the COP in Marrakesh 2016.[88] Among other things, it has the mandate to identify and address the gaps 'both current and emerging, in implementing capacity building in developing country parties and further enhancing capacity-building efforts'.[89] One way of addressing the gap is to make recommendations to the COP which will inform the future architecture of the Paris Agreement.[90]

Some authors expressed optimism that the mentioned provisions would drive the development of endogenous capacity in renewable electricity technologies in developing member states.[91] True to their words, the PCCB has worked assiduously towards building the capacity in developing countries.[92] Some of its notable achievements include sharing information on best practices with respect to capacity building in addressing climate change.[93] This aims to provide the platform for developing countries to extrapolate relevant lessons that will enable them to build the needed capacity.[94] It also brings private and public entities together to collaborate on various projects, culminating in capacity building in developing countries. In the COP 26th Meeting, 2021, the PCCB presented a work plan for the period 2021–2024, which the parties adopted.[95] The work plan reinforces that the PCCB will continue to improve past initiatives in the area of capacity building.[96]

Regardless, the absence of a clear goal on capacity building reduces the certainty as to the extent to which the Paris Agreement would help address the capacity barrier in the OGRE sector in developing countries. There is no clear goal on the extent to which developed member states shall contribute to the development of ESTs such as OGRE in developing countries through this capacity-building support. Commenting on this, Rajamani noted that 'while mitigation and adaptation (albeit qualitative) goals have been identified in the Agreement, there are no clearly identifiable goals in relation to … capacity building. This introduces an element of uncertainty in the assessment of progress'.[97] The implication is that while developed countries have and continue to support capacity building, including developing OGRE in developing countries, there is no obligation on them to provide the degree of support that would build the needed capacity to manufacture and maintain OGRE in developing countries in the light of addressing climate change.

Regrettably, it is not certain that the PCCB will provide the needed clarity on what is the goal of the capacity-building provisions of the Agreement. As reiterated, the PCCB has the mandate to identify the gaps in implementing the current capacity-building arrangements in developing member states. Its 2021–2023 workplan commits to the latter goal and proposes making recommendations to address it.[98] Thus, it is well within the orbits of the mandate of the PCCB to make recommendations to address the absence of a clear goal. On the one hand, it is also not certain that such a recommendation will be made, irrespective of the obvious call to that effect by existing literature. Assuming it is made, there is no guarantee that the COP will adopt the recommendation.

4.4 Conclusion

This work analysed the role of the international climate change regime in driving the development of OGRE. It finds that the Paris Climate Change Agreement has impacted positively on the development of OGRE. Its provisions on the adoption of low greenhouse gas emission strategies have crystallised into the adoption of OGRE strategies by most member states. The obligation on developed member states to provide financial resources to developing member states through the GCF and GEF has also helped address the problem of affordability and accessibility of the capital cost of OGRE projects. The recommendatory provision on the mobilisation of climate finance has galvanised affirmative action from some public and private entities, contributing to the acceleration of the development of OGRE. Since the Paris Agreement, there has been an influx of technology transfer to developing countries. One cannot discountenance that the Agreement has played a role in this development. It is also argued that its provisions are helping in facilitating the development of the capacity to manufacture and maintain OGRE in the context of climate change.

Regardless, its ability to drive the development of OGRE in the context where it is needed is constrained in several respects. The lack of definition of ESTs to mean or prioritise renewable electricity gives the leeway for member states to adopt and give financial assistance to fossil fuel strategies. As such, it is not surprising that the share of renewable electricity, including OGRE, is limited compared to fossil fuels amidst concerns about climate change. Among other things, it was argued that the definition of financial resources to mean concessionary funding had limited its impact in addressing the financial barrier to the development of renewable energy, including OGRE. Lack of a substantial obligation on capable member states to transfer ESTs to member states as well as an absence of a clear goal on capacity building was argued to affect the effectiveness of the Agreement in this context.

The Paris Agreement is a progressive document with many possibilities.[99] In the COP Meeting 2021, some of the mentioned lapses were discussed, like the problem of concessionary funding and the inability to meet the USD 100 billion per year goal for mobilisation of finance. In contrast, one is optimistic that there would be some affirmative actions to address the noted lapses. It is not certain that such actions may be adequate to address the noted gaps to the extent that they will garner the needed support for OGRE. In addition, some of the lapses noted in this work are yet to be identified by the COP like the lack of definition of ESTs to mean or prioritise renewable energy including OGRE. However, one can only express optimism that with time, they will identify and address it.

Notes

1 António Guterres' Address to the 74th Session of the UN General Assembly on 23 September 2019, New York City, USA <https://un.dk/navigation/about-the-un/%E2%80%8B%E2%80%8Bantonio-guterres-address-to-the74th-session-of-the-un-general-assembly> accessed 27 October 2021; U.N. Head Warns Leaders the World is Facing a 'Great Fracture' <https://time.co munited-nations-climate-summit-economy> accessed 27 November 2021.

2 Jonathan M Harris and Brian Roach, *Environmental and Natural Resource Economics: A Contemporary Approach* (4th Edition, Chapter 12-*The Economics of Global Climate Change*, Global Development and Environment Institute, Tufts University 2017); See American Geophysical Union, *Human-Induced Climate Change Requires Urgent Action* (American Geophysical Union 2014) <www.agu.org> accessed 27 October 2021; DR Reidmiller and others (eds), *Impacts, Risks, and Adaptation in the United States: Fourth National Climate Assessment*, Volume II: Report-in-Brief. (U.S. Global Change Research Program (USGCRP), Washington, DC, USA, 2018) p. 24.

3 Umair Shahzad, 'Global Warming: Causes, Effects and Solutions' (2015) 1(4) Durreesamin Journal 2. See also UNFCCC, Caring for Climate: A Guide to the Climate Change Convention and the Kyoto Protocol (United Nations Framework Convention on Climate Change 2003) p. 21 <https://unfccc.int/resource/docs/publications/car-

ing_en.pdf> accessed 9 March 2021; Kyoto Protocol <https://www.wikizero.com/en/Climate_Change-Kyoto_Protocol> accessed 4 November 2021; The United Nations Framework Convention on Climate Change & Kyoto Protocol: An introduction (Friends of the Earth, December 2008) p. 1; Global Warming|Globalization101 <http://www.globaliza tion101.org/global-warming/> accessed 4 November 2021.

4 Filip Johnsson and others, 'The Threat to Climate Change Mitigation Posed by the Abundance of Fossil Fuels' (2017) 19(2) Climate Policy 258.

5 Steven Ferrey, 'The Failure of International Global Warming Regulation to Promote Needed Renewable Energy' (2010) 19(2) Boston College Environmental Affairs Law Review 68.

6 Ngozi Chinwa Ole and Bukola Ruth Akinbola, 'Addressing the Capacity Deficiency in the Nigerian Off-Grid Renewable Electricity: The Place of the International Climate Change Regime' (2019) 2 Redeemer's University Law Journal (RUNLAWJ) 35–58:52–55.

7 Robert Rappier, 'Fossil Fuel Still Supply 84% of the World Energy' (2020) <https://www.forbes.com/sites/rrapier/2020/06/20/bp-review-new-highs-in-global-energy-consumption-and-carbon-emissionsin2019/#:~:text=The%20remainder%20of%20global%20energy,primary%20energy%20consumption%20in%202019.> accessed 1 November 2021. See also Pramod Thakur, *Advanced Mine Ventilation: Respirable Coal Dust, Combustible Gas and Mine Fire Control* (Elsevier 2019) 3.

8 Phebe Asantewaa Owusu and Samuel Asumadu-Sarkodie, 'A Review of Renewable Energy Sources, Sustainability Issues and Climate Change Mitigation' (2016) 3(1) Cogent Engineering 3.

9 Climate Council, '11 Countries Leading the Charge on Renewable Energy' (2019) <https://www.climatecouncil.org.au/11-countries-leading-the-charge-on-renewable-energy/> accessed 3 November 2021.

10 Ali Mostafaeipour and others, 'Identifying Challenges and Barriers for Development of Solar Energy by Using Fuzzy Best-Worst Method: A Case Study' (2021) 226 Energy 120355.

11 Seetharaman and others, 'Breaking Barriers in the Deployment of Renewable Energy' (2019) Heliyon e01166.

12 Theresa Smith, 'Fossil Fuels Still Get More Investment than Renewable Energy' (2020) <https://www.esi-africa.com/industry-sectors/generation/fossil-fuels-still-get-more-investment-than-renewable-energy/> accessed 1 November 2021.

13 Craig A Hart and Dominic Marcellino, 'Subsidies or Free Markets to Promote Renewables?' (2012) 3 RELP 3.

14 Ibid.

15 Ibid.

16 Dorcas Kariuki, 'Barriers to Renewable Energy Technologies Development' (2018) <https://www.energytoday.net/economics-policy/barriers-renewable-energy-technologies-development/> accessed 1 November 2021.

17 John Vogler, 'Energy, Climate Change, and Global Governance: The 2015 Paris Agreement in Perspective' in Debra J Davidson and Matthias Gross (eds), *Energy, Climate Change, and Global Governance: The 2015 Paris Agreement in Perspective* (Oxford University Press 2015) 18–24.

18 The United Nations Framework Convention on Climate Change 1992 (adopted 9 May 1992, entered into force 21 March 1994) FCCC/INFORMAL/84(UNFCCC).

19 The IPCC, *Renewable Energy Sources and Climate Change Mitigation* (IPCC 2011) 7.

20 R Schwarze, *Law and Economics of International Climate Change* (Springer 2013) 1.

21 The UNFCCC 1992 (n 18) Article 4 (1) (c).

22 Eija-Riitta Korhola, *The Rise and Fall of the Kyoto Protocol: Climate Change as a Political Process* (Helsinki 2014) 50.

23 Ngozi Chinwa Ole, 'The Paris Agreement 2015 as a Primer for Developing Nigerian Off-Grid Solar Electricity' (2018) 26(3) African Journal of International and Comparative Law 426–451.
24 Steven Ferrey, 'The Failure of International Global Warming Regulation to Promote Needed Renewable Energy' (n 5) 68.
25 Ibid.
26 Rajamani Lavanya, 'Ambition and Differentiation in the 2015 Paris Agreement: Interpretative Possibilities and Underlying Politics' (2016) 65 ICLQ 493 at 494.
27 The Paris Climate Change Agreement 2015 (adopted 12 December 2015, entry into force 4 November 2016) <https://www.un.org/en/climatechange/paris-agreement> accessed 5 November 2021. See also Melissa Denchak, *Paris Climate Agreement: Everything You Need to Know* (Natural Resources Defence Council (NRDC) 2021) <https://www.nrdc.org/stories/paris-climate-agreement-everything-you-need-know> accessed 12 November 2021.
28 The Paris Climate Change Agreement 2015, Article 2(1).
29 Ibid, Article 2(1)(a).
30 Ibid, Article 4(2).
31 Ibid, Article 2(1)(a).
32 Ibid, Article 4(2); See also Articles 4(3); 4(8); 4(9); 4(19).
33 UNFCCC, 'Technological Innovation for the Paris Agreement: Implementing Nationally Determined Contributions, National Adaptation Plans and Mid-Century Strategies' *Technical Executive Committee (TEC) Brief No.* 10 (United Nations Framework Convention on Climate Change 2017) 5.
34 See Anissa Suharsono, Neil McCulloch, Mostafa Mostafa, Richard Bridle, Lucky Lontoh and Philip Gass, *Getting to 23 Per Cent: Strategies to Scale Up Renewables in Indonesia* (The International Institute for Sustainable Development 2019) 1.
35 UNFCCC, '4 Key Achievements of COP26' (2021) <4 Key Achievements of COP26 | UNFCCC> accessed 29 November 2021.
36 IRENA, *Untapped Potential for Climate Action Renewable Energy in Nationally Determined Contributions* (IRENA 2017) 7.
37 IRENE, *Renewable Energy and Climate Pledges: Five Years after the Paris Agreement* (International Renewable Energy Agency, Abu Dhabi 2019) 2; See UNEP, *Renewable Energy and Energy Efficiency in Developing Countries: Contributions to Reducing Global Emissions* (Third Report, United Nations Environment Programme 2017) 14–15.
38 See Richard Baron, 'Energy Transition after the Paris Agreement: Policy and Corporate Challenges' *Background paper for the 34th Round Table on Sustainable Development*, 28–29 September 2016, Wind Europe Summit, Hamburg.
39 UNFCCC, 'National Determined Contributions' (2021) <https://unfccc.int/process-and-meetings/the-paris-agreement/nationally-determined-contributions-ndcs/nationally-determined-contributions-ndcs> accessed 7 November 2021.
40 Ngozi C Ole. 'Combating Electricity Poverty in Nigeria through Off-Grid Renewable Electricity: The Role of Financial Support under the International Climate Change Regime' (2017) 2(1) International Journal of Innovative Studies in Sociology and Humanities (IJISSH*)* 1–16, 6–8.
41 See, for example, Cervigni Raffaello, John Allen Rogers and Max Henrion (eds), *Low-Carbon Development: Opportunities for Nigeria*. Directions in Development (Washington, DC: World Bank 2013) 87–99; See Andrea A Eras-Almeida and Miguel A Egido-Aguilera, 'What Is Still Necessary for Supporting the SDG7 in the Most Vulnerable Contexts?' (2020) 12 Sustainability 7184.
42 UNFCCC, '4 Key Achievements of COP26' (2021) <4 Key Achievements of COP26 | UNFCCC> accessed 29 November 2021.
43 The Paris Agreement, Article 9(1).

44 See The Paris Decision 2015, FCCC/CP/2015/L.9/Para 59.

45 The GEF was established in 1989 by the World Bank and the International Monetary Fund as a credit-providing facility for environmental projects. See World Bank, *Summary Proceedings of Annual Meetings of the Board of Governors in Washington* (World Bank 1989) 79. It was adopted as the permanent financial mechanism of the UNFCCC by the COP to the UNFCCC in 1998.

46 The Green Climate Fund (GCF) was adopted as a financial mechanism of the UNFCCC in 2011 to support global efforts to respond to the challenge of climate change. GCF helps developing countries limit or reduce their greenhouse gas (GHG) emissions and adapt to climate change. It seeks to promote a paradigm shift to low-emission and climate-resilient development, taking into accounts the needs of developing countries that are particularly vulnerable to climate change impacts. See the Governing Instrument for the Green Climate Fund 2011 (adopted 11 December 2011), Decision 3/CP17/2011/9/Ad d.1 (GCF Instrument 2011), Preamble; See also, GCF in Brief: About the Fund <https://www.greenclimate.fund/sites/default/files/document/gcf-brief-about-fund_0.pdf> accessed 5 November 2021.

47 Baker and Mckenzie, 'The Paris Agreement: Putting the First Universal Climate Change Treaty in Context' (January 2016) <http://www.bakermckenzie.com/en/insight/publications/2016/01/the-paris-agreement--putting-the-first-univer sal__/> accessed 24 November 2021; See The Paris Agreement 2015, Article 9(3). See Daniel Bodansky, 'In Brief: Legal Options for U.S. Acceptance of A New Climate Change Agreement' (2015) <http://www.c2es.org/> accessed 21 November 2021.

48 Global Environment Facility, 'Projects' <https://www.thegef.org/projects> accessed 9 October 2021.

49 GCF, 'Projects' <https://www.greenclimate.fund/projects> accessed 9 November 2021.

50 Michael Westphal, 'Getting to $100 Billion: Climate Finance Scenarios and Projections to 2020' (May 2015) 2 <https://www.wri.org/sites/default/files/getting-to-100-billion-final.pdf> accessed 27 November 2021.

51 UNFCCC-COP Lima Call for Action Decision (adopted 11 December 2014) (Decision 1/cp.20)fccc/cp/2014/10/ Add.1, para 14.

52 Alexander Zahar, 'The Paris Agreement and the Gradual Development of a Law on Climate Finance' (2016) 6 Climate Law 75, 80.

53 The Paris Agreement 2015, Article 13(9).

54 Ibid, Article 9(5).

55 Robinson Meyer, 'A Readers Guide to the Paris Agreement' (December 2015) <http://www.theatlantic.com/science/archive/2015/12/a-readers-guide-to-the-paris-agreement/420345/> accessed 26th November 2021.

56 See The Paris Agreement 2015, Article 9(3).

57 Ibid.

58 The Paris Decision 2015, para 62.

59 World Bank, 'Climate Change' (2021) <https://www.worldbank.org/en/topic/climatechange> accessed 10th November 2021. See also the African Development Bank, 'Climate Change' (2021) <https://www.afdb.org/en/topics-and-sectors/sectors/climate-change> accessed 10 November 2021.

60 World Bank, 'World Bank Group President's Statement on Climate Change Action Plan' (2021) <https://www.worldbank.org/en/news/statement/2021/04/02/world-bank-group-president-statement-on-climate-change-action-plan> accessed 10 November 2021.

61 Ngozi Ole, 'Combating Electricity Poverty in Nigeria through Off-Grid Renewable Electricity: The Role of Financial Support under the International Climate Change Regime' (n 40) 4.

62 UNFCCC Initial Guidance to the GEF 1995 (UNFCCC Decision 11/CP 1) 34–35.

63 Ngozi Ole, 'Combating Electricity Poverty in Nigeria through Off-Grid Renewable Electricity: The Role of Financial Support under the International Climate Change Regime' (n 40) 4.

64 Ibid.
65 Ibid.
66 UNFCCC, 'Glasgow Pact 2021', para 52 (2021) FCCC/PA/CMA/2021/L.16.
67 Ibid.
68 Rajamani Lavanya, 'The 2015 Paris Agreement: Interplay Between Hard, Soft and Non-Obligations' (2016) 28 JEL 331, 343.
69 Daniel Bodansky, 'The Legal Character of the Paris Agreement '(2016) 25(2) RECIEL 142, 145; ODI, 'Climate Fund Updates: May 2016' (May 2016) <https://blog.climate-fundsupdate.org/> accessed 28 November 2021.
70 The Paris Agreement 2015, Article 9(3).
71 UNFCCC, 'Glasgow Pact 2021', para 44 (2021) FCCC/PA/CMA/2021/L.16.
72 Ibid.
73 The Paris Climate Agreement 2015, Article 10(2).
74 Ibid, Article 10(4). See also Technology Framework under Article 10, paragraph 4, of the Paris Agreement, Decision-/CMA.1 <https://unfccc.int/sites/default/files/resource/cp24_auv_cop_4_TF.pdf> accessed 10 November 2021.
75 Ibid, Article 1 (2).
76 Cedric Philibert, 'International Energy Technology Collaboration and Climate Change Mitigation' (2004) <www.oecd.org/env/cc/32138947.pdf> accessed 20 November 2021.
77 See Martina Jung, 'Host Country Attractiveness for CDM Non-Sink Projects' (2006) 34 Energy Policy 2174.
78 The Paris Agreement 2015, Article 10(6).
79 Helen De Conick and Amber Sugar, 'Technology Development and Transfer' in Daniel Klien and others (eds), *The Paris Agreement on Climate Change: Analysis and Commentary* (Oxford University 2017) 267.
80 Ngozi Chinwa Ole, 'The Role of Renewable Energy Law 2005 in Supporting the Development of Off-Grid Renewable Electricity in China' (2019) 5 International Energy Law Review 132.
81 UNFCCC, 'The Global Stocktake' (2021) <https://unfccc.int/topics/global-stocktake> accessed 11 November 2021.
82 Ibid.
83 For this section, see generally, Ngozi Chinwa Ole and Bukola Ruth Akinbola (n 6) 55–56.
84 The Paris Agreement 2015, Article 11(3).
85 Ibid, Article 11(1).
86 Ibid, Article 11(2).
87 Ibid, Article 11(5).
88 UNFCCC, 'Report of the Conference of the Parties on its Twenty-Second Session, held in Marrakech from 7 to 18 November 2016' (2019) FCCC/CP/2019/13/Add.2, Decision 2/CP.22, para 1 <FCCC/CP/2019/13/Add.2 (unfccc.int)> accessed 25 November 2021.
89 Ibid, para 2.
90 Ibid.
91 Susanne Droege and others, 'The Trade System and Climate Action: Ways Forward under the Paris Agreement', October 2016 <www.climate-diplomacy.org/publications/trade-system-and-climate-action-ways-forward-under-paris-agreement> accessed 2 November 2021.
92 UNFCCC, 'The PCCB 5th Annual Report 2021' <5th meeting of the Paris Committee on Capacity-building - report (unfccc.int)> accessed 25 November 2021.
93 Ibid.
94 Ibid.
95 UNFCCC, 'Paris Committee on Capacity Building' (2021) <Workplan of the Paris Committee on Capacity 2021-4.pdf (unfccc.int)> accessed 25 November 2021.
96 Ibid.

97 Lavanya Rajamani, 'Ambition and Differentiation in the 2015 Paris Agreement: Interpretative Possibilities and Underlying Politics' (2016) 65 ICLQ, 493, 504.
98 UNFCCC, 'Paris Committee on Capacity Building' (n 95).
99 Chitzi C Ogbumgbada, 'The Paris Agreement: An Imperfect but Progressive Document' (2016) 8 Int'l Energy L. Rev. 320–323.

Bibliography

Books

Baron R, 'Energy Transition After the Paris Agreement: Policy and Corporate Challenges' *Background Paper for the 34th Round Table on Sustainable Development 28–29 September 2016*, Wind Europe Summit, Hamburg.

De Conick H and Sugar A, 'Technology Development and Transfer' in Daniel K and others (eds), *The Paris Agreement on Climate Change*'. *Analysis and Commentary* (Oxford University 2017) 267.

Harris J M and Roach B, 'Environmental and Natural Resource Economics: A Contemporary Approach' in *Chapter 12-The Economics of Global Climate Change* (4th edn, Global Development and Environment Institute, Tufts University 2017).

IPCC, *Renewable Energy Sources and Climate Change Mitigation* (IPCC 2011) 7.

Korhola E, *The Rise and Fall of the Kyoto Protocol: Climate Change as a Political Process* (Helsinki 2014) 50.

Raffaello C, Rogers J A and Henrion M (eds), *Low-Carbon Development: Opportunities for Nigeria. Directions in Development* (World Bank, Washington, DC 2013) 87–99.

Schwarze R, *Law and Economics of International Climate Change* (Springer 2013) 1.

Suharsono A, McCulloch N, Mostafa M, Bridle R, Lontoh L and Gass P, *Getting to 23 Per Cent: Strategies to Scale Up Renewables in Indonesia* (The International Institute for Sustainable Development 2019) 1.

Thakur P, *Advanced Mine Ventilation: Respirable Coal Dust, Combustible Gas and Mine Fire Control* (Elsevier 2019) 3.

United Nations Framework Convention on Climate Change & Kyoto Protocol: An Introduction (Friends of the Earth, December 2008) 1.

Vogler J, 'Energy, Climate Change, and Global Governance: The 2015 Paris Agreement in Perspective' in Debra J Davidson and M Gross (eds), *Energy, Climate Change, and Global Governance: The 2015 Paris Agreement in Perspective* (Oxford University Press 2015) 18–24.

World Bank, *Summary Proceedings of Annual Meetings of the Board of Governors in Washington* (World Bank 1989) 79.

Articles

Andrea A Eras-Almeida E and Egido-Aguilera M E, 'What Is Still Necessary for Supporting the SDG7 in the Most Vulnerable Contexts?' (2020) 12 Sustainability 7184.

Bodansky D, 'The Legal Character of the Paris Agreement '(2016) 25(2) RECIEL, 142, 145.

Craig A H and Marcellino D, 'Subsidies or Free Markets to Promote Renewables?' (2012) 3 RELP 3.

Ferrey S, 'The Failure of International Global Warming Regulation to Promote Needed Renewable Energy'(2010) 19(2) Boston College Environmental Affairs Law Review 68.

Johnsson F and others, 'The Threat to Climate Change Mitigation Posed by the Abundance of Fossil Fuels' (2017) 19(2) Climate Policy 258.

Lavanya R, 'Ambition and Differentiation in the 2015 Paris Agreement: Interpretative Possibilities and Underlying Politics' (2016) 65 ICLQ 493 at 494.

Lavanya R, 'The 2015 Paris Agreement: Interplay Between Hard, Soft and Non-Obligations' (2016) 28 JEL 331, 343.

Mostafaeipour A and others, 'Identifying Challenges and Barriers for Development of Solar Energy by using Fuzzy Best-Worst Method: A Case Study' (2021) 226 Energy 120355.

Ogbumgbada C C, 'The Paris Agreement: An Imperfect but Progressive Document' (2016) 8 International Energy Law Review 320–323.

Ole N C and Akinbola B R, 'Addressing the Capacity Deficiency in the Nigerian Off- Grid Renewable Electricity: The Place of the International Climate Change Regime' (2019) 2 Redeemer's University Law Journal (RUNLAWJ) 35-58:52-55.

Ole N C, 'Combating Electricity Poverty in Nigeria Through Off-Grid Renewable Electricity: The Role of Financial Support Under the International Climate Change Regime' (2017) 2(1) International Journal of Innovative Studies in Sociology and Humanities (IJISSH) 1-16:6-8.

Ole N C, 'The Role of Renewable Energy Law 2005 in Supporting the Development of Off-grid Renewable Electricity in China' (2019) 5 International Energy Law Review 132.

Ole N C, 'The Paris Agreement 2015 as a Primer for Developing Nigerian Off-Grid Solar Electricity' (2018) 26(3) African Journal of International and Comparative Law, 426–451.

Owusu P A and Asumadu-Sarkodie S, 'A Review of Renewable Energy Sources, Sustainability Issues and Climate Change Mitigation' (2016) 3(1) Cogent Engineering 3.

Rajamani L, 'Ambition and Differentiation in the 2015 Paris Agreement: Interpretative Possibilities and Underlying Politics' (2016) 65 ICLQ, 493, 504.

Zahar A, 'The Paris Agreement and the Gradual Development of a Law on Climate Finance' (2016) 6 Climate Law 75, 80.

Policy Documents/Reports/Statutes

IRENA, *Untapped Potential for Climate Action Renewable Energy in Nationally Determined Contributions* (IRENA 2017) 7.

IRENE, *Renewable Energy and Climate Pledges: Five Years After the Paris Agreement* (International Renewable Energy Agency, Abu Dhabi 2019) 2.

Jung M, 'Host Country Attractiveness for CDM Non-Sink Projects' (2006) 34 Energy Policy 2174.

Kyoto Protocol <https://www.wikizero.com/en/Climate_Change-Kyoto_Protocol> accessed 4 November 2021.

Paris Climate Change Agreement 2015 (adopted 12 December 2015, entry into force 4 November 2016) <https://www.un.org/en/climatechange/paris-agreement> accessed 5 November 2021.

Philibert C, 'International Energy Technology Collaboration and Climate Change Mitigation' (2004) <www.oecd.org/env/cc/32138947.pdf> accessed 20 November 2021.

Reidmiller D R and others (eds), *Impacts, Risks, and Adaptation in the United States: Fourth National Climate Assessment*, Volume II: Report-in-Brief (U.S. Global Change Research Program (USGCRP), Washington, DC 2018) 24.

UNEP, *Renewable Energy and Energy Efficiency in Developing Countries: Contributions to Reducing Global Emissions* (Third Report, United Nations Environment Programme 2017) 14–15.

UNFCCC, '4 Key Achievements of COP26' (2021) <4KeyAchievementsofCOP26|UNFCCC> accessed 29 November 2021.

UNFCCC, 'Paris Committee on Capacity Building' (2021) <WorkplanoftheParisCommitteeonCapacity2021-4.pdf(unfccc.int)> accessed 25 November 2021.

UNFCCC, 'Report of the Conference of the Parties on Its Twenty-Second Session, Held in Marrakech from 7 to 18 November 2016' (2019) FCCC/CP/2019/13/Add.2, Decision 2/CP.22, para 1 <FCCC/CP/2019/13/Add.2(unfccc.int)> accessed 25 November 2021.

UNFCCC, 'The Global Stocktake' (2021) <https://unfccc.int/topics/global-stocktake> accessed 11 November 2021.

UNFCCC, 'National Determined Contributions' (2021) <https://unfccc.int/process-and-meetings/the-paris-agreement/nationally-determined-contributions-ndcs/nationally-determined-contributions-ndcs> accessed 7 November 2021.

UNFCCC, 'The PCCB 5th Annual Report 2021' <5thmeetingoftheParisCommitteeonCapacity-building-report(unfccc.int)> accessed 25 November 2021.

UNFCCC, *Technological Innovation for the Paris Agreement: Implementing Nationally Determined Contributions, National Adaptation Plans and Mid-Century Strategies' Technical Executive Committee (TEC) Brief No. 10* (United Nations Framework Convention on Climate Change 2017) 5.

United Nations Framework Convention on Climate Change 1992 (adopted 9 May 1992, entered into force 21 March 1994) FCCC/INFORMAL/84(UNFCCC).

Websites/Blogs

African Development Bank, 'Climate Change' (2021) <https://www.afdb.org/en/topics-and-sectors/sectors/climate-change> accessed 10 November 2021.

American Geophysical Union, *Human-Induced Climate Change Requires Urgent Action* (American Geophysical Union 2014) <www.agu.org> accessed 27 October 2021.

Baker and Mckenzie, 'The Paris Agreement: Putting the First Universal Climate Change Treaty in Context' (January 2016) <http://www.bakermckenzie.com/en/insight/publications/2016/01/the-paris-agreement--putting-the-first-univer sal__/> accessed 24 November 2021.

Bodansky D, 'In Brief: Legal Options for U.S. Acceptance of A New Climate Change Agreement' (2015) <http:// www.c2es.org/> accessed 21 November 2021.

Climate Council, '11 Countries Leading the Charge on Renewable Energy' (2019) <https://www.climatecouncil.org.au/11-countries-leading-the-charge-on-renewable-energy/> accessed 3 November 2021.

Denchak M, 'Paris Climate Agreement: Everything You Need to Know (Natural Resources Defence Council (NRDC) 2021)' <https://www.nrdc.org/stories/paris-climate-agreement-everything-you-need-know> accessed 12 November 2021.

Droege S and others, 'The Trade System and Climate Action: Ways Forward under the Paris Agreement' (October 2016) <www.climate-diplomacy.org/publications/

trade-system-and-climate-action-ways-forward-under-pari s-agreement> accessed 2 November 2021.

GCF in Brief: About the Fund <https://www.greenclimate.fund/sites/default/files/document/gcf-brief-about-fund_0.pdf> accessed 5 November 2021.

GCF, 'Projects' <https://www.greenclimate.fund/projects> accessed 9 November 2021.

Global Environment Facility, 'Projects' <https://www.thegef.org/projects> accessed 9 October 2021.

Global Warming| Globalization101 <http://www.globalization101.org/global-warming/> accessed 4 November 2021.

Guterres A, 'Address to the 74th Session of the UN General Assembly on 23 September 2019, New York City, USA' <https://un.dk/navigation/about-the-un/%E2%80%8B%E2%80%8Bantonio-guterres-address-to-the74th-session-of-the-un-general-assembly> accessed 27 October 2021.

Kariuki D, 'Barriers to Renewable Energy Technologies Development' (2018) <https://www.energytoday.net/economics-policy/barriers-renewable-energy-technologies-development/> accessed 1 November 2021.

Meyer R, 'A Readers Guide to the Paris Agreement' (December 2015) <http://www.theatlantic.com/science/archive/2015/12/a-readers-guide-to-the-paris-agreement/420345/> accessed 26 November 2021.

ODI, 'Climate Fund Updates: May 2016' (May 2016) <https://blog.climatefundsupdate.org/> accessed 28 November 2021.

Rappier R, 'Fossil Fuel Still Supply 84% of the World Energy' (2020) <https://www.forbes.com/sites/rrapier/2020/06/20/bp-review-new-highs-in-global-energy-consumption-and-carbon-emissions-in2019/#:~:text=The%20remainder%20of%20global%20energy,primary%20energy%20consumption%20in%202019> accessed 1 November 2021.

Seetharaman and others, 'Breaking Barriers in the Deployment of Renewable Energy' (2019) Heliyon e01166.

Smith T, 'Fossil Fuels Still Get More Investment than Renewable Energy' (2020) <https://www.esi-africa.com/industry-sectors/generation/fossil-fuels-still-get-more-investment-than-renewable-energy/> accessed 1 November 2021.

U.N. Head Warns Leaders the World is Facing a 'Great Fracture' <https://time.comunited-nations-climate-summit-economy> accessed 27 November 2021.

Umair Shahzad, 'Global Warming: Causes, Effects and Solutions' (2015) 1(4) Durreesamin Journal 2.

UNFCCC, '4 Key Achievements of COP26' (2021) <4KeyAchievementsofCOP26|UNFCCC> accessed 29 November 2021.

UNFCCC, *Caring for Climate: A Guide to the Climate Change Convention and the Kyoto Protocol* (United Nations Framework Convention on Climate Change 2003) 21 <https://unfccc.int/resource/docs/publications/caring_en.pdf> accessed 9 March 2021.

Westphal M, 'Getting to $100 Billion: Climate Finance Scenarios and Projections to 2020' (May 2015) 2 <https://www.wri.org/sites/default/files/getting-to-100-billion-final.pdf> accessed 27 November 2021.

World Bank, 'Climate Change' (2021) <https://www.worldbank.org/en/topic/climatechange> accessed 10 November 2021.

World Bank, 'World Bank Group President's Statement on Climate Change Action Plan' (2021) <https://www.worldbank.org/en/news/statement/2021/04/02/world-bank-group-president-statement-on-climate-change-action-plan> accessed 10 November 2021.

Part 2

Africa

5 Kenya

Kingsley Osinachi Onu and Samuel Piyanile Lomole

5.1 Introduction

The desire to protect the environment and also attain energy security is the anchor basis for the demand for off-grid renewable electricity (OGRE) development in recent times.[1] OGRE is concerned with 'the production of renewable electricity from smaller electric plants at or near the point of sale, and its distribution is direct to users without connecting to a national grid'.[2] Lack of access to reliable energy has been a major problem in Kenya, especially among rural dwellers, where over two-thirds of those do not have access to electricity; hence they resort to the use of charcoal and wood fuel for their basic energy needs.[3] Research[4] and policy documents[5] have shown that OGRE is the most suitable energy option for these rural dwellers in Kenya. The vast majority of Kenyans dwell in rural communities that are not connected to the grids, which makes OGRE a viable option for Kenyan rural energy demands.[6] Aside from the rural electrification usefulness of OGRE in Kenya, it can also serve as a viable backup for grid electricity in urban areas.

Kenya, like most African countries, has been energy deficit due to increased demand for energy necessitated by urbanisation as well as population and economic growth.[7] This prompted the Kenyan government to seek a lasting solution to the shortage of energy resources, and also pave the way for sustainable energy development in Kenya. The Government of Kenya (GoK) has over the past two decades revolutionised the energy sector through the promulgation of the following laws and policy framework. The legal frameworks are: the Energy Act, 2006; the Least Cost Power Development Plan, 2009; the National Energy Policy, 2015; the Sustainable Energy for All Action Agenda, 2016; and the Energy Act, 2019. These laws and policy frameworks have diversified the energy sources in Kenya and have incorporated renewable energies (REs) as part of Kenya's energy sources. Kenya has a regulated electricity market, with a significant presence of independent power producers (IPPs) following the unbundling of the electricity sector and partial privatisation of Kenya Power and Lighting Company. The country has an installed capacity of 2,819 megawatts (MW) and a current peak demand of 1,912MW. Following the unbundling of the energy sector in 2006, the role of the private sector in power generation has grown significantly. However, the transmission network remains a public monopoly.

DOI: 10.4324/9781003178088-7

Currently, Kenya is regarded as a model for other African countries in terms of energy regulation, for its successful integration of RE into her energy mix.[8] Kenya is also the only African country to meet the World Bank's RE development benchmark.[9]

This chapter will provide an overview of the OGRE landscape in Kenya. In particular, this chapter will set out the policies and legal framework applicable to OGRE projects with a view to establishing the extent to which it is adequate in addressing the barriers to its development.

The chapter is divided into five sections. Section 5.1 is the general introduction. Section 5.2 espouses on the imperatives, opportunities. and barriers to OGRE development in Kenya. Section 5.3 appraises the policy targets, Section 5.4 examines the regulatory framework for OGRE in Kenya. Finally, Section 5.5 contains the conclusion and recommendations.

5.2 Imperatives, Opportunities, and Barriers

Kenya is the economic capital of East Africa and a global tourism nation.[10] This has resulted in a lot of economic activities, population growth, and urbanisation in Kenya with the consequential high demand for energy.[11] In the early 1990s, large hydro plants served as the major source of energy in Kenya, followed by petroleum, geothermal, and small contributions from RE sources and coal.[12] These energy sources posed some serious changes for Kenya. On one hand, electricity supplies from the hydro plants were frequently fluctuating, while petroleum importation was hampered by high importation costs. In addition to these challenges, Kenya's installed electricity capacity could not satisfy her electricity demands, even if they were to function at full capacity. Kenya in most cases suffered a shortage in both energy generation and supply.[13]

Kenya resorted to importing electricity from Tanzania and Uganda as a short-term measure of addressing its energy deficit.[14] This approach paid off for the industrial sector but greatly impacted the residential sector given that most of the country's population are predominantly within the low purchasing power category and could not afford the exorbitant electricity tariffs.[15] These negative outcomes pushed the GoK to opt for a lasting solution to its energy challenge by adopting a long-term policy that favours the diversification of Kenya's electricity sources. The policy acknowledges that the combination of the use of fossil fuels and RE as energy-generating sources will provide the cheapest sustainable energy for the country.[16] The policy using the least cost approach has prioritised the development of geothermal plants, wind energy plants, and solar-fed mini-grids as OGREs rural electrification in Kenya.[17]

The above policy has triggered an evolution and uses RE as Kenya's Sustainable Energy (SE) strategy, and this policy has been vigorously pursued by the GoK. The diversification of Kenya's energy sources is based on the objective of attaining affordable (energy security) and environmentally friendly energy sources. There is no gainsaying that energy security is the condition precedent for the desired economic growth in Kenya, hence its prioritisation in both policies and laws. Energy security connotes the steady availability of energy resources at affordable cost.[18] OGRE will reduce the over-reliance on imported fossil fuel which has been identified as the major reason for the high electricity tariff in Kenya, which has consequently impacted electricity affordability among rural dwellers.[19] OGRE is at the centre of the Kenyan Rural Electrification Programme (KREP), as it has been identified as being essential for enhancing the rural population's access to electricity. The Energy Act, 2019 has repositioned KREP and its implementation in such a manner that it can increase the use of OGRE and as such improve rural access to electricity in Kenya.[20]

The Kenyan economy is driven by climate-sensitive sectors like tourism, rain-fed agriculture, energy, and wildlife which are all vulnerable to the effects of climate change.[21] On the other hand, Kenya's greenhouse gas (GHG) emissions have increased from 56.8 $MtCO^2e$ in 1995 to 93.7 $MtCO^2e$ in 2015, and it is projected to increase further to 143 $MtCO^2e$ by 2030 on a business as usual (BAU) scenario.[22] The integration of OGRE will contribute to energy sustainability in Kenya, and also assist it to meet her environmental protection obligations under international regimes. Kenya has ratified the following international regimes on climate change, which includes the United Nations Framework Convention on Climate Change,[23] the Kyoto Protocol,[24] and the Paris Agreement.[25] Kenya has also submitted her updated Nationally Determined Contribution to the Paris Agreement,[26] and has also designed some domestic policies to enable her to fulfil her obligations under international environmental laws.[27]

These instruments prioritise the diversification of energy sources to pave the way for the inclusion of RE, especially OGRE, for rural electrification. In the same vein Kenya, in line with its environmental protection objective, has promulgated some domestic legislation that is targeted at reducing environmental issues associated with energy generation.[28] Kenya's updated NDC to the Paris Agreement has pledged to reduce GHG emissions by 32% by the year 2030.[29] The above has informed Kenya's development of her OGRE resources.[30] Geothermal accounts for over 50% of Kenya's generated electricity, as it generates 650 MW. It has the potential to generate between 7,000 MW and 10,000 MW of electricity.[31] Electricity prices in Kenya dropped by 30% as an aftermath effect of the commissioning of a 280 MW geothermal plant in 2015 by the government.[32] The country has significant potential for hydropower energy which is estimated to be about 3,000–6,000 MW.[33] Half of this potential stems from small hydro resources which are best suited for rural, small business, and small farm electrification.[34]

Concerning solar, Kenya also has enormous solar energy potential being that she is located close to the equator with 4–6 $kWh/m^2/day$ levels of insolation.[35] Currently, over 200,000 homes use photovoltaic solar home systems basically for powering television sets and lighting. Photovoltaic solar systems can be deployed for rural electrification at a subsidised cost for rural dwellers and farmers. Kenya also has a large potential for wind energy with an average wind speed rate of 6 m/s which can power the whole of Kenya[36]and be used to generate power for rural farmers and their households. OGRE is therefore very fundamental for Kenya's energy diversification plan as it has enormous OGRE potential. The GoK, through the LCPD (as defined below), has prioritised the development of geothermal, wind, and solar energy plants as well as solar-fed mini-grids[37] for rural electrification.[38]

Despite the huge potentials of OGRE in Kenya, there exist challenges to its development. OGRE is mostly used for rural electrification and the project sites are also located in rural communities. These rural communities are predominately inhabited by illiterate citizens who do not understand the need for energy diversification and sustainable development, hence the projects face fierce resistance from the people. For example, the communal crisis halted the construction of Olakaria's geothermal power project and Kinangop's wind power project at some point.[39] Also, the government fails to carry out the needed consultations with the stakeholders before siting some of these projects. They also fail to carry along the people in the decision-making process and execution of projects that are sited within their locality. These gaps in information and inclusion make the people apprehensive of the potential dangers that these projects may elicit. Funding is also a major barrier for OGRE development. It will cost Kenya over two billion dollars to transit completely to clean energy (which is inclusive of OGRE), and Kenya is demanding 87% foreign support and direct foreign investment to achieve this goal.[40] Even when these OGRE sources are integrated into Kenya's energy mix, rural dwellers, who are predominantly poor, will be required to purchase these energies.

Other challenges include administrative bottlenecks in giving approvals for OGRE projects,[41] unavailability of OGRE experts,[42] low budgetary allocation,[43] and inadequate fiscal and regulatory incentives for investors.[44] Where the fiscal incentives are available, they may not directly apply to OGRE.[45]

5.3 Policy Targets

The extant policy on Energy in Kenya is the National Energy Policy (NEP), 2018.[46] The vision of the NEP is to provide 'affordable quality energy for all Kenyans'.[47] Similarly, the central objective of NEP is:

> to ensure affordable, competitive, sustainable and reliable supply of energy at the least cost in order to achieve the national and county development needs, while protecting and conserving the environment for inter-generational benefits.

It is important to note that prior to this era, the Sessional Paper No. 4 of 2004 on Energy guided the GoK's energy sector and informed the enactment of the Energy Act, No. 12 of 2006, the Feed in Tariff (FiT) Policy 2008 (which was revised in 2012), and the enforcement of the Geothermal Resources Act No. 12, of 1982. However, the coming into force of Kenya's Constitution in 2010 and the adoption of Vision 2030 introduced salient provisions on energy governance in the country, hence the need for a new energy policy (NEP, 2018) that will align with the Constitution and the vision. The NEP 2018 has identified geothermal, hydropower, wind, biomass, biofuel, municipal waste, etc. as RE resources with huge potential that can be deployed as both grid-connected electricity and OGRE in Kenya.

Geothermal plants deploy heats and steams from natural underground reservoirs to generate electricity. The Rift valley of Kenya has the occurrence of over 14 high temperature potential sites of geothermal with over 10,000 MWe estimated energy potential.[48] The other locations include the Nyambene Ridges, Chyulu Homa Hills in Nyanza, and the Mwanayamala along the coastal line.[49] The NEP noted that geothermal resources can be used as OGRE for the dairy industry, water heating, space heating and cooling, grain silos, and for industrial purposes.[50]

The GoK's policy targets and strategies for geothermal over the period 2018–2030 include the funding of geothermal prospection and exploitation in order to minimise risks, promotion of capacity building, research and development in geothermal. This will be achieved through fiscal and other incentives, acceleration of geothermal development through the streamlining of allocation and licencing of geothermal blocks with incentives, and also through the promotion and encouragement of the use of geothermal energy through attractive pricing, strict enforcement of regulations, and the promotion of efficient modular geothermal technologies.[51]

However, NEP 2018 identified the site-specific nature of geothermal resources, high investment costs, high exploration and development risks, long duration of development, and land use conflicts, among others as the factors militating against the development of geothermal energy resources for both grid and off-grid utilisation. It also made provisions for hydropower by positing that, as at December 2017, Kenya's estimated hydro potential stood at 6,000 MW, consisting of both large and small hydros. Currently, Kenya has only been able to develop about 17.5 MW of these hydro sources. Hydropower resources, especially small hydros, are a major component of the FiT Policy 2008 which targets to give rural dwellers and small farms access to OGRE for their daily energy needs. The strategy targets for hydropower in the NEP 2018 include to: develop hydro risk mitigation mechanisms and power projects, organise water resources management, fund the conservation of water resource catchment areas, balance the competing interests on water resources, grant incentives to investors, etc.

The NEP also identified the huge energy potential of biomass, biofuel, municipal waste, solar energy, wind energy and specifically made short- (2018–2022), medium- (2018–2026) and long-term (2018–2030) plans and policies for their maximisation in Kenya.[52] It outlined the following policies and strategies to maximise these renewable energy resources for both on-grid and off-grid usages, to wit: development of institutional capacity, continuous review and enforcement of regulations on these RE sources, provide incentives, support and finance hybrid energy generation systems, support transmission efficiency projects, and undertake research development and dissemination (RD&D).[53]

The policy also made provisions for the FiT scheme. The FiT scheme connotes that 'a utility to produce Renewable Energy Sources Generated Electricity (RES-E) and sell the output to a distributor at a pre-determined tariff for a given period of time'.[54] As at 2018, investors have signified interest to invest in grid and OGRE under the FiT policy which will generate 4508.40 MW of energy.[55] The FiT policy is facing the following challenges: lack of awareness among investors, preference of one RE source over others, lack of clear-cut negotiation mechanism, lack of technical and financial support, etc.[56] The NEP 2018 designed the following policies and strategies for FiT optimisation which are: private sector motivation through incentives, formulation and implementation of campaigns for investors' drive, periodic review and implementation of the FiT policy. It also includes the provision for capacity-building programmes and finance, development and review of model power purchase agreements, and improving research.

5.4 Regulatory Framework

5.4.1 Introduction

The Republic of Kenya enacted a constitution in 2010. This Constitution made some striking provisions on RE that were not captured in the Energy Act, 2006. The Constitution obligated the GoK to promote sustainable development of the country's natural resources and the environment, which did not feature in the Energy Act.[57] Although, the Constitution created two control layers for energy resources which were not provided for under the Act.[58] Second, the GoK developed and approved an economic master plan in 2008 titled 'Vision 2030'. It is therefore imperative to overhaul the Act to bring it up to speed with recent policy and constitutional advancements in the energy sector. The foregoing cumulated into the enactment of Kenya's Energy Act, 2019.[59] The Energy Act, 2019 (hereafter referred to as the Act) was assented to by the President of Kenya on the March 12, 2019.[60] The Act consolidated all the laws on energy and obligated the GoK to facilitate the provision of affordable energy services to all persons in Kenya.[61] The Cabinet Secretary is charged with the responsibility of promoting a conducive environment for energy investment through policy designs and implementation.[62] The new Act comprehensively made provisions on RE and OGRE.[63] In line with the Constitution of Kenya, 2010,[64] the Act vests all unexploited RE resources under or in the land in the GoK 'subject to any rights which, by or under any written law, have been or are granted or recognised as being vested in any other person'.[65] In the same vein, the new Act vests all un-extracted geothermal resources in Kenya on the government.[66]

However, this provision is hazy as the phrase 'RE resources under or in the land' is too shallow to encompass all RE resources. RE resources are not limited to resources under or inland, but could also be in form of solar, wind, tide, and hydro which are not affixed to the land and mostly serve as OGRE. The definition of RE resources may be subjected to diverse interpretations. The GoK is consequently responsible for the licencing of operators in the generation, transmission, and distribution of RE in Kenya.[67] The 2019 Act also decentralised electricity distribution in Kenya by permitting the entry of new entities, especially the sale of off-grid electricity.[68] This involvement of the private sector has weakened the monopoly of the KPLC over electricity distribution, hence paving way for competition.

The Energy Act, 2019 established the Energy and Petroleum Regulatory Authority (EPRA) which replaced the defunct ERC as the sole regulator of the energy sector except for nuclear facilities licencing and petroleum refining, exportation, and importation.[69] The GoK, through the Cabinet Secretary, was still saddled with the responsibility of policy formulation for the energy sector and also responsible for the promotion of the development and use of RE technologies in Kenya. This responsibility is to be discharged through the formulation of national research strategies in RE as well as enabling a framework for sustainable development and distribution of RE.[70] The Cabinet Secretary is also mandated to prepare a RE resources inventory and map within 12 months from the commencement of the Act;[71] this RE resources inventory and map is expected to make way for efficient utilisation of RE resources to satisfy the energy needs of people in Kenya.[72]

The 2019 Act also established two new institutions for KREP, which are: the Rural Electrification and Renewable Energy Corporation (REREC);[73] and the Renewable Energy Resources Advisory Committee (RERAC).[74] The coordination of KREP was left within the purview of REREC, which replaced the defunct REA.[75] The new Act conferred expanded powers and functions on REREC to comprehensively oversee the development of RE, inclusive of OGRE for KREP,[76] unlike what was obtainable under the defunct REA. The functions of the REREC also extend to the development and use of OGRE resources. REREC is mandated to 'undertake on-farm and on-station demonstration of wood fuel species, seedling production, and management for energy generation'.[77] It is also mandated to promote, develop, and manage the use of RE resources and technologies such as, but not limited to, biomass (biodiesel, bio-ethanol, charcoal, fuel wood, biogas), municipal waste, solar, wind, tidal waves, small hydropower, and co-generation but excluding geothermal.[78] REREC is also charged to promote the use of quickly maturing trees for energy generation.[79]

Overall, the new Act charged the REREC to develop a sustainable production and utilisation framework for OGRE resources' development. The REREC is mandated to collaborate with county governments in applying the principle of equity in the development of OGRE for use by rural people in Kenya.[80] The new Act expanded the powers of the REREC for effective uptake of RE and OGRE in tackling Kenya's problem of electricity access. The expanded powers also gave room for expanded sources of funding. The funding of the REREC expanded beyond the KREP established under Section 143, to interest from bank deposits, money appropriated by the Kenyan parliament, and revenue from other sources (e.g. donations).[81] Furthermore, the new Act allows the REREC to seek additional funding for OGRE and KREP.[82]

Through the implementation of KREP, over 6 million households have been connected to electricity.[83] Kenya has many OGRE projects intended to boost electricity supply networks in Kenya.[84] The World Bank–funded Kenya Off-Grid Solar Access Project is one such project.[85] The Energy Act, 2019 also establishes the Energy and Petroleum Tribunal (EPT) to hear and determine civil disputes and appeals from the EPRA and any other licencing authority relating to the energy and petroleum sector. The Energy Act provides the jurisdiction and procedure of the EPT.[86] Therefore, any grievances with EPRA may be filed at the EPT and an appeal from the EPT would be made in the High Court.[87]

5.4.2 Licencing Regime for Electricity Projects

The Act saddled EPRA with the responsibility of granting licences in the electricity sector which is inclusive of OGRE.[88] EPRA, in granting or rejecting an application for an electricity licence, is expected to consider certain factors such as impact on social life, need to protect the environment, economic relevance of the venture, and cost implication of the undertaking, among others.[89]

An applicant for an OGRE licence may apply for a commercial use licence, technology-specific clearance, or capacity-specific clearance.[90] An OGRE commercial use licence applicant is expected to secure an environmental impact assessment licence from the National Environmental Management Authority (NEMA).[91] On the other hand, an OGRE technology-specific clearance applicant is required to also secure the Kenya Civil Aviation Authority Clearance, water abstraction permit, geothermal resource licence, and geothermal exploration licence.[92] Such an applicant is also required to deposit an environmental impact assessment (EIA) report to the NEMA prior to the financing and commencement of an OGRE project.[93]

It is submitted that the requirement for licence and compliance with other regulatory frameworks is in line with international best practice as it ensures for efficiency and coordination in OGRE development. However, these requirements may become an unintended clog in the wheel of OGRE development, hence deterring investors. Kenya like must African countries is prone to administrative bottlenecks. The presence of an administrative bottleneck in OGRE licence application and requirement to comply with other regulatory frameworks makes the process cumbersome for investors who will be more interested in a one-stop shop regulatory approach.

EPRA is mandated to communicate its decision of a grant or refusal of licence to the applicant within 60 days of completion of the application.[94] Where the EPRA declines to grant an applicant a licence, it must communicate the reasons for such denial to the applicant.[95] This requirement for feedback to unsuccessful applicants is in tandem with the principles of accountability, transparency, and fairness. However, where such an unsuccessful applicant is not satisfied with the reasons adduced by EPRA, he/she may seek redress before the EPT.[96] For successful applicants, the EPRA is empowered to determine the form of the licence that will be issued,[97] what it shall contain depending on what is applicable.[98]

Further to the terms above, the licence must contain the following conditions: obligation of the licensee to comply with applicable and relevant environmental, health, and safety; liability of the licensee under contract and tort; and obligation for timeous remittance if all necessary fees associated with the licence.[99] An Electricity Generating Permit will be granted to an OGRE undertaking that is between 1 and 3 MW, while an Electricity Generating Licence will be granted to an OGRE undertaking that is above 3 MW.[100] A licence issued under this Act may not be modified, altered, or revised without the consent of the licensee.[101] An operator of a licenced undertaking must ensure that such undertaking for either electricity transmission or distribution is operated within the terms of the EPRA.[102] The Act also prohibits sub-contract or licencing of any undertaking without the prior consent of EPRA being sought and gained.[103] A breach in this term may lead to a revocation of the licence granted.[104]

A licence may be revoked and suspended by the EPRA if the licensee fails to commence operation within 24 months of the granting of the licence.[105] However, EPRA must give a licensee 30 days' notice before effecting the revocation or suspension of the licence;[106] and the notice must explicitly set out the grounds for such revocation or suspension.[107]

5.4.3 Environmental Regulation

The development of OGREs in Kenya with all its benefits may also pose a threat to the environment, hence, the Kenyan parliament has made provisions for it. The first law in this regard is the Constitution of Kenya 2010,[108] which elaborately made provisions on environmental management and protection. Section 42 of the Constitution guarantees the right of Kenyans to a clean and healthy environment. Similarly, Section 69 creates obligations for the GoK concerning the environment. The GoK is obligated by the Constitution to sustainably manage Kenya's natural resources which are inclusive of OGRE for the benefit of all Kenyans.[109] It is also the responsibility of the GoK to establish systems of environmental impact assessment, environmental audit, and monitoring of the environment;[110] and ensure public participation in environmental management.[111] The Constitution also made provision for redressing of breach of the right to a clean and healthy environment. Hence, where a person's right to a clean and healthy environment is being, or is likely to be, infringed on by undertakers of an OGRE project, the person may seek redress before a court.[112]

The second legislative framework is the Environmental Management and Coordination Act (EMCA),[113] which is a comprehensive legislative framework on environmental protection and management in Kenya. EMCA just like the Constitution guarantees the right of Kenyans to a clean and healthy environment and provides for an avenue to seek redress before the Environment and Land Court upon a breach/infringement of that right.[114] The Act also grants Kenyans the right to access information and public participation in the decision-making process that relates to the environment.[115] EMCA established the National Environmental Management Authority[116] and charged it with the duty 'to exercise general supervision and coordination over all matters relating to the environment and to be the principal instrument of Government in the implementation of all policies relating to the environment'.[117]

A proponent of an OGRE undertaking is required to submit an EIA to NEMA before the commencement and financing of such a project.[118] Kenya operates a centralised EIA governance, as such, NEMA[119] centrally controls all EIA processes in Kenya including the issuance of licences. It is only in decision-making that NEMA consults with County Environmental Committees.[120] EPRA is prohibited from issuing an operating licence for an OGRE undertaking if the proponent of such undertaking fails to produce an EIA certificate issued by NEMA.[121]

An OGRE operation licensee whose undertaking is likely to impact wetlands, riverbanks, and seashores is required to adhere strictly to the Environmental Management and Co-Ordination (Wetlands, River Banks, Lake Shores, And Sea Shore Management) Regulations, 2009.[122] Similarly, OGRE undertakings are also prohibited from discharging effluents into the aquatic environment without first procuring a valid effluent discharge licence issued under the EMCA.[123] OGRE operations are also mandated to collect, segregate, and dispose of all waste in the manner prescribed by the EMCA and its regulation.[124] The EMCA established the National Environment Tribunal (NET) to adjudicate disputes that may arise from the operation of the EMCA.[125] It also has jurisdiction to entertain appeals from entities who are not satisfied with the actions of NEMA.[126] Appeals from the NET shall lie before the Environment and Land Court.[127] The EMCA also instituted a National Environment Restoration Fund[128] which operates as supplementary insurance to address environmental degradation where the perpetrator cannot be identified or in exceptional environment degradation cases.[129]

It is pertinent to note that NEMA 'do not have adequate capacity to effectively guide and review EIAs',[130] which underscores the danger in a centralised EIA governance especially for OGRE capacities that are mostly located in rural communities inhabited by the most vulnerable groups of the society.[131] It is important to note that the essence of EIA is to ensure that the public is informed and engaged in information sharing as regards proposed projects in the society in order to decipher their potential benefits and harms to the environment;[132] and collectively devise a means to avert the adverse effect of such harm on the environment.[133] EIA is fundamental for OGRE development, especially for large OGRE facilities. However, the current centralised EIA procedure in Kenya and the lack of adequate facilities for carrying out an EIA process by NEMA makes the process difficult for OGRE developers.

5.5 Pricing Mechanism (Tariff Fixing)

The Energy Act, 2019 is the central regulatory framework on energy in Kenya. The Act expressly made provision for the fixing of electricity energy tariffs in its Section 11(b) that provides that EPRA shall 'set, review and adjust electric power tariffs and tariff structures and investigate tariff charges, whether or not a specific application has been made for a tariff adjustment'. It therefore means that EPRA has the exclusive power to fix, review, and adjust electricity tariffs in Kenya. Tariff review involves some systematic steps.[134]

Similarly, the Energy Act, 2019 mandated EPRA to protect consumers, investors, and other stakeholder interests pursuant to Section 10(hh). The EPRA achieves the above mandate by ensuring fair competition within the energy sector through regular review of the tariff policy framework for energy generation, transmission, and distribution. EPRA also conducts stakeholder engagements to seek views on tariffs charged and also monitor the compliance of approved tariffs.

The Act also established the renewable energy feed-in tariff (ReFIT) system[135] for better pricing and encouragement of investment in mini-grid.[136]

5.5.1 Support Schemes

Although OGRE is efficient, effective, and cheap in the long run, investments in OGRE are expensive. OGRE is expensive and requires financial investment at a pace and scale that imply substantial costs for both the GoK and investors. The GoK has therefore designed some support schemes to cushion the financial effect of OGRE development and also inject ease of doing business for both foreign and local investors in OGRE. First, the Energy Act established a ReFIT system.[137] The essence of the system is to 'facilitate resource mobilisation by providing investment security and market stability for investors in electricity generation from renewable energy sources, and reduce transaction and administrative costs and delays associated with the conventional procurement processes'.[138]

The Cabinet Secretary was empowered by the Energy Act, upon the recommendation of EPRA, to make regulations necessary for the administration and implementation of the feed-in tariff system.[139] The GoK has promulgated the ReFiT policy, 2021, which gives effect to the above statutory provision. The ReFiT system promotes the generation of electricity from renewable energy sources by allowing IPPs to sell generated electricity to the off-taker at a predetermined tariff for a given period. The ReFiT policy, 2021 (unlike the previous FiT, 2012) expressly applies to renewable energy power plants not exceeding 20 MW in biomass, biogas, and small hydro technologies, which mini-grid and OGRE projects may fall under.

Second, the Value Added Tax (VAT) Act 2013 and VAT (Amendment) Act 2014, exempted RE power-generating plant equipment (inclusive of OGRE) importations from VAT and import duties.[140] The economic downturn occasioned by the outbreak of the coronavirus pandemic led to the introduction of 14% VAT on all supplies imported in the sector to cushion the effects of the pandemic.[141] This move would spell hardship for both the investors and consumers of OGRE. The move would have led to higher prices of OGRE (especially PV solar equipment), and the tax pressure transferred by investors to the helpless consumers who are rural dwellers.[142] Higher retail and installation costs would prevent poor and middle-income earners from accessing OGRE resources, and also loss of jobs already created through the distribution of OGRE equipment (especially solar equipment). The GoK reacted swiftly to the outcry of experts and investors, by reverting to VAT and import duty exemptions for RE supplies importation.[143] This move by the GoK will motivate investors to locally set up shops and also reduce the purchase/installation costs of OGRE equipment so that low-income earners can afford it. Third, the GoK adopted the Special Economic Zone (SEZ) Act in 2015,[144] as part of its economic growth and development approach to attract investors into the manufacturing sector, of which OGRE is inclusive.[145] Entities operating within an SEZ shall, from the date of commencement of operation, have a chargeable tax rate of 10% for the first ten years; and subsequently at the rate of 15%.[146] The scheme is a laudable one that has stipulated OGRE-direct foreign investment in Kenya.[147]

In addition, the Public-Private Partnership Act provides for the participation of the private sector in the financing, construction, development, operation, and maintenance of government infrastructure and development projects through concession and other contractual arrangements. The PPP Act, subject to undertaking an analysis of the nature and structure of the project, may be applicable where the intention is to connect the off-grid project to the main grid. Investments in the mini-grid sector are supported by different development agencies including the World Bank, African Development Bank, KfW Development Bank, Deutsche Gesellschaft für Internationale Zusammenarbeit GmbH (GIZ), and Agence Française de Développement. Wako and Nagumo[148] highlighted the other incentives available to RE (inclusive of OGRE) to include exemption of OGRE entities from stamp duties for company registration.[149] Also, instruments used to procure foreign loans for the energy sector are exempted from stamp duties.[150] Kenya also has a Double Taxation Agreement (DTA) with some countries,[151] it, therefore, connotes that the importation of OGRE equipment from such countries will only be taxed once. Kenya does not have any restriction on investment fund's conversion or transfer. Finally, the Foreign Investment Protection Act Chapter 518, Laws of Kenya, guarantees the repatriation and remittance of foreign investors' dividends and interest.

5.6 Conclusion

This chapter has succinctly examined OGRE in Kenya. It is pertinent to note that Kenya has an abundance of OGRE resources, and is also one of the pace-setting states in Africa on the issue of energy policy and governance. Kenya is certainly on the right path to energy efficiency and environmental sustainability through the giant strides it has made in the area of a robust regulatory framework for energy governance. This chapter has examined the opportunities for OGRE investments/development in Kenya and has also examined the challenges currently faced by the sector. Some of these challenges include the high cost of OGRE equipment and installation; illiteracy of most of the end consumers of OGRE who are mostly rural dwellers who lack basic knowledge of the essence of clean energy but are only interested in cheap energy; there is also the lack of consultation and participation of the people in the siting and execution of some these OGRE, for instance, the construction of Olakaria's geothermal power project and Kinangop's wind power project;[152] with the avalanche of incentives in the energy sector in Kenya, funding remains a major drawback as it is relying on 87% foreign support and FDI to attain its sustainable energy development goals;[153] there are still exit administrative bottlenecks and duplicity of functions between EPRA, NEMA, and REREC in OGRE governance.

Based on the foregoing, it is hereby recommended that the GoK should subsidise the cost of OGRE resources for Kenyans, especially the rural dwellers. This will help them to afford such resources for their day-to-day basic energy needs, which will, in turn, assist Kenya in meeting her obligations on carbon emission reduction. The GoK should also intentionally insist on public participation and access to information in the citing and implementation of OGRE plants and installations. Furthermore, mass education on the essence of RE (especially OGRE) should be prioritised in Kenya. Where the purpose of OGRE resources is unknown or unclear to the masses, their abuse or abhorrence becomes inevitable. On the issue of technology, Kenya must look inward and invest in local technology for OGRE development. Over-reliance on FDI and foreign supports may not spell all is well for Kenya, its economy, and people in the long run. Finally, the GoK must quickly address the cumbersome protocols for the securing of OGRE licences. Also, the jurisdictional overlap between the EPRA, NEMA, and REREC should quickly be addressed for smooth OGRE governance in Kenya.

Notes

1 Ngozi Chinwa Ole, 'The Nigerian Electricity Regulatory Framework: Hotspots and Challenges for Off-Grid Renewable Electricity Development' (2020) JENRL <https://doi.org/10.1080/02646811.2020.1771845> accessed 25th March 2022
2 Ibid, 1.
3 K M Lawal, 'The Role of Legislation in Ensuring Sustainable Energy Development in Nigeria: Lessons from Kenya'. LLM dissertation submitted to the School of Law, Adelaide University.
4 Ibid, 87.
5 The Republic of Kenya, *Updated Least Cost Power Development Plan Study Period: 2011–2037* (2018) 49, Retrieved October 19, 2021, <http://gak.co.ke/wp-content/uploads/2019/02/Updated-Least-Cost-Power-Development-Plan-2017-2022-min.pdf>; Republic of Kenya, *Kenya Vision 2030* (2007) 49.
6 Lawal (n 3) 87.
7 Ibid, 13.
8 World Bank, *The World Bank Electricity Expansion (P103037), Energy & Extractives Global Practice Africa Region* (Report No ICR00004496, 2018) 73.
9 Ana Pueyo, 'What Constrains Renewable Energy Investment in Sub-Saharan Africa? A Comparison of Kenya and Ghana' (2018) 109 World Development 85.
10 Lawal (n 3) 89.
11 Ibid.
12 Joseph Kapika and Anton Eberhard, 'Enabling Private-Sector Participation in Electricity Generation' in Joseph Kapika and Anton Eberhard (eds), *Power-Sector Reform and Regulation in Africa: Lessons from Kenya, Tanzania, Uganda, Zambia, Namibia and Ghana* (HSRC Press 2013) 21, 21.
13 Carolyn Fortuna, 'Renewable Energy in Kenya: Meeting the Needs of an Expanding Population' *Clean Technica* (Online) <https://cleantechnica.com/2018/04/16/renewable-energy-in-kenya-meeting-the-needs-of-an-expandingpopulation/> accessed 16th October 2021.
14 Lawal (n 3) 89.
15 Ibid.
16 Republic of Kenya, *Updated Least Cost Power Development Plan Study Period: 2011–2037* (2018).

17 Ibid, 55.
18 Ibid.
19 Kees Mokveld and Steven Von Eije, *Final Energy Report Kenya* (Netherlands Enterprise Agency 2018) 3.
20 The provisions of this Act shall be discussed later in this chapter.
21 The Government of Kenya, 'Kenya's Updated Nationally Determined Contribution (NDC)' (United Nations Framework Convention on Climate Change (UNFCCC), 20 December 2020) <https://www4.unfccc.int/sites/ndcstaging/PublishedSDocuments/Kenya%20First/Kenya%27s%20First%20NDC%20(updated%20version).pdf> accessed 23rd October 2021.
22 Ibid.
23 United Nations Framework Convention on Climate Change, adopted 14 June 1992, UNCED Doc.A/CONF.151/5/Rev.1, (Vol I), Annex I, 13 June 1992, 31 *ILM* 874 (1992).
24 Kyoto Protocol to the United Nations Framework Convention on Climate Change, Report of the Conference of the Parties at its Third Session, 1–11 December 1997, U.N. Doc FCCC/CP/1997/7/Add.1, 18 March 1998, Annex.
25 United Nations Framework Convention on Climate Change, Conference of the Parties, Adoption of the Paris Agreement, Dec. 12, 2015, U.N. Doc. FCCC/CP/2015/L.9/Rev/1 (Dec. 12, 2015).
26 Kenya's Updated Nationally Determined Contribution (n 21).
27 For instance, see Kenya's National Climate Change Action Plan 2015 and the National Climate Change Response Strategy 2010.
28 Lawal (n 3) 92.
29 Kenya's Updated Nationally Determined Contribution (n 21).
30 These include small hydro, wind, solar, biomass, and geothermal.
31 Lawal (n 3) 92.
32 Ibid.
33 Republic of Kenya, Updated Least Cost Power Development Plan Study Period: 2011–2037 (2018).
34 Ibid.
35 Ibid.
36 Kenya Renewable Energy Association, *Wind Energy* <http://kerea.org/renewable-sources/wind-energy/>.
37 Section 2.1 of the Least Cost Development Plan 2021-2030 defines mini-grids as 'a set of electricity generators and energy storage systems interconnected to a distribution network that supplies electricity to a localized group of customers not covered by the interconnected national power grid as approved by EPRA'.
38 Paragraph 4.5, Least Cost Development Plan 2021–2030.
39 Lawal (n 3) 91.
40 Kenya's Updated Nationally Determined Contribution (n 21).
41 Export Solutions, 'Kenya: Electrical Power Systems' <https://www.export.gov/article?id=Kenyaelectricalpower-systems> accessed 16th April 2022.
42 Ibid.
43 Ibid.
44 Lawal, 2020 (supra), 94.
45 Ibid.
46 Republic of Kenya's Ministry of Energy, 'Republic of Kenya's National Energy Policy, 2018' (Kenya Power and Lighting Company, 1 October 2018) <https://kplc.co.ke/img/full/BL4PdOqKtxFT_National%20Energy%20Policy%20October%20%202018.pdf> accessed 24 March 2022.
47 Ibid.
48 Ibid, 25.

49 Ibid.
50 Ibid.
51 Ibid, 26.
52 Ibid, 37.
53 Ibid, 38.
54 Ibid, 40.
55 Ibid, 41
56 Ibid.
57 Kenya's Constitution, 2010, s 69. See also Kenya's Constitution 2010, ch 4, Items 8, 21, 31.
58 Ibid.
59 Energy and Petroleum Regulatory Authority (EPRA), 'Energy Act 2019. *Kenya Gazette Supplement No. 29 (Acts No. 1)*' (EPRA, 12 November 2020 <https://www.epra.go.ke/download/the-energy-act-2019/> accessed 30th November 2021.
60 Ibid.
61 The Energy Act 2019, s 7(1).
62 The Energy Act 2019, s 8(1).
63 The Energy Act 2019, s 43, 74–75.
64 The Energy Act 2019, s 43, 69, and the Constitution of Kenya, 2010, s 260.
65 The Energy Act 2019, s 73.
66 The Energy Act 2019, s 77.
67 Energy Act 2019, s 78.
68 Lawal (n 3) 106.
69 The Energy Act 2019, s 163(1) and s 9(1).
70 Ibid, s 75(1) and (2).
71 Ibid, s 74.
72 Mukha Njau, Shah Paras and Field Rainbow, 'New Energy Act Embraces Renewable Energy' *Bowmans*, 2 December 2021, <https://www.bowmanslaw.com/insights/oil-gas/new-energy-act-embraces-renewable-energy/>.
73 Energy Act, 2019, Section 43.
74 Ibid, s 73.
75 Ibid.
76 Ibid, s 44.
77 Section 44(1)(h).
78 Ibid, (j).
79 Ibid, (n).
80 Ibid, (g) and (i).
81 The Energy Act 2019, s 53(1). See also Lawal, (n 3) 107.
82 The Energy Act 2019, s 44(1)(c).
83 Lily Kuo, 'Kenya's National Electrification Campaign is Taking Less than Half the Time it Took America, *Quartz Africa* (Online), 16 January 2017, <https://qz.com/882938/kenya-is-rolling-out-its-nationalelectricityprogram-in-half-the-time-it-took-america/>.
84 Kees Mokveld and Steven von Eije, *Final Energy Report Kenya* (Netherlands Enterprise Agency 2018) 6.
85 Ibid.
86 The Energy Act 2019, s 36, 38.
87 The Energy Act 2019, s 37(2).
88 The Energy Act 2019, s 117.
89 Ibid.
90 Republic of Kenya's Ministry of Energy, 'Off-Grid Application Flow Chart' (Kenya's Ministry of Energy, 1 November 2021) <https://renewableenergy.go.ke/licensing/clearances-flow-chart/off-grid-applications/> accessed 24th March 2022.

91 Ibid.
92 Ibid
93 Environmental Management and Coordination Act (EMCA) 2018, s 58, 59, and the second schedule of the Act.
94 See broadly The Energy Act 2019, s 121 and 122.
95 Ibid, s 121(3).
96 Ibid, s 121 (4). The EPT is established under Section 25 of the new Act.
97 Ibid, s 122(1).
98 Ibid.
99 Ibid, s 122(2).
100 Republic of Kenya's Ministry of Energy, 'Off-Grid Application Flow Chart' (Kenya's Ministry of Energy, 1 November 2021) <https://renewableenergy.go.ke/licensing/clearances-flow-chart/off-grid-applications/> accessed 24th March, 2022.
101 The Energy Act 2019, s 122(3).
102 The Energy Act 2019, s 122(4).
103 The Energy Act 2019, ss 123 and 124.
104 Ibid.
105 Ibid, s 126(1).
106 Ibid, s 126(2).
107 Ibid, s 126(3).
108 The Government of Kenya, 'Constitution of Kenya, 2010' (Government Printer, 1 May 2011) <http://extwprlegs1.fao.org/docs/pdf/ken127322.pdf> accessed 10 December 2021.
109 The Constitution of Kenya 2010, ss 69(1)(a) and(h); and 260.
110 The Constitution of Kenya 2010, s 69(1)(f).
111 Ibid, (d).
112 The Constitution of Kenya 2010, s 70(1)
113 The Government of Kenya, 'Environmental Management and Coordination Act (EMCA) No. 8, 1999, (as Amended in 2015 and 2018)' (EMCA, 1 June 2018) <https://rise.esmap.org/data/files/library/kenya/Energy%20Access/EA%2021.3A%20Environmental%20Management%20Coordination%20Act,%20No%208%20of%201999.pdf> accessed 10 December 2021.
114 EMCA 2018, s 3.
115 EMCA 2018, s 3A.
116 EMCA 2018, s 7.
117 EMCA 2018, s 9.
118 EMCA 2018, s 58, 59, and the second schedule of the Act.
119 EMCA 2018, s 7 created NEMA, and by Sections 9 and 59, the Act saddled NEMA with the sole governance of EIA procedure in EIA. This function is more detailed in the Environmental (Impact Assessment and Audit) Regulations, 2003 made under the Act.
120 See EMCA 2018, s 9.
121 Environmental (Impact Assessment and Audit) Regulations, 2003, Regulation 4 (2).
122 See Regulation 21.
123 EMCA 2018, s. 75; and Environmental Management and Co-Ordination (Water Quality) Regulations, 2006, regulation 6.
124 Ibid.
125 EMCA 2018 s 12.
126 Ibid, s 129.
127 Ibid, s 130.
128 Ibid, s 25.
129 Jovia Bogere, 'Extractives and the Environment: Kenya's Legal and Institutional Framework' (Justice Project, 12 August 2020) <https://justice-project.org/wp-content

/uploads/2020/08/EXTRACTIVES-AND-THE-ENVIRONMENT-FINAL-PDF .pdf> accessed 10th March 2022, 6.

130 United Nations Environment Programme (UNEP), 'Assessing Environmental Impacts - A Global Review of Legislation, Nairobi, Kenya' (UNEP, 6 February 2018) <https://www.unep.org/resources/assessment/assessing-environment-impacts-global -review-review-legislation> accessed 10th December 2021, 49.

131 Kingsley Onu and Rebecca Sojinu, 'The Quagmire of Environmental Impact Assessment Governance in Nigeria: A Review of the Case of Helios Towers Nig. Ltd V. NESREA' (2021) 20 NIALS ELJ 4.

132 David Hunter, James Salzman and Durwood Zaelke, *International Environmental Law and Policy* 250 (Foundation Press 1998).

133 Kingsley Onu and Rebecca Sojinu, 'The Quagmire of Environmental Impact Assessment Governance in Nigeria: A Review of the Case of Helios Towers Nig. Ltd V. NESREA' (2021) 20 NIALS ELJ 4.

134 EPRA, 'Tariff Setting: Electricity' (EPRA, 12 December 2020) <https://www.epra.go .ke/services/economic-regulation/tarrif-setting/tarrif-setting-electricity/> accessed 25th March 2022.

135 Energy Act, 2019, s 91.

136 This REFiT system is better discussed under the next subheading.

137 Energy Act, 2019, s 91.

138 All Africa, 'The Nation, Kenya: Tax Exemption for Solar Will Woo Investors' (All Africa, 29 June 2021) <https://allafrica.com/stories/202106290154.html> accessed 11 December 2021. It is usually commenced through a demand forecast of both bulk and retail markets based on social, economic and political environment. Secondly, sector revenue requirements will be determined 'based on forecasts of costs likely to be incurred for generation, transmission, distribution and supply of power'. Third, the 'determination of marginal costs of generation, transmission, distribution and retailing; based on approved pricing periods.

Other steps include:

- Allocation of total revenue requirement among approved customer classes on the basis of the marginal costs and price sensitivity
- Computation of initial retail tariff proposals
- Sensitivity analysis of the proposed retail tariffs in order to fine-tune the proposed retail tariffs.
- Public exposure of the proposed tariffs, leading to public debate and hearing
- Determination of the final retail tariffs'.

139 Energy Act 2019, s 91(1).

140 International Energy Agency (IEA), 'Tax Incentives for Renewable Energy' (IEA, 1 June 2016) <https://www.iea.org/policies/6007-tax-incentives-for-renewable -energy> accessed 3rd June 2022.

141 All Africa, 'The Nation, Kenya: Tax Exemption for Solar Will Woo Investors' (All Africa, 29 June 2021) <https://allafrica.com/stories/202106290154.html> accessed 11th December 2021.

142 Ibid.

143 Ibid.

144 Government of Kenya, 'Special Economic Zone Act, 2015 (SEZ)' (Kenya Law, 1 May 2015) <http://kenyalaw.org/kl/fileadmin/pdfdownloads/Acts/2015/Special EconomicZonesNo16of2015.pdf> accessed 25 March 2022.

145 See SEZ Act, 2015, s 15. See also Emmanuel Laryea, Dennis Ndonga and Bosire Nyamori 'Kenya's Experience with Special Economic Zones: Legal and Policy Imperatives' (2020) 20(2) AJICL 171–194, 171. <https://www.euppublishing.com/doi/full/10.3366/ajicl.2020.0309> accessed 12 December 2021.

146 Ibid.
147 Ibid.
148 Julius Wako and Jacinta Nagumo, 'Renewable Energy Law and Regulation in Kenya' (CMS, 12 June 2021) <https://cms.law/en/int/expert-guides/cms-expert-guide-to-renewable-energy/kenya> accessed 12th December 2021.
149 Stamp Duties Act, No. 8, 2021, s 95.
150 Stamp Duties Act, s 117.
151 Ibid.
152 Lawal (n 3) 91.
153 Kenya's Updated Nationally Determined Contribution (n 21).

Bibliography

Books

Hunter D, Salzman J and Zaelke D, *International Environmental Law and Policy 250* (Foundation Press, New York 1998).

Kapika J and Eberhard A, 'Enabling Private-Sector Participation in Electricity Generation' in J Kapika and A Eberhard (eds), *Power-Sector Reform and Regulation in Africa: Lessons from Kenya, Tanzania, Uganda, Zambia, Namibia and Ghana* (HSRC Press 2013) 21.

Articles

Kuo L, 'Kenya's National Electrification Campaign is Taking Less Than Half the Time It Took America' *Quartz Africa (Online)* (16 January 2017) <https://qz.com/882938/kenya-is-rolling-out-its-nationalelectricityprogram-in-half-the-time-it-took-america/>.

Laryea E, Ndonga D and Nyamori B, 'Kenya's Experience with Special Economic Zones: Legal and Policy Imperatives' (2020) 20(2) AJICL 171–194, 171. <https://www.euppublishing.com/doi/full/10.3366/ajicl.2020.0309> accessed 12 December 2021.

Njau M, Paras S and Rainbow F, 'New Energy Act Embraces Renewable Energy' *Bowmans* (2 December 2021) <https://www.bowmanslaw.com/insights/oil-gas/new-energy-act-embraces-renewable-energy/>.

Ole N C, 'The Nigerian Electricity Regulatory Framework: Hotspots and Challenges for Off-grid Renewable Electricity Development' (2020) JENRL <https://doi.org/10.1080/02646811.2020.1771845> accessed 25 March 2022.

Onu K and Sojinu R, 'The Quagmire of Environmental Impact Assessment Governance in Nigeria: A Review of the Case of Helios Towers Nig. Ltd V. NESREA' (2021) 20 NIALS ELJ 4.

Wako J and Nagumo J, 'Renewable Energy Law and Regulation in Kenya' *CMS* (12 June 2021) <https://cms.law/en/int/expert-guides/cms-expert-guide-to-renewable-energy/kenya> accessed 12 December 2021.

Websites/Blogs

Bogere J, 'Extractives and the Environment: Kenya's Legal and Institutional Framework' *Justice Project* (12 August 2020) <https://justice-project.org/wp-content/uploads/2020/08/EXTRACTIVES-AND-THE-ENVIRONMENT-FINAL-PDF.pdf> accessed 10 March 2022, 6.

Export Solutions, 'Kenya: Electrical Power Systems' <https://www.export.gov/article?id=Kenyaelectricalpower-systems> accessed 16 April 2022.

Fortuna C, 'Renewable Energy in Kenya: Meeting the Needs of an Expanding Population' *Clean Technica (Online)* <https://cleantechnica.com/2018/04/16/renewable-energy-in-kenya-meeting-the-needs-of-an-expandingpopulation/> accessed 16 October 2021.

Kenya Renewable Energy Association, *Wind Energy* <http://kerea.org/renewable-sources/wind-energy/>.

Newspapers

All Africa, 'The Nation, Kenya: Tax Exemption for Solar Will Woo Investors' *All Africa* (29 June 2021) <https://allafrica.com/stories/202106290154.html> accessed 11 December 2021.

POLICY DOCUMENT/STATUES

Energy and Petroleum Regulatory Authority (EPRA), 'Energy Act 2019. Kenya Gazette Supplement No. 29 (Acts No. 1)' *EPRA* (12 November 2020) <https://www.epra.go.ke/download/the-energy-act-2019/> accessed 30 November 2021.

Environmental (Impact Assessment and Audit) Regulations, 2003 (Kenya).

Environmental Management and Co-Ordination (Water Quality) Regulations, 2006.

EPRA, 'Tariff Setting: Electricity' *EPRA* (12 December 2020) <https://www.epra.go.ke/services/economic-regulation/tarrif-setting/tarrif-setting-electricity/> accessed 25 March 2022.

Government of Kenya, 'Special Economic Zone Act, 2015 (SEZ)' *Kenya Law* (1 May 2015) <http://kenyalaw.org/kl/fileadmin/pdfdownloads/Acts/2015/SpecialEconomicZonesNo16of2015.pdf> accessed 25 March 2022.

Government of Kenya, 'Constitution of Kenya, 2010' *Government Printer* (1 May 2011) <http://extwprlegs1.fao.org/docs/pdf/ken127322.pdf> accessed 10 December 2021.

Government of Kenya, 'Environmental Management and Coordination Act (EMCA) No. 8, 1999, (as Amended in 2015 and 2018)' *EMCA* (1 June 2018) <https://rise.esmap.org/data/files/library/kenya/Energy%20Access/EA%2021.3A%20Environmental%20Management%20Coordination%20Act,%20No%208%20of%201999.pdf> accessed 10 December 2021.

International Energy Agency (IEA), 'Tax Incentives for Renewable Energy' *IEA* (1 June 2016) <https://www.iea.org/policies/6007-tax-incentives-for-renewable-energy> accessed 3 June 2022.

Kyoto Protocol to the United Nations Framework Convention on Climate Change, Report of the Conference of the Parties at its Third Session, 1–11 December 1997, U.N. Doc FCCC/CP/1997/7/Add.1, 18 March 1998, Annex.

Mokveld K and Von Eije S, *Final Energy Report Kenya* (Netherlands Enterprise Agency 2018).

Pueyo A, 'What Constrains Renewable Energy Investment in Sub-Saharan Africa? A Comparison of Kenya and Ghana' (2018) 109 World Development 85.

Republic of Kenya, *Updated Least Cost Power Development Plan Study Period: 2011–2037* (2018) 49, Retrieved October 19, 2021, <http://gak.co.ke/wp-content/uploads/2019/02/Updated-Least-Cost-Power-Development-Plan-2017-2022-min.pdf>; Republic of Kenya, Kenya Vision 2030 (2007) 49.

Republic of Kenya, *Updated Least Cost Power Development Plan Study Period: 2011–2037* (2018).

Republic of Kenya's Ministry of Energy 'Off-Grid Application Flow Chart' *Kenya's Ministry of Energy* (1 November 2021) <https://renewableenergy.go.ke/licensing/clearances-flow-chart/off-grid-applications/> accessed 24 March 2022.

Republic of Kenya's Ministry of Energy, 'Republic of Kenya's National Energy Policy, 2018' *Kenya Power and Lighting Company* (1 October 2018) <https://kplc.co.ke/img/full/BL4PdOqKtxFT_National%20Energy%20Policy%20October%20%202018.pdf> accessed 24 March 2022.

Stamp Duties Act, No. 8, 2021 (Kenya).

The Government of Kenya, 'Kenya's Updated Nationally Determined Contribution (NDC)' *United Nations Framework Convention on Climate Change (UNFCCC)* (20 December 2020) <https://www4.unfccc.int/sites/ndcstaging/PublishedSDocuments/Kenya%20First/Kenya%27s%20First%20NDC%20(updated%20version).pdf> accessed 23 October 2021.

United Nations Environment Programme (UNEP) 'Assessing Environmental Impacts- A Global Review of Legislation, Nairobi, Kenya' *UNEP* (6 February 2018) <https://www.unep.org/resources/assessment/assessing-environment-impacts-global-review-review-legislation> accessed 10 December 2021, 49.

United Nations Framework Convention on Climate Change, Conference of the Parties, Adoption of the Paris Agreement, Dec. 12, 2015, U.N. Doc. FCCC/CP/2015/L.9/Rev/1 (12 December 2015).

World Bank, *The World Bank Electricity Expansion (P103037), Energy & Extractives Global Practice Africa Region* (Report No. ICR00004496 2018) 73.

Thesis

Lawal K M, 'The Role of Legislation in Ensuring Sustainable Energy Development in Nigeria: Lessons from Kenya' LL.M Dissertation submitted to the School of Law, Adelaide University.

6 Nigeria

Eti Best Herbert and Peter Kayode Oniemola

6.1 Introduction

Renewable energy sources, such as biomass, hydropower, solar, and wind energy, among others, have been identified as viable off-grid electricity sources, as distinct from the traditional mode of feeding electricity generated directly to the national grid for onward transmission and distribution to consumers.[1] The Nigerian electricity sector relies heavily on energy sourced from oil and gas and, more recently, renewable energy to a lesser extent. The sole reliance on the grid network for electricity supply across the country has placed Nigeria as a country with a centralised energy system. Despite several decades of operation, the national grid for electricity supply in Nigeria has not been able to find answers to the question of erratic power supply. The grid expansion drive has taken place at a snail's pace, as a large part of the country, particularly rural areas, is yet to be covered by the grid.[2]

Notwithstanding the vast land mass and population of Nigeria, the entire country can come within the grid network, as a country like America with a greater land mass and size boasts 100% electricity coverage through the length and breadth of the country.[3] The transmission network is notorious for being the weak link in the electricity value chain of Nigeria.[4] There have been reports of system breakdowns on several occasions.[5] The transmission lines cannot carry and wheel a load of power captured by electricity generation companies, as only 3,879 MW out of the 12,522 MW generated power are transmitted while 2,519 MW reaches the final consumers as a result of huge system loss.[6] Experts have argued that there is no way Nigeria can experience a constant power supply if there is a continuous total dependence on the transmission network.[7]

The foregoing challenges necessitate the need for alternative energy sources for Nigeria. Off-grid electricity readily fits as a viable means to complement grid electricity towards meeting the energy needs of the teeming unserved and underserved population in the country. Renewable energy also serves as a viable energy source for off-grid platforms. Several reasons account for this: renewable energy is easily sourced from the environment and it is closer to the point of use. Also, given the global drive towards green technology and phase-out of fossil-based energy sources due to their environmentally harmful properties, it is only reasonable that Nigeria starts making developmental moves in that direction. The reform

DOI: 10.4324/9781003178088-8

of the power sector envisages the emergence of independent power producers, deployment of off-grid electricity solutions, and utilisation of renewable energy sources. Therefore, given the enabling regulatory environment, there are available investment opportunities in off-grid electricity generation and distribution in Nigeria.

In view of the foregoing, this chapter shall, therefore, examine the imperatives for developing renewable energy potential in Nigeria, as well as the opportunities and barriers that affect such development. It will examine the major policies and targets supporting the development of off-grid renewable energy; the regulatory environment in which off-grid renewable energy will operate will be analysed to determine the extent to which they are viable for the deployment of off-grid renewable energy in Nigeria.

6.2 Imperatives, Opportunities, and Barriers to Off-Grid Renewable Electricity

Nigeria is rich in renewable energy sources, as well as other sources of energy. Virtually the whole country is endowed with abundant solar energy and biomass for the generation of electricity. The potential of renewable energy to contribute to the power generation capacity of Nigeria is enormous. Due to its massive land mass and tropical vegetation, northern Nigeria enjoys an abundance of wind supply which can be a useful energy source. It is estimated that should 5% of the land mass of northern Nigeria be set out for the development of solar-based power generating plants; it can hypothetically contribute 42,700 MW to the power generation capacity of the country.[8] The rainforests of the southern part of Nigeria have a large body of water, which can serve as hydropower sources for the construction of hydroelectric dams. Also, by virtue of being a coastal state with a southern boundary contiguous to the Atlantic Ocean, the tidal waves from the Atlantic can be utilised as an energy source.

It has to be noted that a major criticism against renewable energy has been the question of variable weather and climate conditions. The changing seasons of the year implies that the different forms of renewable energy would have an on- and off-peak period, depending on the season of the year. During such off-peak periods, electricity tends to be in short supply. Hybrid renewable energy options can be used to circumvent this challenge. For instance, an off-grid platform may be powered by a mix of solar and wind, solar and hydro, wind and hydro, or hydro and biomass hybrid systems, in anticipation of the changing weather conditions associated with different seasons.

It is true that Nigeria has a high deficiency in rural electrification. An estimate of 77 million Nigerian rural dwellers lack access to modern electricity.[9] While this is a not such pleasant news, the flipside is that, it confers on Nigeria the status of being a potential hub for off-grid renewable energy investment. Given the renewable energy potentials which have been highlighted above, these renewable energy sources can be deployed by both government and private investors in the development of off-grid rural electrification. Despite its ambition to ensure fresh 130 million grid connections by 2025, the African Development Bank recognised the need to ensure 75 million off-grid connections if the goal of universal energy access is to be attained in the set date of 2030. This is because regions which are not easily accessible by the grid network can only be electrified with the aid of off-grid renewable energy.[10] Indeed, government and investors have taken notice of this potential and keyed into this opportunity. This explains the federal government of Nigeria's recent launch of an off-grid renewable energy electricity project, code named 'Solar Power Naija'. The project, which is a measure of the Economic Sustainability Plan (ESP), has the objective of establishing 5 million solar facilities targeted at meeting the energy needs of 25 million dwellers in rural communities that are located off the grid network.[11] Green Village Electricity (GVE) Ltd has embarked on 24 kW–500 kW capacity mini-grid electricity projects in 72 communities which cut across seven states in the country. The project is targeted at addressing the energy needs of about 73,500 students.[12]

Most businesses and manufacturers depend on the power supply to carry out their business operations. The lack of a sufficient power supply has driven these business owners to resort to electric generators powered by diesel and premium motor spirits. Apart from the environmental hazards arising from noise and fume emission, it increases the production and overhead costs of business owners. The corresponding effect is the increase in the price of goods and services which are borne by final consumers. Off-grid power solutions can be harnessed to address this shortfall of power supply for business owners. By so doing, more persons would be encouraged to invest in business, the cost of goods and services would become affordable, economic productivity would peak up, thereby leading to growth increase in gross domestic product.[13] A practical instance is the test run of the GVE projects which was found to have resulted in a decrease in electricity-related expenses by 40% and increased productivity for agro-processing enterprises and other businesses.[14]

Access to an affordable electricity supply is one of the identified measures of ensuring poverty alleviation and guaranteeing a better standard of living not just for rural dwellers but urban dwellers also. On the contrary, inadequate power supply stands as a major barrier to development, given the limitation it places on energy requirements for meeting critical services in homes as well as agricultural, educational, and health services.[15] Using these renewable off-grid platforms to solve electricity challenges in rural areas would bolster the growth of small and medium range enterprises operating within those regions. This could well serve the purpose of poverty alleviation for residents of those areas.

Cost is a major concern against the deployment of renewable energy in Nigeria. Most rural dwellers, who should be the major beneficiaries of off-grid energy, may not be able to pay for the cost of electricity. This is challenging to investors, who may not likely recoup their investments. This fact would result in investors' reluctance to embark on off-grid energy development, as they would not want to incur huge financial risk without a guarantee to break even. The financing is not readily available from domestic financiers because they regard off-grid electricity endeavours as an untested investment domain with incomprehensible risk to financiers.

Apart from creating better self-reliance in energy generation for domestic and industrial needs, in line with the needs of the country, a renewable energy based off-grid power system will also create a varied and decentralised energy supply option.[16] Off-grid enabled decentralised energy options would ensure less reliance on the national grid, thereby reducing excessive electricity loads placed on the grid network, which also contributes to its periodic breakdown.[17] However, the federalist structure of the country does not make it any easier for the country to attain a decentralised energy system in terms of energy sources and governance.[18] The implication of having electricity under the concurrent legislative list[19] implies that the federal and state government can legislate on the same subject matter. The enactment of the Electricity Power Sector Reforms Act (EPSRA) 2005 by the National Assembly suggests that the federal legislation has covered the field on this subject.[20] As such, this puts the Lagos State Electric Power Sector Reform Law 2018 subservient to the EPSRA. Also, the creation of federal institutions such as the Nigerian Electricity Regulatory Commission (NERC) and Rural Electrification Agency places the federal government in pole position to control and determine the pace of off-grid electricity development in the country. However, it has been argued that the question is not so much about state governments having the constitutional powers to legislate on the subject matter of off-grid electricity; but that most states of the federation do not possess the economic prowess to embark on such capital-intensive projects, due to the lopsided federalism practised by Nigeria.[21]

6.3 Policy Targets and Frameworks for Off-Grid Renewable Electricity

Before 2003, the Nigerian government had not established any known policy that is receptive to renewable energy applications in the energy sector. Hence, there was a lack of coordinated renewable energy engagement in Nigeria. At best, interest and undertakings in renewable energy were only implemented by a few individuals, companies, or communities. The existing national policies of that time had directed its attention to centrally coordinated conventional energy sources. Government incentives were only available for the benefit of entities which are involved in or utilise these conventional energy sources as a means of stimulating investment in this area.[22]

Nigeria's high endowment and dependence in petroleum resources is also reflected in the choice of gas as the major energy source in the power sector. However, the Nigerian government has, through its more recent policies, recognised the need to diversify the energy sources of the electricity sector to achieve an energy mix of renewable and non-renewable energy. Even the Nigerian National Petroleum Policy (NNPP) identifies that one of the strategic policy objectives is 'managing the balance between depleting oil resources vs renewable energy'. The NNPP states that 'the government is aware though that renewable energy is increasingly gaining ground, worldwide and in Nigeria'. It stated further that, 'as renewable energy costs continue to come down, over the long and even medium term, renewable energy is expected to, and should be encouraged to, become a significant part of a diversified energy mix'. If the government is to be faithful to these policies, it is expected that oil and gas companies can be carried along in the maximisation of the energy mix of the country and engaging them in off-grid solutions.

It is not strange that a fossil fuel-based policy would embrace renewable energy. This turn of events is not synonymous with government policies alone. Examples abound where oil and gas companies now venture into renewables as a means of energy diversification. Generation Oil has plans to deploy wind and solar technology to operate their offshore oil platforms.[23] Shell has announced its intention to pack up all its onshore oil platforms in Nigeria in its transition bid towards non-emission of carbon by 2050.[24] On the part of Nigerian National Petroleum Corporation, apart from its drive towards the attainment of clean development mechanisms, it has commenced the development of off-grid energy retail mega-stations and other formations with the aid of solar photovoltaic technology.[25]

The National Energy Policy (NEP) 2003 expresses support for renewable energy. The plan of the NEP 2003 is to harness the various forms of renewable energy sources for development of robust energy needs.[26] The National Renewable Energy and Energy Efficiency Policy (NREEEP) was initiated by the Nigerian government in 2015. Contained in the policy are frameworks for integration of renewable energy into the energy mix, including off-grid renewable electricity promotion. Part of its key objectives is to use hydro, non-wood fuel biomass, solar, and wind energy sources for rural and off-grid electricity development.[27] The target is to employ an increased energy mix to guarantee off-grid electricity supply to remote areas of Nigeria in tandem with the ECOWAS Renewable Energy Policy target[28] of attaining a 22% and 25% increment in the number of rural dwellers that are served with off-grid renewable electricity by 2020 and 2030, respectively.[29] The proposed strategy for achieving this outcome is to create market incentives for private sector engagement in off-grid undertakings, while federal government performs the role of framework formulation and offeror of financial guarantees for its implementation.[30]

The World Bank's Regulatory Indicators for Sustainable Energy Report of 2016 indicates that, Nigeria has a colossal energy deficit due to an underdeveloped policy environment for the sustenance of energy access.[31] So, therefore, it is not the case that Nigeria lacks policies on renewable energy; but that the renewable energy resource prospects of Nigeria remain under-utilised, largely owing to the shortfall of investment in this regard. Among the several factors responsible for this turn of events include policymakers' lack of commitment towards implementation; communication breach between policymakers and regulators; lack of enabling atmosphere for policies to thrive.[32] There are no obvious attempts to integrate renewable energy sources into the power sector as envisioned under the NEP 2003 and other policies on renewable energy.

Policy statements and objectives of the government is not within the hierarchy of law; hence, lack imperative status to command compliance. Policies should essentially culminate into legislative enactments, which in turn gives coercive impetus to policy objectives.[33] The National Assembly, which is conferred with law-making powers, has failed to give the NREEEP and similar policies the required legal impetus as a substantive body or as a subsidiary legislation by the relevant administrative organ.[34] This would imply that reasonable steps should be taken to not only enact legal support for renewable energy infusion to the power sector, but the government must also be committed to implement the laws.[35] At present, the government lacks the political will to adopt a regulatory framework that will enable it to make a significant change and/or improvement in the nation's electricity usage from conventional to renewable energy sources. This view suggests that the Nigerian government is not doing enough towards transitioning from the use of conventional energy to renewable energy.[36]

6.4 Regulatory Environment

As discussed above, the Nigerian electricity sector is highly centralised. Though the Constitution of Nigeria conceives that both the federal and state government can be involved in the generation of electricity, the federal government has made laws that are regarded as all-encompassing and much overbearing on states. Thus, states are left with no powers to regulate but to merely initiate renewable off-grid electricity projects. The EPSR Act 2005 is the principal law for regulation of electricity in Nigeria. The law establishes NERC as the apex body conferred with regulatory powers in the power sectors.[37] These powers include monitoring, setting out market rules, tariff regimentation, etc.[38]

6.4.1 Licensing Regime

The advent of privatisation as an offshoot of power reforms has liberalised the power sector to enable private investors to operate electricity generation, distribution, and transmission undertakings as licensees of NERC.[39] However, it is noted that electricity transmission is yet to be liberalised, as only Transmission Company of Nigeria (a company wholly owned by the federal government of Nigeria) has been issued with an operating licence. Though there are 11 distribution companies, these companies solely cover franchise areas spanning about three to four states in most cases. It is also appreciated that the generation segment of the Nigerian power sector is not that restricted. Indeed, it is the most liberal unit of the power sector. With the emergence of independent power producers and in furtherance of the objectives of the power sector reform in Nigeria, investors can consider investing in the Nigerian energy sector. The advantage this state of affairs serve for off-grid investors is that, since most off-grid platforms can be operated with the aid of electricity transmission and distribution companies, the bottleneck associated with procuring the services or getting licences for electricity transmission and distribution companies has been avoided.

The EPSRA does not really have provisions that directly and expressly addresses off-grid regulation. However, pursuant to its law-making powers as contained in Sections 45(2) and 96(1) of the EPSRA 2005, NERC has exerted its authority as an off-grid electricity regulator by the establishment of the following regulations: NERC (Permits for Captive Power Generation) Regulation 2008; NERC (Embedded Generation) Regulation 2012; NERC (Independent Electricity Distribution Network) Regulation 2012 and NERC Mini-grid Regulation 2016. These are the major regulations applicable to off-grid electricity undertaking.

It is important to emphasise that, licence is not required for electricity generation undertaking that is less than a cumulative of 1 MW in a particular location.[40] Off-grid developers must not generate power of more than 1 MW in a specific site, else they would be required to obtain NERC electricity generation licence as on-grid electricity generators would do.[41] This implies that the developer would have to pay for operating licence and licence application fees to onward licence processing by NERC.[42] The cost implication thereof is US$2,500 and N50,000, respectively.[43] This would further compound the challenge of capital financing for developers and, by extension, consumers of off-grid electricity projects. However, captive power generators and developers of mini-grid within the range of 1 MW are required to obtain a permit from NERC before they can undertake any of these endeavours.[44] Supply of excess power not in use by any of the permit holders would also require written consent had, and obtained, from NERC.[45] There are other regulatory considerations which off-grid developers must bear in mind. The following section of this chapter shall address some of these issues and the challenges associated therewith.

6.4.2 Environmental Regulations

Renewable energy is generally referred to as clean energy. However, it is not without environmental implications. The process of electricity generation may lead to pollution or spillages. Given the global sustainable development drive, it is widely accepted that developmental projects should not be carried out at the detriment of the environment. Hence, the regulation of the environment is very important. This is the essence of Environmental Impact Assessment (EIA) Act 1992. The EIA Act compulsorily require that EIA should be carried out before embarking on projects which portends environmental risks. Electricity generation is identified as an example of projects that require compliance with EIA, hence electricity licensees of renewable off-grid projects are bound by this provision.[46] The EIA Act expressly mentions power projects such as: dam, steam, hydro and combined cycle power plants.

Although EIA is generally under the regulatory control of National Environment Management Authority (NEMA), NEMA usually rely on the expertise of industry specific regulators as 'lead agencies'[47] to give their expert evaluation on their industry related projects. NERC stands the opportunity to function as a lead agency on proposed power projects, whether off-grid or renewable energy based. Neither the EPSRA 2005 nor any NERC regulation require EIA clearance as a condition for issuance of electricity licences to applicants. In line with sustainable development requirements and its obligation as a lead agency, NERC is expected to establish an industry requirement for EIA in order to gain some certainty and transparency in the environmental requirements for power projects.

The Penal Code and the Criminal Code criminalises the release of harmful substances into the environment. Section 247 Criminal Code 1916 regards acts which are likely to upset air quality as public health offences. Section 245 Criminal Code prohibits acts which foul water reservoir courses, otherwise making them less fit for ordinary usage. Both offences are misdemeanours punishable with six months' imprisonment. Section 13(1) Public Health Act 1917 contains similar provision. It is also important that relevant provisions of the National Environmental Standards and Regulations Enforcement Agency (NESREA) Act 2007 and regulations on environmental safety should be taken into account. Hence, developers of hydro- and biomass-based off-grid platforms must, respectively, take cognisance of these penal provisions. The tort of trespass to land, private and public nuisance can also apply in the case of noise pollution and release of toxic substances to the environment, in the course of developing and operating renewable energy off-grid electricity facilities. Whereas developers accumulate substances, in the non-natural use of the land, the rule in *Ryland v. Fletcher*[48] apply to occasion civil liabilities on the owner of such facilities, if the substance escapes and causes damage to adjoining land.

6.4.3 Pricing Mechanisms

The pricing of electricity is a critical issue in Nigeria. The tariff for electricity in Nigeria is fixed by NERC.[49] While the Constitution empowers state governments to undertake off-grid power projects, they cannot fix the tariff regime for such a project. By virtue of being a federal body, NERC enjoys the constitutional right of regulating electricity undertakings, including tariff fixing. Thus, any state that purports to fix tariffs would be acting unconstitutionally and such action would be void to the extent of its inconsistency.[50]

NERC has adopted the Multi Year Tariff Order (MYTO), as the methodology for fixing on-grid electricity pricing. Electricity producers and distributors cannot fix tariffs outside the NERC approval price.[51] Even off-grid producers may need to have approval for their electricity prices. In the case of *Funke Adekoya SAN v. VGC Management & Maintenance Co. Limited & Eko Electricity Distribution Company*,[52] it was held that a NERC licence and approved price regime is required of an off-grid electricity facility in the respondents' estate, as the aggregate generated power exceeded 1 MW. The MYTO regime is mostly applicable to on-grid electricity supply.

NERC has created the feed-in tariff, as a special regime applicable to renewable energy, through the Regulation for Feed-in Tariff of Renewable Energy Sourced Electricity in Nigeria (REFIT) 2015. The REFIT Regulation set out an electricity pricing scheme for a fixed period for renewable energy sourced electricity. The aim is to guarantee renewable energy developers' profitability from their investment.[53] The tariff would serve as an incentive for companies to invest in renewable energy. However, off-grid electricity is expressly excluded from the REFIT pricing regime.[54]

The various off-grid regulations contain separate regimes on how prices of off-grid electricity are determined. For isolated mini-grid, the registered developers are required to execute the 'tripartite contract' between the mini-grid investor, the distribution company, and host community.[55] Consensus is reached between these parties in order to determine the price of off-grid electricity. This transaction is flexible and subject to the negotiating power of the parties concerned. Hence, the pricing is flexible, as different communities may be paying different costs for electricity, depending on the peculiarity of their situation. In the case of interconnected mini-grid, apart from the three parties reaching a consensus, MYTO methodology also applies, as approved by NERC.[56]

6.4.4 Support Schemes

The REFIT Regulation 2015 provides incentives for investors who develop renewable energy sourced electricity. REFIT assures renewable energy developers of access to the grid network for a 20-year duration for solar, biomass, and small-hydro power plant life-span.[57] These incentives are meant to support the deployment of renewable energy for electricity in Nigeria. As noted earlier, these incentives are only available to renewable energy served into the national grid network, excluding off-grid. Providing subsidy for grid power without a corresponding incentive for off-grid generators makes off-grid renewable electricity investment less attractive for investors. The lack of an equal playing field for undertakings form a daunting obstacle for off-grid development.[58] Even the lofty provision of the REFIT Regulation is yet to be actualised. Although Nigerian Bulk Electricity Trader NBET had signed a power purchase agreement (PPA) with 14 on-grid renewable energy developers, the companies later pulled out of the deal on the ground that the agreed tariff plan of US$0.115/kWh was unilaterally reduced to US$0.75/kWh by NBET.[59]

Renewable energy investors require support in form of tax incentives, aid, grants, subsidy or duty importation of renewable energy technology, loan support at interest free or low rates,[60] and feed-in tariff for off-grid renewable electricity. The Finance Act 2020 amended the Industrial Development (Income Tax Relief) Act. By this amendment, a maximum 2.5% withholding tax cap has been placed on power projects.[61] Developers of renewable energy power projects can benefit from this provision, as value added tax for renewable energy technology for power projects must not exceed 2.5%. Duties have been removed for importation of materials used for solar power plants. However, this incentive has been said to be merely theoretical. Light Up Africa, a solar energy vendor, experienced a situation where Nigerian customs officials regarded integrated solar lights to be lanterns and included other charges which raised the importation cost to 100%.[62]

Funding for off-grid renewable electricity rural electrification projects can be obtained through the Rural Electrification Fund (REF) created under the EPSRA.[63]

The REF is under the management and administration of the Rural Electrification Agency (REA) which the Act created.[64] The REF has been conceived to provide funding for electrification of the country by extending electricity to rural areas not covered by the national grid. The specific objectives of the fund are: attainment of equitable spread of electricity to the various regions of the country; maximise a sustainable use of rural electrification subsidies; ensure grid expansion and off-grid development; encourage innovative deployment of rural electrification, which could encompass renewable energy sources.[65] The EPSRA provided for the strict application of the REF for the purpose of its creation, and therefore, restrained the misappropriation of the REF for payment of electricity consumption subsidies.[66]

The REF derives its funding from various sources, including budgetary allocation by the National Assembly; surpluses from NERC budget;[67] fines exacted on defaulting licencees;[68] gifts, grants, or loans received from government, local communities, corporate entities, and international bodies; financial contributions made by eligible customers;[69] savings interest and investment returns of the REF. Although NERC may attach a certain percentage of licensees' annual income for REF funding,[70] there is no record of this discretion being exercised. The REA has set aside US$150 million to fund 'minimum subsidy tender' and 'performance-based grant program' for mini-grid projects.[71] Given the magnitude of rural electrification development required in the country, the REF certainly needs an expanded and larger funding base. Other suggested forms of funding include introduction of a renewable energy development charge, REF tax; corporate social responsibility obligations directed at rural electrification; crowd-funding and private equity funding for renewable energy based rural electrification.[72]

Also, in addressing the funding challenge, the African Development Bank has been one of the foremost financiers for entities interested in venturing into off-grid electricity. In a collaborative effort with the Nigerian government, the National Electrification project was launched in 2020, towards the development of mini-grid facilities. The African Development Bank and the Africa Growth Together Fund have jointly earmarked the sum of US$200 million to fund private undertakings in the Nigerian off-grid power sector.[73] This has been followed by a recent collaborative effort wherein the African Development Bank is to contribute US$10 million, while All On is to contribute US$5 million towards the creation of the Nigeria Energy Access Fund to offer financial assistance to solar off-grid development companies.[74] Another off-grid renewable energy project has been embarked upon by Havenhill Synergy Limited, receiving Nigeria Infrastructure Debt Fund (NIDF) support of up to N1.8 billion from Chapel Hill Denham. The target is to bankroll the erection of 22 mini-grid facilities which will connect 70,000 residents of the beneficiary communities to a power supply when the project is concluded.[75] With the availability of the foregoing funds and others, developers of off-grid platforms can get new a lease of life for solving the nagging problem of financing or the huge interest rates associated with financing offers from local commercial banks.[76]

6.4.5 Purchasing Arrangements

The purchase of electricity in Nigeria is based on vesting contract. Generators can enter into PPA with distribution companies to directly provide electricity for transmission in the distribution network or with NBET.[77] Ordinarily, party autonomy prevails in contractual agreements. But the special contractual nature of PPA does not allow it to go without regulatory surveillance. Hence, parties to PPA can only exercise their freedom of contract to the extent that the law and regulatory stipulations allow. For instance, a power purchaser cannot execute a PPA without submitting the tender report to NERC to confirm that there was a bid for the purchase and that a successful bidder had emerged.[78] Also, the terms of the PPA must not exceed the stipulated tariff term set out by NERC. These make it necessary for NERC to approve PPA before it takes effect.[79]

The REFIT Regulation provides a partial risk guarantee for renewable energy power to be channelled into the grid. Out of the 2,000 MW renewable energy target of the Regulation, NBET statutorily undertake to purchase 50% from the developers, while the remainder is left for the distribution companies to procure. This offers a partial risk guarantee for renewable energy developers, in order to encourage investment in this regard. As noted earlier, currently, there is no purchase guarantee for off-grid electricity developers. Off-grid power is purchased based on agreement reached by the developer and consumers. If an electricity distribution company is in the picture, it may make a lump sum payment to the developer, while it progressively realises its cost from the consumer. However, distribution companies do not find it attractive to invest in this kind of scheme. This is because off-grid electricity is usually required in unserved rural areas largely occupied by poor rural dwellers, who may lack the ability to pay for a power supply. This is one of the reasons why there is a dearth of financing in this area. Hence, off-grid developers have to rely on several funding incentives if they must take up such an endeavour.

6.5 Conclusion

Nigeria is rich in renewable energy sources. Exploring the deployment of renewable energy for off-grid solutions is a viable option that should be explored. Despite the potential of renewable off-grid electricity options to solve the energy poverty situation of the country, its successful deployment has continued to be hindered by several obstacles highlighted in this chapter. These include incoherent policies and institutional framework, weak policy implementation, financing, power purchase, and pricing challenges. The current legal and institutional framework will require reform to provide adequate support for off-grid renewable energy. This implies the need to improve the regulatory environment if off-grid renewable energy deployment is to create a significant impact in resolving Nigeria's energy deficiency.

Notes

1 EB Herbert, 'Developing a Renewable Energy Based Off-Grid Electricity Solution for Nigeria' (2021) 2(2) Global Energy Law and Sustainability 182, 200.

2 VR Osu, 'A Critical Evaluation of the Prospects for a Transition Towards Ocean Based Renewable Energy Development in Nigeria' (PhD Thesis, Robert Gordon University, 2017) 95.

3 L Pellegrini and L Tasciotti, 'Rural Electrification Now and Then: Comparing Contemporary Challenges in Developing Countries to the USA's Experience in Retrospect' (2012) Forum for Development Studies 1, 1.

4 'Why Transmission is Weakest Link in Power Chain' *The Guardian*, 25 August 2016, <https://m.guardian.ng/news/why-transmission-is-weakest-link-in-power-chain/> accessed 22 June 2021.

5 There are record of grid breakdown on more than 126 occasions since 2013 till date. See, K Jeremiah, 'Power Grid Fails 126 Times in 7 Years', *The Guardian*, 8 December 2020, <https://m.guardian.ng/news/power-grid-fails-126-times-in-7-years/> accessed 22 June 2021.

6 GC Adeyanju, OA Osobajo, A Otitoju and O Ajide, 'Exploring the Potentials, Barriers and Option for Support in the Nigeria Renewable Energy Industry' (2020) 1(7) Discover Sustainability 1, 5.

7 'Nigeria Will Never Have Stable Power Using Main Grid, Ex-Power Minister' *Thisday*, 16 July 2018, <http://energymixreport.com/nigeria-will-never-have-stable-power-using-main-grid-ex-power-minister/> accessed 22 June 2021.

8 C Newsom, 'Renewable Energy Potential in Nigeria: Low-Carbon Approaches to Tackling Nigeria's Energy Poverty' International Institute for Environment and Development the SUNGAS project (2012) 15, <http://www.iied.org_G03512.pdf> accessed 22 June 2021.

9 MY Roche, H Verolme, C Agbaegbu, T Binnington, M Fischedick and EO Oladipo, 'Achieving Sustainable Development Goals in Nigeria's Power Sector: Assessment of Transition Pathways' (2020) 20(7) Climate Policy 846, 846.

10 RJ Heffron, HI Hussein, CH Yang and N Sun, 'The Global Future of Energy Law: 2017 Review' (2017) 8 International Energy Law Review 291, 293.

11 'Nigeria Announces New Energy Access Project 'Solar Power Naija' Industry Sectors Future Energy' News, 7 December 2020, <https://www.esi-africa.com/industry-sectors/renewable-energy/nigeria-announces-new-energy-access-project-solar-power-naija/> accessed 22 June, 2021.

12 'GVE Nigeria', <https://repp.energy/project/gve-nigeria/> accessed 22 June 2021.

13 E Ejoh, 'How Off-Grid Devt'll Rescue Nigeria Power Sector', <https://rea.gov.ng/off-grid-devt-ll-rescue-nigeria-power-sector/> accessed 22 June 2021.

14 'GVE Nigeria', n. 12.

15 YO Akinwale and IO Ogundari, 'Exploration of Renewable Energy Resources for Sustainable Development in Nigeria: A Study of the Federal Capital Territory' (2017) 7(3) International Journal of Energy Economics and Policy 240, 241.

16 UB Akuru and OI Okoro, 'Renewable Energy Investment in Nigeria: A Review of the Renewable Energy Master Plan' (2014) 25(3) Journal of Energy in Southern Africa 67, 71.

17 Y Oke, 'Beyond Power Sector Reforms: The Need for Decentralised Energy Options ('DEOPs') for Electricity Governance in Nigeria' in Yemi Oke (ed.) *Essays on Nigerian Electricity Law* (Princeton Publishing 2016) 120.

18 *Ibid.*

19 Item 13 and 14, Part II of the 2nd Schedule to the Constitution of the Federal Republic of Nigeria 1999 (as amended).

20 PK Oniemola, 'Developing a Legal Framework for Promoting Investment in Renewable Energy in the Nigerian Power Sector: An Analysis of the Design and Implementation Challenges' (PhD Thesis, University of Aberdeen, 2014) 19.

21 EB Herbert, 'Application of Electricity Federalism in Nigeria: Drawing Inspiration from America' (2021) 29(2) African Journal of International and Comparative Law 223, 241.

22 H Ajaelu and R Okereke, 'Assessment of Nigerian Renewable Energy Potentials and Energy Poverty: Challenges and Prospects' (2020) 9(6) PM World Journal 1, 10.
23 RJ Johnson, R Blackmore and R Bell, 'The Role of Oil and Gas Companies in the Energy Transition' *Atlantic Council* (2020) 32, <https://www.atlanticcouncil.org/in-depth-research-report/the-role-of-oil-and-gas-companies-in-the-energy-transition/> accessed 30 June 2021.
24 L Hurst, 'Shell in Talks with Nigerian Government to Exit Onshore Oil fields as Part of Green Push' *World Oil* (2021) <https://www.worldoil.com/news/2021/5/18/shell-in-talks-with-nigerian-government-to-exit-onshore-oil-fields-as-part-of-green-push> accessed 30 June 2021.
25 Nigerian National Petroleum Corporation, 'Renewable Energy Division' (2020) <https://www.nnpcgroup.com/ventures/Pages/RenewableEnergy.aspx 18> accessed 30 June 2021.
26 Energy Commission of Nigeria, *National Energy Policy* (2003) 23–34, <http://fao.org/docs/pdf/ins64284.pdf> accessed 22 October 2021.
27 Federal Republic of Nigeria, 'National Renewable Energy and Energy Efficiency Policy (NREEEP)' (2015) 11, 13, 14, 16, <http://admin.theguides.org/Media/Documents/NREEE%20POLICY%202015-%20FEC%20APPROVED%20COPY.pdf> accessed 22 October 2021.
28 *Ibid.* 20.
29 ECOWAS Center for Renewable Energy and Energy Efficiency, 'ECREEE Rural Electrification Programme', <http://www.ecowas.int/wp-content/uploads/2015/11/ECREEE-FLYER-TEMPLATES_VERSION_RURAL-ELECTRIFICATION_09112015.pdf> accessed 22 October 2021.
30 Federal Republic of Nigeria n. 27.
31 Heffron, n. 10.
32 S Duru and A Oni, 'A Succinct Guide to Investing in Nigeria's Renewable Energy Sector' *Banwo & Ighodalo*, <https://www.banwo-ighodalo.com_6cbd0c5b298daf9fe0754887c856e429.pdf> accessed 22 June 2021.
33 Y Omorogbe, *Why We Have No Energy* (University of Ibadan Press 2008) 21.
34 CV Basil, 'Renewable Energy Deployment in Nigeria', African Energy and Mineral Management Initiative 10, <https://africanenergyminerals.org-Chioma-V-Basil-MAIN-ARTICLE-ON-ENERGY-POLICY-DEPLOYMENT-IN-NIGERIA-converted.pdf> accessed 22 June 2021.
35 MF Akorede, O Ibrahim, SA Amuda, AO Otuoze and BJ Olufeagba, 'Current Status and Outlook of Renewable Energy Development in Nigeria' (2017) 36(1) Nigerian Journal of Technology 196, 210.
36 There are records of grid breakdown on more than 126 occasion since 2013. See K Jeremiah, 'Power Grid Fails 126 Times in 7 Years' *The Guardian*, 8 December 2020, <https://m.guardian.ng/news/power-grid-fails-126-times-in-7-years/> accessed 22 June 2021.
37 Section 31 (1) EPSRA 2005.
38 Section 32 (2) EPSRA 2005.
39 OC Tasie and PK Oniemola, 'Legal Analyses of Liberalisation and Privatisation of State-Owned Companies in the Nigerian Power Sector' (2016) 27(5) International Company & Commercial Law Review 141–147.
40 Section 62 (1) and (2) EPSRA 2005.
41 Section 8 (b) NERC Regulations for the Granting of Permits for Captive Power Generation 2008.
42 See Regulation 6 NERC Regulation for the Application for Licence (Generation, Transmission, System Operations Distribution and Trading) 2010.
43 Resolution Law Firm, 'How to Obtain Power Generation License in Nigeria' <https://www.resolutionlawng.com/how-to-obtain-power-generation-license-in-nigeria/#:~

:text=The%20prescribed%20generation%20licence%20fees,a%20licence%20fee%20of%20%242%2C500.00> accessed 22 October 2021.

44 See, Section 3 NERC Regulations for Permits for Captive Power Generation 2008 and Section 6 (2) NERC Mini-grid Regulation 2016.

45 Section 8 (a) NERC Regulations for the Granting of Permits for Captive Power Generation 2008.

46 Item 13 Section 64 EIA Act 1992.

47 See, Sections 2 and 6 (2) EIA Act 1992.

48 (1866) L.R. EX 265.

49 Section 32 (1) d. (2) c. EPSRA 2005.

50 PK Oniemola and OC Tasie, 'Incorporating Off-Grid Renewable Energy Options under the Framework for Rural Electrification in Nigeria' (2016) 34(8) International Energy Law Review 327, 333.

51 See, *Petadis Enterprise v HFP Properties Limited* NERC//10/0011/08.

52 NERC/H/061.

53 N Saifuddin, S Bello, S Fatihah and KR Vigna, 'Improving Electricity Supply in Nigeria-Potential for Renewable Energy from Biomass' (2016) 11(14) International Journal of Applied Engineering Research 8322, 8332.

54 Section 6 (b) REFIT Regulation 2015.

55 Section 6 (2) Mini-grid Regulation 2016.

56 *Ibid.*

57 SG Dauda and SO Idehen, 'An Examination of the Implementation of Existing Policies on Renewable Energy in Nigeria: How Effective' (2021) 9 Journal of Power and Energy Engineering 104, 112.

58 H Ajaelu and R Okereke, 'Assessment of Nigerian Renewable Energy Potentials and Energy Poverty: Challenges and Prospects' (2020) 9(6) PM World Journal 1, 10.

59 F Adeniyi, 'Overcoming the Market Constraints to On-Grid Renewable Energy Investments in Nigeria' (2019) Oxford Institute for Energy Studies Paper EL 37, <https://doi.org/10.26889/9781784671495> accessed 1 July 2021.

60 OO Ajayi, PA Aderonmu and OO Ajayi, 'Nigeria's Energy Policy: Issues of Sustainable Energy Development in Nigeria', 9, <http:eprints.covenantuniversity.edu.ng_Nigeria's_Energy_Policy_Issues_of_Sustainable_Energy_Development_in_Nigeria.pdf> accessed 22 June 2021.

61 KPMG, 'Finance Act, 2020: Impact Analysis' (2020) 15, <https://assets.kpmg-pdf-tax-finance-act-2020.pdf> accessed 1 July 2021.

62 Newsom, n. 8.

63 See, Section 88 (13) EPSRA 2005.

64 Section 88 (1) EPSRA 2005.

65 Section 67 (13) EPSRA.

66 Section 67 (13) (d) EPSRA.

67 See Section 53 EPSRA.

68 This should extend to regulations made by NERC pursuant to EPSRA.

69 See Section 90 EPSRA.

70 See, generally, Section 88 (12) EPSRA.

71 Rural Electrification Agency, 'Nigerian Electrification Project (NEP) Solar Hybrid Mini Grids Component' <https://rea.gov.ng/minigrids/> accessed 1 July 2021.

72 SM Ohiare, 'Financing Rural Energy Projects in Developing Countries: A Case Study of Nigeria (PhD Thesis, Leicester De Montfort University, 2014), xx+272.

73 African Development Bank, 'African Development Bank, to Provide Underserved Communities in Nigeria with Sustainable Energy Solutions' (2020) <https://www.afdb.org/en/news-and-events/press-releases/african-development-bank-provide-underserved-communities-nigeria-sustainable-energy-solutions-34855> accessed 28 June 2021.

74 African Development Bank, 'Nigeria - Energy Access Fund - SEFA Project Summary Note' 3 March 2021, <https://www.afdb.org/en/documents/nigeria-energy-access-fund-sefa-project-summary-note> accessed 28 June, 2021.

75 Off Grid Energy Independence, '22 Mini Grids for Nigeria Electrification Project' (2021) <https://www.offgridenergyindependence.com/articles/23274/22-mini-grids-for-nigeria-electrification-project> accessed 22 June 2021.

76 African Development Bank, 'Local Currency Financing for Off-Grid Energy Solutions in Africa Limited, Needs Scaling up- African Development Bank Report' (2020) <https://www.afdb.org/en/news-and-events/press-releases-local-currency-financing-grid-energy-solutions-africa-limited-needs-scaling-african-development-bank-report-39698> accessed 28 June 2021.

77 PK Oniemola, SJ Ajayi and EB Herbert, 'Contractual Architecture for Wholesale Electricity Purchase in Nigeria' (2019) 5(2) Journal of Commercial Law 728, 729.

78 Section 10.5 NERC Regulation for the Procurement of Generation Capacity 2014.

79 Section 12 (b) REFIT Regulation 2015.

Bibliography

'GVE Nigeria' <https://repp.energy/project/gve-nigeria/> accessed 22 June 2021.

'Nigeria Announces New Energy Access Project "Solar Power Naija" Industry Sectors Future Energy' *News* (7 December 2020) <https://www.esi-africa.com/industry-sectors/renewable-energy/nigeria-announces-new-energy-access-project-solar-power-naija/> accessed 22 June 2021.

'Nigeria Will Never Have Stable Power Using Main Grid, Ex-Power Minister' *Thisday* (16 July 2018) <http://energymixreport.com/nigeria-will-never-have-stable-power-using-main-grid-ex-power-minister/> accessed 22 June 2021.

'Why Transmission is Weakest Link in Power Chain' *The Guardian* (25 August 2016) <https://m.guardian.ng/news/why-transmission-is-weakest-link-in-power-chain/> accessed 22 June 2021.

Adeniyi F, 'Overcoming the Market Constraints to On-grid Renewable Energy Investments in Nigeria' (2019) Oxford Institute for Energy Studies Paper EL 37 <https://doi.org/10.26889/9781784671495> accessed 1 July 2021.

Adeyanju GC, Osobajo OA, Otitoju A and Ajide O, 'Exploring the Potentials, Barriers and Option for Support in the Nigeria Renewable Energy Industry' (2020) 1(7) Discover Sustainability 1.

African Development Bank, 'African Development Bank, to Provide Underserved Communities in Nigeria with Sustainable Energy Solutions' (2020) <https://www.afdb.org/en/news-and-events/press-releases/african-development-bank-provide-underserved-communities-nigeria-sustainable-energy-solutions-34855> accessed 28 June 2021.

African Development Bank, 'Local Currency Financing for Off-grid Energy Solutions in Africa Limited, Needs Scaling Up- African Development Bank Report' (2020) <https://www.afdb.org/en/news-and-events/press-releases-local-currency-financing-grid-energy-solutions-africa-limited-needs-scaling-african-development-bank-report-39698> accessed 28 June 2021.

African Development Bank, 'Nigeria - Energy Access Fund - SEFA Project Summary Note' (3 March 2021) <https://www.afdb.org/en/documents/nigeria-energy-access-fund-sefa-project-summary-note> accessed 28 June 2021.

Ajaelu H and Okereke R, 'Assessment of Nigerian Renewable Energy Potentials and Energy Poverty: Challenges and Prospects' (2020) 9(6) PM World Journal 1.

Ajayi OO, Aderonmu PA and AjayiOO, 'Nigeria's Energy Policy: Issues of Sustainable Energy Development in Nigeria' 9 <http:eprints.covenantuniversity.edu.ng_Nigeria's_Energy_Policy_Issues_of_Sustainable_Energy_Development_in_Nigeria.pdf> accessed 22 June 2021.

Akinwale YO and Ogundari IO, 'Exploration of Renewable Energy Resources for Sustainable Development in Nigeria: A Study of the Federal Capital Territory' (2017) 7(3) International Journal of Energy Economics and Policy 240.

Akorede M F, Ibrahim O, Amuda SA, Otuoze AO and Olufeagba BJ, 'Current Status and Outlook of Renewable Energy Development in Nigeria' (2017) 36(1) Nigerian Journal of Technology 196.

Akuru UB and Okoro OI, 'Renewable Energy Investment in Nigeria: A Review of the Renewable Energy Master Plan' (2014) 25(3) Journal of Energy in Southern Africa 67.

Basil CV, 'Renewable Energy Deployment in Nigeria' *African Energy and Mineral Management Initiative* 10 <https://africanenergyminerals.org-Chioma-V.-Basil-MAIN-ARTICLE-ON-ENERGY-POLICY-DEPLOYMENT-IN-NIGERIA-converted.pdf> accessed 22 June 2021.

Dauda SG and Idehen SO, 'An Examination of the Implementation of Existing Policies on Renewable Energy in Nigeria: How Effective' (2021) 9 Journal of Power and Energy Engineering 104.

Duru S and Oni A, 'A Succinct Guide to Investing in Nigeria's Renewable Energy Sector' *Banwo & Ighodalo* <https://www.banwo-ighodalo.com_6cbd0c5b298daf9fe0754887c856e429.pdf> accessed 22 June, 2021.

ECOWAS Center for Renewable Energy and Energy Efficiency, 'ECREEE Rural Electrification Programme' <http://www.ecowas.int/wp-content/uploads/2015/11/ECREEE-FLYER-TEMPLATES_VERSION_RURAL-ELECTRIFICATION_09112015.pdf> accessed 22 October 2021.

Ejoh E, 'How Off-grid Devt'll Rescue Nigeria Power Sector' <https://rea.gov.ng/off-grid-devt-ll-rescue-nigeria-power-sector/> accessed 22 June 2021.

Energy Commission of Nigeria, National Energy Policy (2003) 23–34 <http://fao.org/docs/pdf/ins64284.pdf> accessed 22 October 2021.

Federal Republic of Nigeria, 'National Renewable Energy and Energy Efficiency Policy (NREEEP)' (2015) 11, 13, 14, 16 <http;//admin.theguides.org/Media/Documents/NREEE%20POLICY%202015-%20FEC%20APPROVED%20COPY.pdf> accessed 22 October 2021.

Heffron RJ, Hussein HI, Yang CH and Sun N, 'The Global Future of Energy Law: 2017 Review' (2017) 8 International Energy Law Review 291.

Herbert EB, 'Application of Electricity Federalism in Nigeria: Drawing Inspiration from America' (2021) 29(2) African Journal of International and Comparative Law 223.

Herbert EB, 'Developing a Renewable Energy Based Off-grid Electricity Solution for Nigeria' (2021) 2(2) Global Energy Law and Sustainability 182.

Hurst L, 'Shell in Talks with Nigerian Government to Exit Onshore Oil Fields as Part of Green Push' *World Oil* (2021) <https://www.worldoil.com/news/2021/5/18/shell-in-talks-with-nigerian-government-to-exit-onshore-oil-fields-as-part-of-green-push> accessed 30 June 2021.

Jeremiah K, 'Power Grid Fails 126 Times in 7 Years' *The Guardian* (8 December 2020) <https://m.guardian.ng/news/power-grid-fails-126-times-in-7-years/> accessed 22 June 2021.

Johnson RJ, Blackmore R and Bell R, 'The Role of Oil and Gas Companies in the Energy Transition' *Atlantic Council* (2020) 32 <https://www.atlanticcouncil.org/in

-depth-research-report/the-role-of-oil-and-gas-companies-in-the-energy-transition/> accessed 30 June 2021.

KPMG, 'Finance Act, 2020: Impact Analysis' (2020) 15 <https://assets.kpmg-pdf-tax-finance-act-2020.pdf> accessed 1 July 2021.

Newsom C, 'Renewable Energy Potential in Nigeria: Low-carbon Approaches to Tackling Nigeria's Energy Poverty' *International Institute for Environment and Development The SUNGAS Project* (2012) 15 <http://www.iied.org_G03512.pdf> accessed 22 June 2021.

Nigerian National Petroleum Corporation, 'Renewable Energy Division' (2020) <https://www.nnpcgroup.com/ventures/Pages/RenewableEnergy.aspx18> accessed 30 June 2021.

Off Grid Energy Independence, '22 Mini Grids for Nigeria Electrification Project' (2021). <https://www.offgridenergyindependence.com/articles/23274/22-mini-grids-for-nigeria-electrification-project> accessed 22 June 2021.

Ohiare SM, 'Financing Rural Energy Projects in Developing Countries: A Case Study of Nigeria' (PhD Thesis Leicester De Montfort University, 2014).

Oke Y, 'Beyond Power Sector Reforms: The Need for Decentralised Energy Options ("DEOPs") for Electricity Governance in Nigeria' in Yemi Oke (ed.), *Essays on Nigerian Electricity Law* (Princeton Publishing 2016).

Omorogbe Y, *Why We Have No Energy* (University of Ibadan Press, 2008) 21.

Oniemola PK, 'Developing a Legal Framework for Promoting Investment in Renewable Energy in the Nigerian Power Sector: An Analysis of the Design and Implementation Challenges' (PhD Thesis University of Aberdeen, 2014).

Oniemola PK, Ajayi SJ and Herbert EB, 'Contractual Architecture for Wholesale Electricity Purchase in Nigeria' (2019) 5(2) Journal of Commercial Law 728.

Oniemola PK and Tasie OC, 'Incorporating Off-grid Renewable Energy Options Under the Framework for Rural Electrification in Nigeria' (2016) 34(8) International Energy Law Review 327.

Osu VR, 'A Critical Evaluation of the Prospects for a Transition Towards Ocean Based Renewable Energy Development in Nigeria' (PhD Thesis, Robert Gordon University, 2017) 95.

Pellegrini L and Tasciotti L, 'Rural Electrification Now and Then: Comparing Contemporary Challenges in Developing Countries to the USA's Experience in Retrospect' (2012) Forum for Development Studies 1, 1.

Resolution Law Firm, 'How to Obtain Power Generation License in Nigeria' <https://www.resolutionlawng.com/how-to-obtain-power-generation-license-in-nigeria/#:~:text=The%20prescribed%20generation%20licence%20fees,a%20licence%20fee%20of%20%242%2C500.00> accessed 22 October 2021.

Roche MY Verolme H, Agbaegbu C, Binnington T, Fischedick M and Oladipo EO, 'Achieving Sustainable Development Goals in Nigeria's Power Sector: Assessment of Transition Pathways' (2020) 20(7) Climate Policy 846.

Rural Electrification Agency, 'Nigerian Electrification Project (NEP) Solar Hybrid Mini Grids Component' <https://rea.gov.ng/minigrids/> accessed 1 July 2021.

Saifuddin N, Bello S, Fatihah S and Vigna KR, 'Improving Electricity Supply in Nigeria-Potential for Renewable Energy from Biomass' (2016) 11(14) International Journal of Applied Engineering Research 8322.

Tasie OC and Oniemola PK, 'Legal Analyses of Liberalisation and Privatisation of State-owned Companies in the Nigerian Power Sector' (2016) 27(5) International Company & Commercial Law Review 141.

7 Uganda

Olugbenga Oke-Samuel, Kamoru T. Lawal, and Tajudeen Sanni[3]

7.1 Introduction

The Ugandan economy is one of the poorest in the world. The country has been experiencing relatively slow economic growth since the 1960s. The lack of access to a stable and low-cost power supply has been the bane of Uganda's economic development. Biomass (comprising of charcoal, wood, and agricultural wastes) and fossil fuels are Uganda's main sources of energy supply with more than 92% of the primary energy supply coming from renewable energy.[1] Electricity, a major driver of economic development, has been a source of concern to policymakers. With a population of over 45 million, and not less than 70% of the population living in rural areas,[2] only 28% of Uganda's population has access to electricity.[3] The rural areas, where there is a large concentration of the population, have the lowest rate of electricity access. Only a few households and businesses are connected to the grid. According to a report, it is only urban households that are within a range of 20 km and rural households within a 22 km distance from national grid coverage are connected to the grid.[4] Others have had to rely on alternative sources for electricity access. The challenge of electricity access, as will be seen later in this chapter, is caused by a combination of supply side (absence of grid coverage) and the demand side (poor/no uptake of electricity) factors.[5] The two factors are the major reason Uganda has been recording a poor progress rate in rural electrification access for some decades now.[6]

Faced with the challenges of electricity access, the Government of Uganda (GoU) launched a number of programmes. One of such programme is the Vision 2040 that was launched in 2013 with the sole aim of transforming Uganda from a low-income economy to 'a competitive upper middle income country'.[7] The Vision 2040 program is meant to serve as a springboard for the actualisation of the objectives of the National Development Plan.[8] At the heart of Vision 2040 is the provision of energy that can propel industrialisation. The GoU considers, in addition to grid electricity, a system of electricity generation and distribution that will be independent of the grid network but can support population growth and sustainable economic development. Off-grid renewable electricity (OGRE) in Uganda, thus, emerged out of the necessity to increase the rate of electricity access proportionate to population growth

DOI: 10.4324/9781003178088-9

and economic development.[9] The choice of OGRE is based on the recognition of the link between energy and sustainable economic development, and the role of OGRE in providing access independent of the grid.

The term 'OGRE' is used to describe 'the production of renewable electricity from smaller electric plants at or near the point of sale, and its distribution is direct to users without connecting to a national grid'.[10] It is a system of electricity generation and distribution that relies on '(semi) autonomous capacity to satisfy electricity demand through local power generation'.[11] Globally, OGRE has served as a means of providing low-cost reliable electricity access to rural households and communities that are isolated or far away from the electricity grid.[12] According to the IEA, more than 70% of the lack of access to rural electrification can be addressed through either mini-grid or standalone systems.[13]

As a major step towards the development of OGRE technologies in Uganda, the GoU first embarked on master planning processes to identify areas that are appropriate for grid extension, and the areas that are better suited for OGRE. The GoU also introduced programmes to promote and encourage the development of OGRE. Among these programmes, three, that is, the rural electrification programme, the Energy for Rural Transformation, and the Sustainable Energy for All Initiative, are relevant to the discussions in this chapter.[14] The Uganda Electricity Act of 1999 is the main enabler of OGRE. The Act mandates the GoU to promote grid and off-grid electricity access for an equitable regional distribution of electricity. The Electricity Act provides thus:

> The Government shall undertake to promote, support and provide rural electrification programmes through public and private sector participation in order to –
>
> (a) achieve equitable regional distribution access to electricity;
> (b) maximise the economic, social and environmental benefits of rural electrification subsidies;
> (c) promote expansion of the grid and development of off-grid electrification; and
> (d) stimulate innovations within suppliers.[15]

In 2002, the Ugandan Cabinet approved the Energy Policy as the policy document for the sector. The Energy Policy sets out the GoU's commitment to the development and utilisation of renewable energy for the economic development of the country. In furtherance of the objectives in the 2002 Policy, the Renewable Energy Policy was approved in March 2007 to drive the development of renewable energy. The electrification objectives of the law and policy, as well as in the selected programmes, create opportunities for OGRE, especially among unserved and the underserved sections of the population. However, despite the opportunities for OGRE, there are still a number of challenges.

The aim of this chapter is to examine the extent to which the law and policy support the development of OGRE in Uganda. It will examine factors that have contributed to the growth and development of OGRE, particularly in rural areas where the issue of electricity access has remained a major challenge. In this chapter, reference to OGRE will mean the generation and distribution of renewable-sourced electricity via standalone systems, pico-, mini-, and micro-grids.[16] This chapter is structured into five sections. Section 7.1, this introduction, sets the context for OGRE in Uganda. Section 7.2 will analyse electricity generation in Uganda with a view to identifying the opportunities and challenges of OGRE. Discussions in Section 7.2 will incorporate an overview of selected programmes. In this part, the chapter will examine the rural electrification programme, the Energy for Rural Transformation, and the Sustainable Energy for All Action Agenda. Section 7.3 will analyse the policy framework for OGRE in Uganda in order to establish the extent of its support for the latter. Specifically, the 2002 Energy Policy and the Rural Electrification Strategy and Plan (RESP) will be analysed. The 2002 Energy Policy, though, remains the primary policy document for the promotion of OGRE in Uganda, and does not contain specific directions on how to drive OGRE.[17] For instance, there is no financing structure to assist OGRE developers get access to credit facilities. While the local financial institutions do not have the means to provide the needed funds, their international counterparts often impose conditions which the OGRE developers find difficult to fulfil. Furthermore, the absence of subsidies to support end-users of OGRE is an on-going challenge in Uganda. The non-availability of end-user subsidies is affecting the affordability of OGRE technologies. The GoU has considered a revision of the 2002 Energy Policy to address some of the challenges of OGRE. The proposed revised Energy Policy will address the 'limited consideration for off-grid energy solution' in the 2002 Energy Policy.[18] Section 7.4 will analyse the regulatory framework for OGRE based on the 1995 Constitution of Uganda and the Electricity Act of 1999. In Section 7.5, this chapter will conclude and make recommendations on the future development of OGRE in Uganda.

7.2 Opportunities and Challenges of OGRE

7.2.1 Case for OGRE

Since economic liberalisation in 1987, the electricity sector has been playing an important role in Uganda.[19] Revenue from electricity-related activities (such as the imposition of value-added tax on electricity consumption, licence fees and royalties, and levies on transmission of bulk purchases of electricity) has served as a source of income for the GoU.[20] Given the place of electricity, and the requirements of the Electricity Act, the GoU takes steps to not only increase electricity generation capacity but also ensure equitable regional access. Over the years, the total installed electricity capacity has increased to 1,179 MW from the 872 MW it was in 2012.[21] The average electricity generation has also increased to around 850 MW while the average electricity demand is about 600 MW.[22] In terms of generation and demand, Uganda is generating surplus electricity for its population. The state of surplus electricity generation is aptly captured in the Draft Energy Policy:

> The electricity sub-sector has shifted from generation capacity shortages between 2002 and 2012 to the current anticipated surplus of power generation compared to demand.[23]

Ironically, Uganda has one of the lowest rates of electricity access and consumption in the world.[24] Two main factors are responsible for the poor state of electricity access.

First is the non-coverage of the electricity grid (that is, the supply side limitation). The existing grid network does not cover the whole of Uganda. Although extension of the grid network is one of the options for addressing the lack of electricity access, the extension comes with high construction costs. As a result of the high costs, grid extension is considered as not economically feasible. Apart from the high costs of extending the grid, there are also high connection costs. The average electricity connection fee for households is UGX359,000, and the households in urban areas pay UGX53,000 above the average fee.[25] A programme, Electricity Connections Policy (ECP), which was introduced to connect more households and businesses to the national grid, has not recorded a complete success. Under the ECP, the GoU plans to connect at least 300,000 customers annually by subsidising electricity connection costs. Although the ECP has succeeded in removing the high electricity connection cost (which is estimated at US$165) and other sundry charges, convincing people to embrace the programme has remained a challenge. As a result, only a few households have so far been connected under the ECP.[26]

The second factor is the low uptake of electricity (that is, the demand side limitation) caused by high electricity tariffs. In Uganda, the average residential electricity tariff is US$0.21/kWh, and only a few households can afford this tariff.[27] As a result of the high tariff, electricity consumption is low especially in rural areas. In addition, the volume of electricity demand is grossly insufficient to justify the costs of extending the national grid.[28] In most cases, electricity companies find it difficult to recover the fixed costs of connecting more areas to the grid.[29]

As a result of the non-coverage of the grid and the low uptake of electricity, OGRE becomes a veritable means of providing electricity access independent of the national grid. In Uganda, mini-grids that rely on renewable energy sources offer opportunities for off-grid electrification (that is, OGRE).[30] A major boost for the policy shift in favour of OGRE is the need to address the challenges of grid electricity. The GoU recognises the role of off-grid technologies in providing low-cost electricity in communities that are dispersed and isolated from the grid. Moreover, the percentage of the rural population without access to grid electricity makes a 100% reliance on grid electricity untenable. With OGRE, Uganda has been able to harness its renewable energy resources for electricity generation. Uganda is one of the few African countries that are blessed with abundant renewable energy resources. The country's hydro-electricity potential is estimated to be more than 2,000 MW (large-scale, micro-, and mini-scale).[31] The average solar radiation is approximately 5.2 kWh per square metre. The solar irradiation is between 1,825 kWh/m^2 and 2,500 kWh/m^2 per year.[32] The high speed of solar insolation has made solar energy an important renewable energy source with high potential in Uganda. Out of the estimated 200 MW solar capacity, the total contribution of solar is 50 MW (4.2%). The GoU has set an ambitious target of generating 5,000 MW from solar by 2040.[33] Uganda has invested in policies and programmes to attract investors that can develop the country's solar energy. With the aid of donor funds, the solar industry has recorded significant improvements since it started in 1980. The first solar PV in Uganda was installed under the United Nations Development Programme (UNDP) funded rural electrification project, that is, the Uganda Photovoltaic Pilot Project for Rural Electrification (UPPPRE).[34] There is now mass production of solar panels in countries like China for markets in Uganda. The price of a solar panel is now about 80% cheaper than when it was first introduced.[35] There is also a World Bank funded Solar Energy Development ERT programme under which over 1,000 solar PV systems have so far been installed in more than 50 Ugandan districts.[36] There are also community-based mini-grid and solar PV in dispersed and isolated communities.

The growth of OGRE has been further supported through programmes introduced by the GoU. Among these are the REP, the Energy for Rural Transformation, and the Sustainable Energy for All. Following the enactment of the Rural Electrification Act of 1936, REP was introduced as a means of providing electricity to farms and villages.[37] OGRE has served as a vehicle through which Uganda has benefitted from internationally sponsored rural electrification.[38] As part of the REP strategies, OGRE is deployed in areas where grid extension is not feasible. Solar is the only energy resource that is deployed for electrification in the rural areas.[39] The availability of solar energy is a major reason why Uganda's energy policy favours its adoption over other forms of renewable energy for electricity access in the rural areas. Sunlight, which can be converted to solar energy, is abundantly and evenly distributed in every part of Uganda.[40] The country enjoys more solar on account of the country's location on the equator.[41] Given the abundance of solar energy, OGRE technologies that rely on solar energy (such as solar PV) have become the preferred choice when it comes to the electrification of rural areas. The flexibility in the mode of payment and the uniqueness of solar PV makes it a preferred choice over other OGRE technologies.[42] Under the REP, 140,000 new customers will be connected annually using OGRE (solar PV, micro hydro, and community-based mini-grid). Within a 10-year period, approximately 1.42 million households in rural areas will have electricity access through OGRE (standalone, community-based mini-grid, solar PV, and solar home systems).[43] The idea being that the electrification of rural areas will facilitate the realisation of the Vision 2040 electrification targets.[44]

The Energy for Rural Transformation (ERT) is another programme that has aided the growth of OGRE. ERT, a World Bank supported programme, is a multi-sector programme with the overall goal of increasing electricity access in rural areas.[45] The programme is a major promoter of solar PV connected mini-grids in Uganda. The First Phase of the program (ERT I) started in 2001 with a 10%

electricity access target by 2012.[46] ERT I sought to put in place a functioning and conducive environment for the sustainable delivery of rural electrification and communication technologies but could not achieve this by the cut-off date.[47] The Second Phase (that is, ERT II) targeted electricity access beyond grid extension areas. In support of the target, there was a proposal for accelerated investment for increased national electricity coverage and provision of credit facilities to facilitate energy access.[48] Currently, ERT is in the Third Phase (2016–2021) (ERT III).[49] The goal of ERT III is increased rural electricity access through grid extensions and the exploitation of renewable energy off-grid.[50] The inclusion of off-grid renewable energy as a key strategy of ERT III is a booster for OGRE. Under ERT III, solar PV will be deployed to selected areas (e.g., public institutions) in rural areas and for business development support.[51] The Ministry of Energy and Mineral Development (MEMD) coordinates, monitors, and evaluates the performance of ERT III to ensure that the objectives of the programme are realised.[52] Last but not the least among the programmes is the Sustainable Energy for All (SE4All) Action Agenda. SE4All, an initiative that was conceived in 2015, is the GoU's platform for coordinating universal energy access. SE4All is Africa's sustainable energy initiative aimed at ensuring that African countries meet their obligations under the United Nations' sustainable energy programme. SE4All is an innovative empowerment programme that seeks to advance gender equality.[53] Electricity generation, as well as universal electricity access, is one of the prioritised action areas in the SE4All. The initiative targets more than 98% electricity access and at least 90% renewable energy share in final energy by 2030.[54] However, the universal access target in the SE4All initiative is not likely to contribute to increased deployment of OGRE given that the initiative will rely, in the main, on on-grid electricity services.[55]

7.2.2 Opportunities for OGRE

The OGRE market in Uganda is growing. Over the past few years, markets for solar PV have grown steadily with new players (foreign investors and private developers) entering the OGRE markets. OGRE solutions offer great potential for electrification, especially among rural households. Currently, OGRE via solar PV provides electricity access to about 4.9% of households in Uganda, and the projection is that this figure could grow to around 33% of households by 2030.[56] Looking at some factors, there are opportunities for OGRE in Uganda. First, a significant percentage of the Ugandan population does not have access to (grid) electricity means there are still largely untapped markets for OGRE. According to a report on the 2018 Energy for Rural Transformation (ERT III), many households are not connected to any source of electricity.[57] Findings, based on categorisation of the population along the demand side and the supply side limitations, reveal that 60% of households lack electricity access due to non-coverage while the remaining 40% have grid access but lack access due to no uptake.[58] The section of the population without grid access make use of either dry cell batteries or other off-grid electricity which does not qualify as Tier 1.[59] In either case, there are opportunities (that is, untapped markets) that can be explored by investors who are interested in OGRE technologies.

Second, climate and environmental considerations in energy policy are a source of opportunities for OGRE. Uganda is committed to reducing greenhouse gas (GHG) emissions under the Paris Agreement.[60] Globally, studies have found the energy sector to be a major source of GHG emissions. So, there are on-going campaigns to promote the use of technologies that rely on renewable energy sources as a way reducing GHG emissions.[61] Hydro, the main source of electricity generation in Uganda, is vulnerable to changes in the climate. Apart from the fluctuation in the water level, hydro also emits carbon dioxide (CO_2), a major source of GHG emissions.[62] Based on this climate commitment, Uganda has taken steps to ensure reduction of GHG emissions. The GoU energy policy in this regard targets the promotion of renewable energy technologies such as solar PV.[63]

Third, the development in mobile telecommunication in Uganda means more demand for OGRE. Recently, Uganda witnessed an increase in the deployment of mobile technologies. There are now many mobile payment platforms, telecommunication masts, and mobile monitoring of systems. The bulk of the mobile telecommunication platforms rely on OGRE for power.[64] So, with the GoU's plan to deploy more mobile telecommunication platforms to facilitate economic development, there is going to be additional demand for OGRE solutions.

Overall, the opportunities for OGRE in Uganda continue to evolve. Given that the deployment of solar PV is a key strategy of the REP coupled with the projected population of 61.3 million by 2040, OGRE will continue to be key to electrification in Uganda.[65]

7.2.3 Challenges of OGRE

Notwithstanding the opportunities for OGRE, there are challenges, which, if not addressed, may impede the growth of OGRE in Uganda. A major challenge of OGRE in Uganda is the lack of access to capital.[66] OGRE is a capital-intensive venture that requires huge investment outlays. However, OGRE investors do not have access to capital to build their business due to limited financing options. Many OGRE companies are experiencing financial difficulties due to lack of public funds. The available public-funded efforts are concentrated more on grid expansion with little on OGRE.[67] Given that it is a new business area, and in view of the fact that policies are still evolving, banks are hesitant in granting loans to companies that intend to invest in OGRE.

Local banks in Uganda lack the capacity to offer long-term financing. The challenge of financing is particularly felt by local developers who are yet to have records of accomplishment. Local developers, in most cases, do not have the necessary documents to access finance from international banks. As a result, many developers are unable to access long-term international financing.[68] The only fund in the Electricity Act, the Rural Electrification Fund, does not provide financing support for electricity projects that are not connected with rural electrification.[69] While OGRE is key to rural electrification, it is not limited to rural areas. There are many unserved areas in Uganda including urban poor and the rural communities. Moreover, the REF is meant to subsidise electricity access in rural areas and not to provide loans to OGRE developers.[70] Another challenge is the absence of support for OGRE consumers. The cost of OGRE technologies is relatively high. Only a few households and businesses can afford the OGRE technologies without any form of financial assistance. However, there is neither subsidies (particularly end-user subsidies) nor specific financial support for OGRE users (especially for solar PV). The existing supports are grossly 'inadequate and poorly streamlined'.[71] In cases where loans are available, the interest rate is usually very high.[72] As a result of this lack of access to financing support for OGRE consumers, affordability of OGRE technologies has become an issue.[73] This is because the majority of Ugandans are living below the poverty line and will, therefore, find it difficult to pay for OGRE in the absence of GoU support.

Absence of direction on how to drive OGRE is another major challenge. The market for solar PV, for instance, has been driven mainly by the lack of electricity access arising from non-coverage of the grid. There is no empirical evidence to support the actual potential of solar PV and the section of the country that will be better suited to electrification based on solar PV. In other words, there is no data to ascertain the actual number of operational solar PV systems in Uganda.[74] So, investors and developers are not clear as to the actual potential of solar PV within the household sector to be able to make future projections.[75] This is evident in the electricity targets in Vision 2040. By the Vision 2040, Uganda plans to generate 5,000 MW from solar. However, is not clear how much of this solar target will be generated and distributed through off-grid media.[76] Moreover, given that investors will want to have an idea about the market opportunities before investing their money, the lack of empirical data regarding OGRE potential may affect the volume of investments in solar PV. The actual market potential may also be the basis for which banks and other financial institutions will make funding available for OGRE.

The absence of mini-grid regulation is also a challenge. In Uganda, there are many entities that now own and operate mini-grids (micro-grids) that rely on renewable energy sources. However, despite the number of mini-grid operators there is no regulation on the modes of operating them.[77] As a result of the absence of specific regulation regarding the operation of mini-grids, stakeholders are raising concerns about the tariff structure for OGRE via these mini-grids.[78] The renewable energy feed-in tariff (ReFiT), set by the Ugandan electricity regulator, does not apply to OGRE.[79] There is also the concern about the affordability of OGRE through mini-grids in view of the fact that the latter are relatively expensive compared to the cost of constructing and connecting traditional grids.[80] Last, the reliability of OGRE technologies is still a challenge. Generally, the quality of an energy technology is a factor for its acceptance.[81] In Uganda, the costs of OGRE technologies are relatively high. These high costs have prompted some developers to manufacture lower qualities as alternatives.[82] While the low-cost technologies are mostly affordable, these technologies do not offer the same level of reliability. As a result, many OGRE users view some of the technologies with mistrust. According to a report, the inability of battery or solar devices to power appliances for a longer period of time is a major issue that has affected the acceptability of solar products by some sectors of the population.[83]

7.3 Policy Framework for OGRE

7.3.1 Energy Policy 2002

The policy landscape for OGRE evolved with the release of the 2002 Energy Policy. The Energy Policy sets out the direction on the sustainable way of satisfying the energy needs of the people for social and economic development. On the strength of the provisions of the 2002 Energy Policy, Uganda has been able to drive the country's electricity generation and access.[84] However, looking at the development of OGRE, the 2002 Energy Policy has failed to drive OGRE, as it does not provide for a clear-cut direction for the promotion of OGRE. The failure of the 2002 Energy Policy to drive OGRE is captured in the Draft Energy Policy as follows:

> The 2002 policy needed a stronger focus on biomass as the primary energy source in order to curb inefficient use and related environmental degradation. The policy also had limited consideration for off-grid energy solutions.[85]

The Draft Energy Policy identifies the absence of a strong regulatory environment as one of the major impediments to the development of off-grid electricity access in Uganda, especially in rural areas.[86] In the Draft Energy Policy, the GoU sets out priority areas one of which is increased access to modern and reliable energy services through scaling up of 'off-grid renewable energy supply systems'.[87] The GoU plans to maintain the current state of the grid expansion programme, and promote the use of mini-grid and off-grid installations until electricity demands are industrially justified. Looking at the two provisions of the Draft Energy Policy (that is, scaling up of off-grid renewable energy and the deployment of off-grid installations), there is no doubt that this will aid the deployment of OGRE. Grid expansion projects in many parts of Uganda have suffered on account of poor demand for electricity.[88] So, halting grid expansion to promote OGRE is a welcome development in the Draft Energy Policy.

The OGRE strategy in the Draft Revised Energy Policy also covers the agricultural sector whereby the GoU will promote the deployment of OGRE to reduce reliance on fossil fuels.[89] The promotion of OGRE to reduce reliance on fossil fuels, no doubt, is going to be a major boost for OGRE development. Another area in which the Draft Revised Energy Policy seeks to promote the development of renewable energy technologies is through the establishment of credit facilities, pricing policies, and tax incentives.

However, while the Draft Revised Energy Policy suggests a brighter future for OGRE its provisions remain inoperative until the GoU approves the proposed draft. In other words, the 2002 Energy Policy remains the extant policy framework for OGRE in Uganda notwithstanding its inadequacies.

In furtherance of the renewable energy objectives, the Ugandan Cabinet approved the Renewable Energy Policy in March 2007. The Renewable Energy Policy is a high-level policy document on the development of the latter. The RE Policy provides a legal and institutional framework for investments in renewable energy in a bid to increase its contribution to the energy mix. It is on the basis of these provisions of the RE Policy that the GoU upholds the need to address the challenges of electricity access in Uganda through OGRE,[90] and increase the use of modern renewable energy.[91] The GoU has further set ambitious targets for renewable energy. Hydropower, for instance, will contribute not less than 90% electricity generation by 2030.[92] In terms of number, 19 out of the 35 hydro plants in Uganda are small hydro (OGRE). However, in terms of capacity, large hydro accounts for more than 84% of the total hydro capacity (that is, 855 MW out of 1011 MW).[93] Moreover, the RE Policy does not specify any target for hydro-based OGRE beyond the general target for hydropower. What this means is that large hydropower has been and will continue to dominate electricity generation in Uganda, while small hydro contribution remains marginal despite the number of plants. Therefore, except there is a specific policy strategy to small hydro-based OGRE, the 90% target will not bring about any significant increase in OGRE. By not having a specific target there is no basis for pursuing OGRE or for measuring performance. According to Kieffer and Couture (2015), having a renewable energy target serves as a guide as well as a motivation for the implementation of a policy.[94] This important objective of having a target applies also to OGRE.

7.3.2 Rural Electrification Strategy and Plan (RESP)

The Minister of Energy developed the Rural Electrification Strategy and Plan (2013–2022) as a policy direction on the implementation of the REP. The RESP is a 10-year rural electrification strategy designed to address the unequal electricity access between urban and rural areas in Uganda. The strategy is based on the understanding that communities and beneficiary populations play important roles in the provision of electricity services by making use of local models such as cooperatives.[95] The current RESP was approved in 2013 following the expiration of the First Phase in 2012. The primary objective of the RESP is 'to achieve an accelerated pace of electricity access and service penetration to meet national development goals during the planning period and beyond'.[96] Other objectives include the facilitation of access to modern energy as well as other forms of traditional cooking and heating.[97] The RESP highlights the policy direction and programme on how the GoU will accelerate the rate of electricity access in rural areas. Based on the policy objectives of the RESP, the GoU has set a 26% rural electrification access for the 10-year period.[98] The good thing about the rural electrification target in the RESP is that it has prompted the GoU to prioritise OGRE given that electricity access through the grid is still very low. According to the RESP:

> Off-grid electrification development based primarily on the current target market approach for solar electrification development, as may be modified

> to draw solar PV services more directly under REA's service territory concession planning and service provider customer service model to gain program implementation and cost efficiencies, will continue unabated during the RESP performance period.[99]

The GoU plans to make loans available for off-grid electrification and provide new enabling legal and regulatory frameworks for renewable energy to ensure smooth implementation of the programme.[100] The proposed off-grid financing scheme approach in the RESP is a major improvement from previous phases. A major factor that was responsible for the failure of the First Phase of RESP is the huge investment capital but limited financing schemes. The grid extension projects involve huge initial capital outlay whereas the initial demand for electricity in rural areas is abysmally low.[101] The private sector could not provide the needed investments for grid extension projects. In the end, the programme achieved less than 5% electrification access rate as opposed to the 10% target that was set for the period.[102] In a bid to avoid the pitfalls of the First Phase, the current RESP seeks to entrench sustainable capital financing by proposing investments that are based on a 10-year business plan for each service territory.[103] The financing option, if properly implemented, can assist OGRE developers to overcome the challenges of accessing long-term financing which has been a major impediment to its development. The RESP also introduces a change to the role of the GoU in the financing of rural electrification projects. Under the new financing arrangement in the RESP, the GoU will assume greater responsibility in the planning, management, and financing of rural electrification projects while the private sector will play a complementary role.[104] The GoU will absorb much of the financial risk associated with OGRE projects.[105] That the GoU is absorbing some of the financial risks is good news as banks may be prepared to provide additional finance to OGRE developers. The financial risk has been a major reason why banks are hesitant to finance OGRE. Based on the provisions of RESP, licensed concession holders are permitted to oversee the management of rural electrification infrastructures and services for efficient service delivery. What this means is that the GoU, acting through the REA, will monitor rural electrification activities in view of the fact that planning and management of projects are now centralised in the REA.

Another area of interest in the current RESP is the electrification strategy. Unlike the First Phase where reliance was placed on grid expansion, the current RESP relies mainly on off-grid electricity technologies (solar PV and community-based mini-grids) which are independent on the national grid. As stated in the RESP:

> Off-grid electrification services comprising energy service technologies not dependent on the national grid shall, preferably, be planned, offered and furnished to eligible consumers in the service territories in tandem with on-grid electrification services.[106]

The off-grid electricity strategy, which relies on renewable energy, will further encourage the deployment of OGRE technologies. The GoU is prepared to facilitate investment in small-distributed power generation systems that will boost OGRE.

The policy framework, though not providing specific guidance on the development of OGRE, has nevertheless aided the growth of the latter in Uganda.[107] The policy approach to solar PV, for instance, has largely been one of non-interference from the GoU. The non-interference approach has helped in promoting increased participation of the private sector in OGRE through the stimulation of private markets.[108] The ERA, Uganda's electricity regulator, also does not regulate the costs and tariffs of solar PV products but rather sets a cost-reflective tariff through feed-in tariffs. Solar PV equipment also enjoys tax and tariff exemptions.[109] What the GoU needs to do is to align different policy objectives as a way of addressing the perceived inconsistency and/or overlap of electrification targets. In the SE4ALL, for instance, the GoU sets at least 98% electrification targets by 2030,[110] whereas, in Uganda's Vision 2040 the plan is to achieve 80% grid access by 2040.[111] The SE4ALL is a sector-wide target that seeks to capture the sustainable energy agenda under three headings: universal access to modern energy, doubling of the rate of energy efficiency, and the attainment of twice the share of renewable energy in the energy mix.[112] Vision 2040, on the other hand, symbolises the fundamentals of the Ugandan economy in which electricity will play a key role. The implication is that there is no coherent and unified approach to the development of OGRE, and this will impact negatively on its future development.

7.4 Regulatory Framework for OGRE

7.4.1 1995 Constitution

The 1995 Constitution is the foundation of energy policies in Uganda. The Constitution mandates the GoU to 'promote and implement energy policies that will ensure that people's basic needs and those of environmental preservation conditions are met'.[113] Although the Constitution does not expressly refer to OGRE, the requirement of environment preservation and sustainable economic growth creates a strong impetus for the development of OGRE for equitable access to clean electricity in Uganda. The latter is attributed to the nexus between renewable energy and sustainable economic growth cum environmental preservation. OGRE will not only ensure equitable electricity access in areas that are not connected to grid electricity but will also help in preserving the environment. Sustainable economic growth and widespread access continue to be the basis of every electricity law and policy in Uganda.

7.4.2 Electricity Act of 1999

Following the approval of the Power Sector Reform and Privatization Strategy by the Ugandan Cabinet, the Electricity Act of 1999 was enacted for the purposes of driving policy change in the electricity sector.[114] The Electricity Act provided the necessary legal framework for reforms in Uganda's power sector. The Electricity

Act enabled a departure from a state-dominated electricity structure, and introduced a system that permits the active participation by the private sector in core electricity functions. The Electricity Act is important as it sets out the fundamental obligations of the GoU, which include the attainment of equitable regional distribution of access to electricity, the expansion of grid and the development of off-grid electricity.[115] Based on the provision of the Act, the GoU is obligated to promote and support the REP. As a way of supporting the REP, the Minister of Energy is mandated by the Act to set up the Rural Electrification Fund as a vehicle for achieving equitable electricity access in that country. The Fund will apply to subsidise the development costs of electrification projects in rural areas based on the criteria within the Act.[116] REA has been the REF to provide financial support for mini-grid operators through either direct funding or reimbursement for costs incurred.[117]

It is important to note that the REF does not specifically target OGRE. REF applies to every electricity project that is connected to rural electrification, and which projects fulfil the conditions within the Electricity Act. The non-dedication of the REF or a part of the fund to OGRE projects means that it is not every OGRE project that will qualify for subsidies under the REF. What is, however, a plus to OGRE is the requirement of the Act whereby the Minister of Energy is required every year to submit a report showing the installation of solar PV in isolated areas where grid-based electrification is not economically feasible.[118] The Ministry of Energy has also, in the exercise of the supervisory powers contained in the Electricity Act, approved the RESP as a policy framework for undertaking the REP, and the deployment of solar is a key strategy in the RESP.[119]

Further, and in compliance with the regulatory requirements in the Electricity Act, the GoU in 2000 established the ERA as an independent corporate body with powers over electricity matters. The ERA regulates the electricity industry independent of the Ministry of Energy, and supervises key electricity functions (that is, generation, distribution, and transmission).[120] By law, the ERA issues electricity licences, sets tariffs for electricity, and develops and enforces standards within the electricity industry. The power of the ERA over electricity licensing covers grid electricity and OGRE.[121] It is the responsibility of the ERA to ensure that licence holders comply with the conditions of the licence. The ERA also exercises regulatory control over mini-grids. It grants approval, sets retail tariffs for OGRE, and enforces standards over mini-grids.[122] Every developer of OGRE that relies on mini-grids needs to obtain an approval from the ERA. The ERA is in the process of establishing a one-stop shop to facilitate a process whereby a staff member of the ERA can coordinate licensing processes and liaise with mini-grid developers.[123]

The ERA exempts electricity generation plants with capacity below 2 MW from obligation to obtain a licence.[124] By exempting plants of less than 2 MW capacity the developers will not have to go through the expensive licensing process. Electricity plants of between 0.5 MW and 2 MW capacity are also not required to pay the annual licence fees.[125] Currently, the ERA charges US$3,500 as an application fee for a licence to undertake an electricity project.[126] Looking at the costs, it is clear that the grant of exemption will serve as a form of incentive to OGRE in terms of cost saving given that OGRE technologies are not subject to feed-in tariff.

In addition to electricity sector approval, some OGRE projects require non-electricity sector approval. One such non-electricity sector approval is by the National Environmental Management Authority (NEMA). By law, NEMA is required to approve an environmental impact assessment report for every mini-grid.[127] Where the mini-grid is in respect of a hydro-based electricity plant, the approval of the Directorate of Water Resources Management is also required.[128] Subjecting OGRE projects to non-electricity sector approval may negate the very reason Uganda opted for OGRE, most especially the equitable access to national electricity. Different considerations and laws guide different regulators, and some of the considerations may not accord with the electricity objectives. The electric sector and non-electric sector approvals create a complex and lengthy approval regime for OGRE in addition to costly feasibility studies. In the end, it takes several months to complete the approval process.[129]

In addition to regulatory functions, the ERA develops and enforces performance standards for electricity distribution including OGRE.[130] The ERA, in conjunction with the Uganda National Bureau of Standard (UNBS), develops, monitors, and enforces standards for renewable energy technology and electricity equipment.[131] The UNBS is a regulatory body empowered by law to develop and monitor standards.[132] The UNBS sets prices for product certification the payment of which is only valid for a period of 12 months.[133] The certification fee is an additional cost to companies involved with OGRE, and this may be passed on to end-users as additional cost.

7.4.3 Statutory Instrument 2001 No. 75 Establishing the Rural Electrification Agency

As well as the ERA, there is also the Rural Electrification Agency. The REA is a semi-autonomous body that serves as a secretariat to the Rural Electrification Board.[134] The REA was established pursuant to the Statutory Instrument No. 75 of 2001 to assist the Rural Electrification Board in the discharge of rural electrification projects in Uganda. One of the functions of the Agency is to advise the Rural Electrification Board on the most efficient way to apply the Rural Electrification Fund.[135] The REA receives and processes applications for financial support from the Fund.[136] The REA is required by law to facilitate the goal of rural electrification in Uganda. As a matter of procedure, the REA will initiate electrification projects and later pass the projects to the ERA for the purposes of advertisement and licensing. The Board is the highest decision-making body on the matter of rural electrification projects. The Board plays a major role in the management of the REF for equitable electricity access as well as the attainment of the goals of the REP.[137] The Board approves projects based on certain set parameters one of which is the impact of the project on the economic development of the rural people. The involvement of the Board in OGRE is in the area of project approval and the provision of subsidies for projects.[138] Going by the RESP, the Board's involvement in OGRE will increase further in view of the proposed financing templates in the RESP which include REA-backed grant funding and medium-term loans for rural electrification projects.[139]

7.5 Conclusions and Recommendations

OGRE in Uganda will continue to evolve as the GoU continues to address legal and regulatory issues. With the large-scale adoption of on- and off-grid solar PV for rural electrification, the contribution of hydropower will decrease significantly in the future. The decrease may pave the way for Uganda to pursue 100% electricity access based on OGRE. What the GoU needs to do is to provide support for OGRE, and address all the identified challenges. It is important to provide financing support for OGRE developers. The financing support can take the form of a long-term loan. In this regard, the GoU can adopt the long-term financing option in the RESP and amend the Electricity Act accordingly. Regarding the affordability of OGRE, the GoU can also introduce innovative financial support schemes that will specifically address end-users' access to OGRE. Furthermore, the issue of reliability of OGRE can be addressed by constant engagement with relevant regulators and key stakeholders. It is important for the GoU to promote greater awareness about OGRE among entities that are directly involved or will benefit from OGRE. The awareness is necessary to realise the full potential of OGRE. Above all, the GoU should introduce a coherent strategy for undertaking OGRE. The introduction of OGRE-focused strategy will go a long way in addressing the challenges of OGRE. On-grid and OGRE are electrification options, each relies on different considerations and objectives. Therefore, only a tailor-made policy can address the challenges of OGRE. With only about 28% out of the country's 45 million population having electricity access, Uganda will continue to provide opportunities for OGRE.

Notes

1 Sylvia Manjeri Aarakit, Vincent Fred Ssennono, and Muyiwa S. Adaramola, 'Estimating Market Potential for Solar Photovoltaic Systems in Uganda' (2021) 9 Frontiers in Energy Research 1, 1; International Renewable Energy Agency (IRENA), 'Energy Profile: Uganda' (IRENA 2021) 1, <https://www.irena.org/IRENADocuments/Statistical_Profiles/Africa/Uganda_Africa_RE_SP.pdf> accessed 23 September 2021.

2 Riley Abbott, 'Scaling Off-Grid Energy Access in Uganda: A Mid-Level Landscape Analysis of Issues and Stakeholders', Catalyzing Partnerships for Scale (FHI 360, Washington, DC, October 2017) 7; The World Bank, 'Population: Uganda, The World Bank Data', <https://data.worldbank.org/indicator/SP.POP.TOTL?locations=UG> accessed 17 October 2021.

3 Manjeri Aarakit et al. (n 1) 1. As at 2019 the rate of electricity access was 26.7% of the population. See, Reuben Gad Mugagga and Hope Baxter Nqcube Chamdimba, 'A Comprehensive Review on Status of Solar PV Growth in Uganda' (2019) Journal of Energy Research and Reviews 1.

4 Ministry of Energy and Mineral Development, 'National Electrification Report for Energy for Rural Transformation-ERT III Baseline Survey 2018' (January 2020) 33.
5 Manjeri Aarakit et al. (n 1) 1.
6 Veit Raisch, 'Financial Assessment of Mini-Grids Based on Renewable Energies in the Context of the Ugandan Energy Market Energy' (2016) 93 Procedia 174, 175.
7 Republic of Uganda, 'Uganda Vision 2040' (Government of Uganda 2013) IV.
8 Ibid 3 and 4; Menno Jan van der Ven, 'An Overview of Recent Developments and the Current State of the Ugandan Energy Sector' (International Growth Centre (IGC) Working Paper June 2020, Ref. No: E-20046-UGA-1) 13.
9 Manjeri Aarakit et al. (n 1) 1.
10 Ngozi Chinwa Ole, 'The Nigerian Electricity Regulatory Framework: Hotspots and Challenges for Off-Grid Renewable Electricity Development' [2020] Journal of Energy & Natural Resources Law 1, 2.
11 International Renewable Energy Agency (IRENA), 'Off-Grid Renewable Energy Systems: Status and Methodological Issues' (IRENA Working Paper 2015) 5.
12 Ibid.
13 United States Agency for International Development (USAID), 'Practical Guide to the Regulatory Treatment of Mini-Grids' (Meister Consultants Group, Inc. November 2017) 9.
14 The three programs will be discussed later in this chapter.
15 Electricity Act 1999, s. 62; See also, Republic of Uganda, Constitution 1995 (as Amended), National Objectives and Directive Principles of State Policy; Menno Jan van der Ven (n 8) 6.
16 Ministry of Energy and Mineral Development, *Renewable Energy Policy*, Republic of Uganda 2007, 12.
17 Ministry of Energy and Mineral Development, *Draft National Energy Policy*, Republic of Uganda 2019, 9.
18 Ibid.
19 Robert Tumwesigye, Paul Twebaze, Nathan Makuregye and Ellady Muyambi, 'Key Issues in Uganda's Energy Sector', Pro-Biodiversity Conservationists in Uganda (PROBICOU 2011) 6.
20 Ibid.
21 Ronald Makanga Kakumba, 'Despite Hydropower Surplus, Most Ugandans Report Lack of Electricity', Afrobarometre No. 441 [April 2021] 1, 1.
22 Africa Electricity Portal, 'Uganda: Demand for Electricity Declines by 10 Per Cent' (AEP 04 May 2020) <http://africa-energy-portal.org/news/uganda-demand-electricity-declines-10-cent> accessed 21 September 2021.
23 *Draft National Energy Policy* (n 17) 9.
24 Ibid 10; Michael Grimm, Luciane Lenz, Jörg Peters and Maximiliane Sievert, 'Demand for Off-Grid Solar Electricity – Experimental Evidence from Rwanda' (2020) 7(3) Journal of the Association of Environmental and Resource Economists 417.
25 National Electrification Report for Energy for Rural Transformation (n 4) 34.
26 Abbott (n 2) 7.
27 Ibid.
28 Dalberg Global Development Advisors, *Improving Access to Electricity Through Decentralised Renewable Energy: Policy Analysis from India, Nigeria, Senegal and Uganda* (2017) 13.
29 Moussa P. Blimpo and Malcolm Cosgrove-Davies, 'Electricity Access in Sub-Saharan Africa: Uptake, Reliability, and Complementary Factors for Economic Impact' International Bank for Reconstruction (Washington, DC 2019) 72.
30 Raisch (n 6) 176.
31 Tumwesigye et al. (n 19) 6 and 30.

32 Manjeri Aarakit et al. (n 1) 1.
33 Menno Jan van der Ven (n 8) 21; Kees Mokveld and Steven von Eije, *Final Energy Report Uganda* (Netherlands Enterprise Agency December 2018) 8.
34 Japan International Cooperation Agency (JICA), 'Dissemination of Renewable Energy into Rural Communities: Study on Photovoltaic and Small-Hydro Projects in East Africa' (Proact International, Inc., Summary Report October 2008) 13 <https://openjicareport.jica.go.jp/pdf/11900685.pdf> accessed 4 February 2022; Umaru Lule, 'Bringing Electricity to the Rural Bringing Electricity to the Rural Areas: Designing Laws and Areas: Designing Laws and Policies that Work' <https://www.un.org/esa/sustdev/sdissues/energy/op/parliamentarian_forum/lule_rural_areas.pdf> accessed 26 October 2021.
35 Gad Mugagga and Nqcube Chamdimba (n 3) 2.
36 Lule (n 34).
37 Brandon McBride, 'Celebrating the 80th Anniversary of the Rural Electrification Administration', USDA Rural Utilities Service (February 2017) <https://www.usda.gov/media/blog/2016/05/20/celebrating-80th-anniversary-rural-electrification-administration> accessed 7 November 2021.
38 Elizabeth Kaijuka, 'GIS and Rural Electricity Planning in Uganda' (2007) Journal of Cleaner Production 203, 205.
39 Mokveld and von Eije (n 33) 8; Ministry of Energy and Mineral Development, *Rural Electrification Strategy and Plan (2013–2022)* (Republic of Uganda 2013) 8.
40 Japan International Cooperation Agency (n 34) 1; Sylvia M. Aarakit, Joseph M. Ntayi, Francis Wasswa, Muyiwa S. Adaramola and Vincent F. Ssennono, 'Adoption of Solar Photovoltaic Systems in Households: Evidence from Uganda' (2021) 329 Journal of Cleaner Production 1, 2.
41 M. Aarakit et al. (n 40) 2.
42 Ibid. The uniqueness of the technology is based on the fact that each household can generate and self-consume electricity at a minimal maintenance costs.
43 Standalone will be deployed in 'dispersed and isolated areas' while mini-grids will be used in concentrated rural areas. See, Menno Jan van der Ven (n 8) 19.
44 Rural Electrification Strategy and Plan (n 39); IRENA (n 11) 17.
45 The World Bank, 'Energy for Rural Transformation III', The World Bank (2015) <https://projects.worldbank.org/en/projects-operations/project-detail/P133312> accessed 13 September 2021; International Energy Agency, 'Energy for Rural Transformation' (IEA, December 2013) <https://www.iea.org/policies/4130-energy-for-rural-transformation-ert> accessed 15 October 2021.
46 International Energy Agency (n 45); Kaijuka (n 38), 205; The World Bank (n 45).
47 National Electrification Report for Energy for Rural Transformation (n 4) 1.
48 Kaijuka (n 38) 205; The World Bank (n 45); National Electrification Report for Energy for Rural Transformation (n 4) 1.
49 The Republic of Uganda, Terms of Reference for ERT III' Ministry of Energy and Mineral Development (2016) <https://www.energyandminerals.go.ug/site/assets/files/1270/terms_of_reference_for_ert_iii_documentary.pdf> accessed 8 September 2021.
50 National Electrification Report for Energy for Rural Transformation (n 4) 1.
51 The World Bank (n 45). During the Second Phase of the program around 3, 799 solar home systems were deployed. See, International Energy Agency, 'Energy for Rural Transformation' (IEA, December 2013) <https://www.iea.org/policies/4130-energy-for-rural-transformation-ert> accessed 8 September 2021.
52 Kaijuka (n 38) 206.
53 Neha Misra, 'Sustainable Energy for All: Empowering Women' *United Nations* <https://www.un.org/en/chronicle/article/sustainable-energy-all-empowering-women> accessed 31 October 2021.
54 Ministry of Energy and Mineral Development (MEMD), *Uganda's Sustainable Energy for All (SE4ALL) Initiative Action Agenda* (Government of Uganda 2015)

26; Mathias Gustavsson, Mark Hankins and Karin Sosis, 'Energy Report for Uganda: A 100% Renewable Energy Future by 2050' WWF-World Wide Fund for Nature - Uganda Country Office (Formerly World Wildlife Fund) (Kampala, 2015) 14.

55 *Uganda's Sustainable Energy for All Initiative Action Agenda* (n 54) 30.

56 Dalberg Global Development Advisors (n 28) 13.

57 National Electrification Report for Energy for Rural Transformation (n 4) 32.

58 Manjeri Aarakit et al. (n 1) 1.

59 National Electrification Report for Energy for Rural Transformation (n 4) 32. The report categorises household electricity supply using Multi-Tier Frame (MTF). Based on the MTF, the best form of electricity is Tier 5 full access) while the lowest is Tier 0 (no access). See, National Electrification Report for Energy for Rural Transformation (n 4) 29.

60 Uganda Vision 2040 (n 7) 101; Kaijuka (n 38).

61 See, Johnson & Johnson, *Climate Friendly Energy Policy*, World Energy Management <https://www.rspo.org/file/acop/johnson-johnson/M-GHG-Grower-Emissions-Report.pdf> accessed 17 October 2021.

62 Gad Mugagga and Nqcube Chamdimba (n 3) 2; The Guardian, Hydroelectric dams emit a billion tonnes of greenhouse gases a year, study finds <https://www.theguardian.com/global-development/2016/nov/14/hydroelectric-dams-emit-billion-tonnes-greenhouse-gas-methane-study-climate-change> accessed 31 October 2021; Scott Foster and David Elzinga, 'The Role of Fossil Fuels in a Sustainable Energy System' *UN Chronicle* <https://www.un.org/en/chronicle/article/role-fossil-fuels-sustainable-energy-system> accessed 13 September 2021.

63 Renewable Energy Policy (n 16) 12.

64 Abbott (n 2) 10 and 23.

65 Uganda Vision 2040 (n 7) 88 and 89; Mokveld and von Eije (n 33) 8.

66 Abbott (n 2); Power Africa, 'Power Off-Grid Productive Use of Energy', March https://powerafrica.medium.com/exploring-off-grid-productive-use-of-energy-in-uganda-91d5c4aa3bf4> accessed 13 September 2021.

67 Renewable, Innovative, and Sustainable Electrification (RISE), Electricity for Integrated Rural Development: The role of businesses, the public sector and communities in Uganda and Zambia, Project RISE practitioner report 2019, 7.

68 Open Capital Advisors, Scaling Off-Grid Energy Access in Uganda: A Political Landscape Analysis, United States Agency for International Development (USAID) 9 <https://pdf.usaid.gov/pdf_docs/PA00W5JV.pdf> accessed 6 January 2022.

69 Energy Act 1999, s. 65.

70 Renewable, Innovative, and Sustainable Electrification (RISE), Electricity for Integrated Rural Development: The Role of Businesses, the Public Sector and Communities in Uganda and Zambia, Project RISE Practitioner Report 2019, 30.

71 Dalberg Global Development Advisors (n 28) 87. Government recently stopped Value-Added-Tax exemption for renewable energy products. See, Mokveld and von Eije (n 33) 5.

72 In most cases the average interest rate is as high as 20%. Mokveld and von Eije (n 33) 20.

73 Abbott (n 2) 10.

74 Gustavsson et al. (n 54) 87.

75 Manjeri Aarakit et al. (n 1) 2; Gad Mugagga and Nqcube Chamdimba (n 3) 1.

76 Menno Jan van der Ven (n 8) 21.

77 Abbott (n 2) 18.

78 Ibid.

79 Electricity Regulatory Authority (ERA), *Uganda Renewable Energy Feed-in Tariff (REFIT)*, Phase 3 Guideline (Revised) June 2019) s. 4.1.

80 Abbott (n 2) 18.

81 Benjamin K. Sovacool, Anthony L. D'Agostino and Malavika Jain Bambawale, 'The Socio-technical Barriers to Solar Home Systems (SHS) in Papua New Guinea: "Choosing Pigs, Prostitutes, and Poker Chips over Panels' (2011) 39(3) Energy Policy 1532.
82 Aleid C. Groenewoudt, Henny A. Romijn and Floor Alkemade, 'From Fake Solar to Full Service: An Empirical Analysis of the Solar Home Systems Market in Uganda' (2020) 58 Energy for Sustainable Development 100.
83 National Electrification Report for Energy for Rural Transformation (n 4) 53.
84 Draft National Energy Policy (n 17) 7; Gad Mugagga and Nqcube Chamdimba (n 3).
85 Draft National Energy Policy (n 17) 9.
86 Ibid 29.
87 Ibid 14.
88 Dalberg Global Development Advisors (n 28) 13.
89 Draft National Energy Policy (n 17) 34.
90 Renewable Energy Policy (n 16) 27.
91 Ibid 16 and 25.
92 International Hydropower Association (IHA), Country Profile: Uganda <https://www.hydropower.org/country-profiles/uganda> accessed 5 January 2022.
93 Vincent Katutsi, Milly Kaddu, Adella G. Migisha, Muhumuza E. Rubanda and Muyiwa S. Adaramola, 'Overview of Hydropower Resources and Development in Uganda' (2021) 9(6) AIMS Energy 1299, 1304.
94 Ghislaine Kieffer and Toby D. Couture, 'Setting Renewable Energy Targets', International Renewable Energy Agency (IRENA 2015) 32.
95 Rural Electrification Strategy and Plan (n 39) 1 and 2.
96 Ibid 7.
97 Ibid.
98 Ibid 9.
99 Rural Electrification Strategy and Plan (n 39) vi.
100 Ibid; Renewable Energy Policy (n 16) 27.
101 Lule (n 34). The ERT I only succeeded in connecting 7,000 households as opposed to the 80,000 projected solar PV installations. See also, Rural Electrification Strategy and Plan (n 39) 10.
102 Rural Electrification Strategy and Plan (n 39) 7.
103 Ibid 8.
104 Menno Jan van der Ven (n 8) 6.
105 Rural Electrification Strategy and Plan (n 39) iii.
106 Ibid iv.
107 Dalberg Global Development Advisors (n 28) 87.
108 Prior to this time, the development of solar PV was driven largely by the Government of Uganda and donor-funded programmes. See, Ulrich Elmer Hansen, Mathilde Brix Pedersen and Ivan Nygaard, 'Review of Solar PV Policies, Interventions and Diffusion in East Africa' (2015) ResearchGate 11 and 12 <https://www.researchgate.net/publication/273790226_Review_of_solar_PV_policies_interventions_and_diffusion_in_East_Africa> accessed 13 December 2021.
109 Ibid 93.
110 Rural Electrification Strategy and Plan (n 39) 25–26; Gustavsson et al. (n 54) 14.
111 Uganda Vision 2040 (n 7) 14; Menno Jan van der Ven (n 8) 6.
112 Rural Electrification Strategy and Plan (n 39) 25.
113 Constitution 1995 (n 15) s. XXVII (III).
114 Lule (n 34).
115 Electricity Act, 1999, s. 62.
116 Electricity Act 1999, s. 64(3).
117 United States Agency for International Development (n 13) 72.
118 Electricity Act 1999, s. 63(2)(d).

119 Energy Act 1999, s. 63
120 Electricity Act 1999, s. 10(a); Electricity Regulatory Authority, 'Uganda's Electricity Sector Overview' (ERA November 2020) <https://www.era.go.ug/index.php/sector-overview/uganda-electricity-sector> accessed 24 September 2021.
121 Electricity Act 1999, s. 10(a).
122 United States Agency for International Development (n 13) 49.
123 Ibid.
124 The licence envisaged here is a combined licence for generation, distribution and sales. See, Electricity (Licence Fees) (Amendment) (No. 3) Regulation 2014, s. 3.
125 Electricity Act 1999, s. 113; Abbott (n 2) 12. For electricity plants of less than 0.5 MW capacity the exemption is automatic while the developers will have to apply for the grant of an exemption in the case of electricity plants of between 0.5 and 2 MW capacities. See, Electricity Regulatory Authority (ERA), 'Developments and Investment Opportunities in Renewable Energy Resources in Uganda' (ERA June 2013) 40.
126 Electricity Regulatory Authority (n 125) 38.
127 United States Agency for International Development (n 13) 49.
128 Ibid 99.
129 Renewable, Innovative and Sustainable Electrification (RISE), 'Electricity for Integrated Rural Development: The Role of Businesses, the Public Sector and Communities in Uganda and Zambia', Project RISE practitioner report 2019, 30.
130 Electricity Act 1999, s. 10.
131 Electricity Act 1999, s. 10; Abbott (n 2) 12; Tumwesigye et al. (n 19) 43.
132 Renewable Energy Policy (n 16) 79; Electricity Regulatory Authority (n 125) 4; Tumwesigye et al. (n 19) 43.
133 Uganda's National Bureau of Standards, UNBS Reduces Cost of Product Certification (Q-Mark) to Support BUBU <https://www.unbs.go.ug/readmore-slider.php?sl=299&banner> accessed 5 January 2022.
134 United States Agency for International Development (n 13) 207.
135 Lule (n 34).
136 Ibid.
137 Ministry of Energy and Mineral Development, 'Electricity Access Scale-Up Project (EASP): Resettlement Policy Framework (RPF)' (Republic of Uganda 2020) 18.
138 Electricity Regulatory Authority (n 125) 3.
139 By the new financial scheme, the Board is required to provide loans to service providers for rural electrification projects regardless of whether the service providers are solar (OGRE) companies or on-grid service providers. See, Rural Electrification Strategy and Plan (n 39) 13.

Bibliography

Aarakit S M, Ntayi J M, Wasswa F, Adaramola M S and Ssennono V F, 'Adoption of Solar Photovoltaic Systems in Households: Evidence from Uganda' (2021) 329 Journal of Cleaner Production 1.

Aarakit S M, Ssennono V F and Adaramola M S, 'Estimating Market Potential for Solar Photovoltaic Systems in Uganda' (2021) 9 Frontiers in Energy Research 1.

Abbott R, 'Scaling Off-grid Energy Access in Uganda: A Mid-Level Landscape Analysis of Issues and Stakeholders' *Catalyzing Partnerships for Scale* (FHI 360, Washington DC October 2017).

Africa Electricity Portal, 'Uganda: Demand for Electricity Declines by 10 Per cent' (AEP 04 May 2020) <http://africa-energy-portal.org/news/uganda-demand-electricity-declines-10-cent>.

Blimpo M P and Cosgrove-Davies M, 'Electricity Access in Sub-Saharan Africa: Uptake, Reliability, and Complementary Factors for Economic Impact' *International Bank for Reconstruction* (Washington DC 2019).

Dalberg Global Development Advisors, Improving Access to Electricity Through Decentralised Renewable Energy: Policy Analysis from India (Senegal and Uganda, Nigeria 2017).

Electricity Regulatory Authority (ERA), 'Developments and Investment Opportunities in Renewable Energy Resources in Uganda' (ERA June 2013).

Electricity Regulatory Authority (ERA), *Uganda Renewable Energy Feed-in Tariff (REFIT)*, Phase 3 Guideline (Revised) June 2019).

Electricity Regulatory Authority, 'Uganda's Electricity Sector Overview' (ERA November 2020) <https://www.era.go.ug/index.php/sector-overview/uganda-electricity-sector>.

Foster S and Elzinga D, 'The Role of Fossil Fuels in a Sustainable Energy System, UN Chronicle' <https://www.un.org/en/chronicle/article/role-fossil-fuels-sustainable-energy-system>.

Grimm M, Lenz L, Peters J and Sievert M, 'Demand for Off-grid Solar Electricity – Experimental Evidence from Rwanda' (2020) 7(3) Journal of the Association of Environmental and Resource Economists 417.

Groenewoudt A C, Romijn H A and Alkemade F, 'From Fake Solar to Full Service: An Empirical Analysis of the Solar Home Systems Market in Uganda' (2020) 58 Energy for Sustainable Development 100.

Guardian, 'Hydroelectric Dams Emit a Billion Tonnes of Greenhouse Gases a Year, Study Finds' <https://www.theguardian.com/global-development/2016/nov/14/hydroelectric-dams-emit-billion-tonnes-greenhouse-gas-methane-study-climate-change>.

Gustavsson M, Hankins M and Sosis K, *Energy Report for Uganda: A100% Renewable Energy Future by 2050, WWF-World Wide Fund For Nature - Uganda Country Office (Formerly World Wildlife Fund)* (Kampala 2015).

Hansen U E, Pedersen M B and Nygaard I, 'Review of Solar PV Policies, Interventions and Diffusion in East Africa' *ResearchGate* (2015) 11 and 12 <https://www.researchgate.net/publication/273790226_Review_of_solar_PV_policies_interventions_and_diffusion_in_East_Africa>.

International Energy Agency, 'Energy for Rural Transformation' (IEA, December 2013) <https://www.iea.org/policies/4130-energy-for-rural-transformation-ert>.

International Hydropower Association (IHA), 'Country Profile: Uganda' <https://www.hydropower.org/country-profiles/uganda>.

International Renewable Energy Agency (IRENA), 'Off-Grid Renewable Energy Systems: Status and Methodological Issues' (IRENA Working Paper 2015).

International Renewable Energy Agency (IRENA), 'Energy Profile: Uganda' (IRENA 2021) 1 <https://www.irena.org/IRENADocuments/Statistical_Profiles/Africa/Uganda_Africa_RE_SP.pdf>.

Japan International Cooperation Agency (JICA), 'Dissemination of Renewable Energy into Rural Communities: Study on Photovoltaic and Small-Hydro Projects in East Africa' (Proact International, Inc., Summary Report October 2008) 13 <https://openjicareport.jica.go.jp/pdf/11900685.pdf>.

Johnson & Johnson, 'Climate Friendly Energy Policy, World Energy Management' <https://www.rspo.org/file/acop/johnson-johnson/M-GHG-Grower-Emissions-Report.pdf>.

Kaijuka E, 'GIS and Rural Electricity Planning in Uganda' (2007) Journal of Cleaner Production 203.

Kakumba R M, 'Despite Hydropower Surplus, Most Ugandans Report Lack of Electricity, Afrobarometre No. 441 (April 2021)' 1.

Katutsi V, Kaddu M, Migisha A G, Rubanda M E and Adaramola M S, 'Overview of Hydropower Resources and Development in Uganda' (2021) 9(6) AIMS Energy 1299.

Kieffer G and Couture T D, 'Setting Renewable Energy Targets' (International Renewable Energy Agency IRENA 2015) 32.

Lule U, 'Bringing Electricity to the Rural Bringing Electricity to the Rural Areas: Designing Laws and Areas: Designing Laws and Policies that Work' <https://www.un.org/esa/sustdev/sdissues/energy/op/parliamentarian_forum/lule_rural_areas.pdf>.

McBride B, 'Celebrating the 80th Anniversary of the Rural Electrification Administration' (USDA Rural Utilities Service February 2017) <https://www.usda.gov/media/blog/2016/05/20/celebrating-80th-anniversary-rural-electrification-administration>.

Ministry of Energy and Mineral Development, *Renewable Energy Policy* (Republic of Uganda 2007).

Ministry of Energy and Mineral Development, *Rural Electrification Strategy and Plan (2013–2022)* (Republic of Uganda 2013).

Ministry of Energy, Electricity (Licence Fees) (Amendment) (No. 3) Regulation 2014.

Ministry of Energy and Mineral Development (MEMD), *Uganda's Sustainable Energy for All (SE4ALL) Initiative Action Agenda* (Government of Uganda 2015).

Ministry of Energy and Mineral Development, *Draft National Energy Policy* (Republic of Uganda 2019).

Ministry of Energy and Mineral Development, *Electricity Access Scale-Up Project (EASP): Resettlement Policy Framework (RPF)* (Republic of Uganda 2020).

Ministry of Energy and Mineral Development, *National Electrification Report for Energy for Rural Transformation- ERT III Baseline Survey 2018* (January 2020) 1.

Misra N, 'Sustainable Energy for All: Empowering Women' (United Nations) <https://www.un.org/en/chronicle/article/sustainable-energy-all-empowering-women>.

Mokveld K and Eije S, 'Final Energy Report Uganda' (Netherlands Enterprise Agency December 2018).

Mugagga R G and Chamdimba H B N 'A Comprehensive Review on Status of Solar PV Growth in Uganda' (2019) Journal of Energy Research and Reviews 1.

National Electrification Report for Energy for Rural Transformation (n 4) 32. The report categorises household electricity supply using Multi-Tier Frame (MTF).

Ole N C, 'The Nigerian Electricity Regulatory Framework: Hotspots and Challenges for Off-grid Renewable Electricity Development' [2020] Journal of Energy & Natural Resources Law 1.

Open Capital Advisors, Scaling Off-Grid Energy Access in Uganda: A Political Landscape Analysis, United States Agency for International Development (USAID) <https://pdf.usaid.gov/pdf_docs/PA00W5JV.pdf>.

Power Africa, 'Power Off-Grid Productive Use of Energy', March https://powerafrica.medium.com/exploring-off-grid-productive-use-of-energy-in-uganda-91d5c4aa3bf4>.

Raisch V, 'Financial Assessment of Mini-Grids Based on Renewable Energies in the Context of the Ugandan Energy Market Energy' (2016) 93 Procedia 174.

Renewable, Innovative, and Sustainable Electrification (RISE), 'Electricity for Integrated Rural Development: The Role of Businesses, the Public Sector and Communities in Uganda and Zambia', Project RISE practitioner report 2019.

Republic of Uganda, Constitution 1995 (as amended).
Republic of Uganda, Electricity Act 1999.
Republic of Uganda, 'Uganda Vision 2040' (Government of Uganda, 2013).
Republic of Uganda, 'Terms of Reference for ERT III' (Ministry of Energy and Mineral Development 2016) <https://www.energyandminerals.go.ug/site/assets/files/1270/terms_of_reference_for_ert_iii_documentary.pdf>.
Sovacool B K, D'Agostino A L and Bambawale M J, 'The Socio-technical Barriers to Solar Home Systems (SHS) in Papua New Guinea: "Choosing Pigs, Prostitutes, and Poker Chips Over Panels"' (2011) 39(3) Energy Policy 1532.
Tumwesigye R, Twebaze P, Makuregye N and Muyambi E, 'Key Issues in Uganda's Energy Sector' (Pro-Biodiversity Conservationists in Uganda PROBICOU 2011).
Uganda's National Bureau of Standards, 'UNBS Reduces Cost of Product Certification (Q-Mark) to Support BUBU' <https://www.unbs.go.ug/readmore-slider.php?sl=299&banner>.
United States Agencssssy for International Development (USAID), 'Practical Guide to the Regulatory Treatment of Mini-Grids' (Meister Consultants Group, Inc. November 2017).
Ven M J, 'An Overview of Recent Developments and the Current State of the Ugandan Energy Sector' International Growth Centre (IGC) Working Paper June 2020 (Ref. No: E-20046-UGA-1).
World Bank, 'Energy for Rural Transformation III' (The World Bank 2015) <https://projects.worldbank.org/en/projects-operations/project-detail/P133312>.
World Bank, 'Population: Uganda' (The World Bank Data) <https://data.worldbank.org/indicator/SP.POP.TOTL?locations=UG>.

Part 3

Asia

8 China

Ngozi Chinwa Ole, Opeyemi Omotuyi, and Imam Mulyana

8.1 Introduction

The challenge of lack of access to reliable and clean electricity is the rationale for developing OGRE in China. In the words of Ole, 'For China, the imperatives for the optimal development of OGRE are also associated with the twin issues of energy security and environmental protection'.[1] In the 1990s, over 4.58 million households did not have electricity in China's rural areas.[2] Remarkably, the lack of access to electricity has been addressed as China now has 100% access to electricity.[3] OGRE has played a role in the provision of access to electricity to rural China[4]The term OGRE ordinarily encompasses electricity generated from renewable energy sources using decentralised technologies that supply directly to users without connecting to the national grid.[5] In China, it has been defined to incorporate the feeding in of excess electricity generated from such decentralised technologies into the national grid.[6] Thus, justifying its role in the electrification of rural areas whether remote or not.[7]Notwithstanding, there is the recurring problem of power blackouts from the national grid, which continues to undermine the reliability of the power supply.[8] The latter is attributed to the fact that the current electricity supply is well below demand.[9] As a result, major manufacturing companies have been called to reduce their production level to ease the incessant power blackouts. Overall, the power blackouts arereducing the productivity of the country.[10] It is apt to mention that OGRE can remedy the problem of power blackouts. First, it can serve as a back-up allowing both households and companies to meet their energy demand irrespective of the limited supply from the grid.[11] As such, economic activities will continue irrespective of the blackouts. The use of OGRE alongside electricity from the grid will help reduce the power blackouts occasioned by excessive demand.[12]

Another impetus for the development of OGRE is environmental protection especially climate change. China is currently 'the world's largest carbon polluter' and, as such, a significant contributor to the exacerbation of the climate change problem.[13] The reason for its lead in this regard is because 57% of the energy used in the country is from coal.[14] Unsurprisingly,there is a strong global pressure on them to take measures to mitigate climate change.[15]Besides, the country is a signatory to the international climate change instruments, including the Paris

DOI: 10.4324/9781003178088-11

Climate Change Agreement 2015.[16] Under the obligations stipulated in the Paris Agreement, the country has pledged to increase renewable energy development as their low greenhouse emission mitigation strategy.[17]The development of OGRE like biogas is part of their rural electrification strategy.[18]The OGRE development is also claimed to be one of the effective mechanisms, not only to promote energy saving and increase energy efficiency, but also to control GHG emissions and increase equitable access to energy for the Chinese people.[19] Thus, the need for the optimal development of OGRE in China.

What is more, the coronavirus (COVID-19) pandemic amplifies the need to intensify efforts towards the accelerated development of OGRE. Indomitably, COVID-19 has affected the global economy negatively. China is not exempted from the mentioned negative impact. Recovery from COVID-19 entails radical deployment of key infrastructures, including the energy sector.[20] It is stipulated that renewable energy development, including OGRE, should play a role.[21]Thus, there is an increased need to develop OGRE. It has also been identified that COVID-19 has hampered the realisation of government policies and objectives in the renewable energy sector.[22] Thus, intensified efforts are needed to get the sector, including OGRE, on track to achieve energy security and climate change mitigation potential.

In light of the above background, it is not surprising that China's 14th five-year plan (2021–2025) recognises that the development of environmentally friendly energy technologies, including electricity, is crucial to achieving a carbon-neutral economy by 2060.[23]In China's Action Plan for Carbon Dioxide Peaking before 2030, it was recognised that vigorously pursuing renewable energy, including OGRE, is pertinent to achieving the noted carbon target.[24]Creating the proper foundation for the accelerated development of OGRE is contingent on the effectiveness of existing laws and policies[25]Against the above background, this chapter contains an analysis of the extent of existing laws supporting OGRE development in China. It builds on the original work of the first author, which examines the role of the China Renewable Energy Law 2009 in supporting the development of OGRE.[26] However, the mentioned study is outdated and incomprehensive. For one, it does not factor in recent barriers, imperatives, relevant government policies, and regulations. The study is also incomprehensive, focusing only on the effectiveness of the China Renewable Energy Law 2005 in supporting the development of OGRE. Even at that, it does not look at how the Renewable Energy Law interacts with recent existing barriers to the sector's development.

The chapter is divided into five sections. The introduction is followed by a discussion of the pressing barriers to the development of OGRE in Section 8.2. Section 8.3 analyses the extent of the policy in supporting the sector's development. Section 8.4 contains an analysis of the regulatory environment for OGRE in the context of the central research issue. Section 8.5 is the concluding section.

8.2 Barriers to the Development of OGRE

Initially, the typical barrier that impeded the development of OGRE in China before 2015 was the problem of accessing and affording its initial capital cost.[27]

The latter was exacerbated by the absence of a level investment platform existing between OGRE and competing fossil fuel electricity technologies.[28] The reason has been summed by Hart and Marcellino, who opine that the latter has retained 'the accumulated advantages of decades of public support in the form of subsidies, loans, guarantees, various favourable laws…'.[29] What is more, there was the absence of needed technological capacity in the sector.[30]Residents were also not aware of the benefits of renewable energy in comparison to the already established one for fossil fuel.[31] As such, the development of OGRE was constrained.

However, some progress has been recorded in addressing the mentioned barriers such that some of them are non-existent or minimised. For one, some renewable electricity technologies have become costcompetitive with fossil fuel electricity technologies such that the problem of the high initial capital cost isgreatly reduced.[32] Some regulatory measures introduced by the Renewable Energy Law of China 2009 is credited with having played a role in addressing the mentioned financial barrier.[33] The latter will be discussed in detail, i.e.,the section on regulatory support. China has become the hub of renewable energy development such that residents are increasingly aware of the need for OGRE development.[34] China has also overcome the absence of technological capacity in the sector.[35]As a fact, they are now leading in the exportation of OGRE know-how to other countries.[36]

Regardless, some factors continue to impede the development of OGRE. For one, there is still the absence of a level playing field between it and competing fossil fuel strategies to the extent that the latter is still maintaining the lead in the sector at an alarming margin.[37]Fossil fuel continues to enjoy support from the government despite the accumulated decades of the same.Even though in April 2021 the government officially announced that China will strictly control coal-fired power generation projects and limit the increase in coal consumption over the 14th five-year plan period (2021–2025),[38]in November 2021, the Chinese government was reported to have invested US$31 billion to support the development of coal electricity.[39] The latter is compounded by reduced subsidies for renewable electricity projects, including OGRE.[40] Another challenge is that OGRE developers are struggling to feed it into the grid because of the intermittent nature of renewable energy.[41] Given this, they are mandated to get storage facilities at a high cost.[42] Some residents are still oblivious of the full benefits of renewable energy, including OGRE, to the extent that it is still constraining its widespread use and acceptability.[43] All the abovementioned factors cumulatively constrain the development of the sector such that it is not what it should and can be.

8.3 Policies for OGRE

The Renewable Energy Law 2005 (RE Law 2005, as amended in 2009) is the primary law supporting the OGRE sector's development. It has the objective 'to promote the development and utilisation of renewable energy, improve the energy structure, diversify energy supplies, safeguardenergy security, protect the environment, and realise the sustainable development of the economy and society'.[44] In the above context, it recognises that 'the Government supports the construction of independent renewable power systems in areas not covered by the power grid to provide power service for local production and living'.[45] It further enjoins private entities to install OGRE technologies in their workplaces and homes.[46] The mentioned provisions capture the energy security and climate change need for the development of OGRE.[47] For one, some rural populations did not have access to electricity at that time.[48] Even the ones that did have access to fossilfuel electricity options were also not constant. Thus, the clear context for OGRE is commendable because it is the foundation upon which an effective law can be built.[49]Without an accurate and precise definition of the problem that law purports to address, it will be ineffective.[50]

Noteworthy, the abovementioned provisions do not capture the COVID-19 imperatives for the development of OGRE. This is not unexpected, given that the Renewable Energy Law (REN Law) 2005 was made several years before the COVID-19 pandemic. The Chinese Government Work Report on COVID-19 targetsinfrastructural development, including energy, as part of the recovery pathway for the country.[51] It doesn't expressly mention the development of renewable energy for this purpose,however, some authors opine that it impliesrenewable energy development, including OGRE.[52] However, the report expressly targets the development of fossil fuel options.[53] In the words of Gosen and Jotzo:

> Analysis of China's recent Government Work Report suggests that while a repeat of recovery measures focused on high-emissions infrastructure following the 2008 global recession is not in the cards, a Chinese Green New Deal is not in sight either. Much investment flows to fossil fuel industries, while support policies for renewable energy industries are absent from Beijing's recovery programme.[54]

The absence of an express commitment to acknowledge the role of renewable energy, including OGRE in the 2021 report, may mean that it can never fully attain its potential in this context. The development of fossil fuel strategies is a step backwards in addressing the global problem of climate change.

Against the background of the need for OGRE, the REN Law empowers the National Development and Reform Commission (NDRC) to adopt targets for developing various forms of renewable energy.[55] Flowing from which, it sets a specific target for the development of solar, hydro, and wind OGRE in a Medium and Long-Term Development Plan for Renewable Energy 2007.[56]However, the absence of a sectoral target generally for OGRE was criticised because it implies that there is no benchmark for adopting measures for addressing the mentioned barriers to its development.[57] It is also impossible to fully ascertain the extent of the progress made without a comprehensive target.[58] Regardless, the end of 2020 saw the accelerated development of renewable energy, including OGRE.[59] While the development of OGRE cannot be fully measured without an initial baseline before the REN Law and a comprehensive target, progress has been made in the sector.

As reiterated, the NDRC medium- and long-term plan for renewable energy ended in 2020. China's 14th five-year plan, 2021–2025, sets a target to achieve carbon neutrality by 2060 and radically reduce carbon emissions by 2030.[60] It mandates that carbon intensity must be reduced by 13.5% in the energy sector.[61]Even though the target is not yet stipulated in law, it provides the basis for adopting policies and implementation guidelines to give flesh to the mentioned provisions.[62] Consequently, the NDRC has adopted the Action Plan for Carbon Dioxide Peaking before 2030. It provides that achieving the 14th five-year plan entails 'commitment to cutting carbon emissions in a safe manner by vigorously promoting substitution of renewable sources of energy under the condition that energy security is ensured'.[63]It proposes the development of OGRE in several respects to attain the proposed carbon neutrality.[64]It proposes reviewing relevant policies and laws, including the REN Law, to tune with the mentioned aspirations.[65]

However, the NDRC has not adopted renewable energy and OGRE targets. This is totally expected given that adequate time is needed to prepare an effective long- and medium-term plan for renewable energy development, including OGRE, in line with the 14thfive-year plan. However, there is no certainty that the plan will be a progression of the previous one by accommodating the COVID-19 imperatives of OGRE and providing comprehensive sectoral targets.

8.4 Regulatory Environment

8.4.1 Introduction

While the REN Law supports various forms of renewable energy, the Electricity Law 1996 is the principal law that governs every spectrum of operations in the OGRE sector.[66]We also have the Administrative License Law of the People's Republic of China 2004[67] which governs the issuance of permits for various activities which ordinarily, without it, would be impermissible. Together, these laws create the regulatory environment for OGRE. Besides the NDRC, the Electricity Law provides the legal basis for adopting the National Energy Administration(NEA) as the primary regulator in the electricity sector with the power to issue licences and oversee environmental impact assessment.[68]The regulatory environment created by the mentioned will be analysed in line with the central research question.

8.4.2 Licensing

The Electricity Law of 1996 provides that 'the administrative department of electric power under the State Council shall be responsible for the supervision and control of the electric power industry in the whole country'.[69] At the provincial, autonomous, and municipal levels, the administrative department of electric power shall have similar control within their jurisdictions.[70] As a result, the State Electricity Regulatory Commission (SERC) was vested with the power to license every spectrum of electricity activities, including OGRE, subject to the law.[71] SERC has been replaced by the NEA and the latter currently performs its functions as per licence.[72] The Administrative Regulation on Power Industry Business Permits 2005 has been adopted to govern the licensing of electricity activities.[73] It provides that any enterprise that wants to undertake any business in any sphere of electricity (including renewable energy) within China shall obtain a licence.[74]The requirements for issuing the licence include the payment of processing fees and demonstrating the financial capacity to undertake the specific business.[75]

The point has been made that fossil fuel electricity options maintain the upper hand as far as competition is concerned with renewable energy, given the accumulated benefits of a decade of continuous subsidies and experience with usage. Even though renewable energy has become costcompetitive and enjoys government support, it is still more expensive as per its high initial capital. Thus, demonstrating financial capacity and paying licensing fees will weigh more heavily on OGRE developers than competingfossil fuel options. Investors are generally motivated by profit and lower investment costs rather than a common good like climate change. The latter will still make fossil fuels more attractive to investors than OGRE.It is apt to mention that the regulation does not apply to individuals generating OGRE solely for their consumption.[76] The latter is good practice in light of what is obtainable in other jurisdictions. For instance, in Nigeria, OGRE below 1MW/h is exempted from licensing[77] provided it is not for trading purposes.[78]

8.4.3 Environmental Impact Assessment

The Electric Power Law emphasises the issue of environmental impact assessment and environmental protection more generally. It provides that:

> the construction, production, supply, and utilisation in relation to electric power shall abide by the principles of protecting the environment according to law, adopting new technology, decreasing the discharge of harmful substances, and preventing pollution and other public hazards. The state *encourages and supports the use of renewable and clean energy resources for electricity generation.*[79]

However, it is noteworthy that environmental impact assessment is one of the internationally accepted principles of environmental law.[80] In addition to being conducive to environmental protection, the law also requires planning for electric power development to take cognisance of the rational use of energy resources.[81]Renewable energy, including OGRE, is scattered in every nook and cranny of China in contrast to fossil fuel energy-depleting sources. As such, the rational use of power should entail developing replenishing energy sources such as renewables. Likewise, the law requires ancillary issues such as environmental protection to be designed and put into operation together with electricity-generating projects.[82] Arguably, therefore, the Electric Power Law does support the development of renewable energy, including OGRE, through its numerous provisions on environmental impact assessment. However, the Electric Power Law does not exclude or attempt to diminish the development of competing fossil fuel options, including in the long run. The development of fossil fuel options is allowed and encouraged under its provisions.

The Renewable Energy Law also gives cognisance to the issue of environmental impact assessment in article 9. According to this provision, in working out a plan for the development and utilisation of renewable energy resources, the relevant authority must take cognisance of local conditions, consider all factors, and stick to orderly development. Such factors that must be considered include; the development targets, significant tasks, regional layout, key projects, progress, construction of associated power network, service systems, and safeguards, among others. It further requires the relevant department to work out the plan for developing renewable energy resources to conduct a scientific assessment. These ensure that the development and utilisation of renewable energy resources is implemented, not detrimental to the environment and society at large. The provision further recognises the significance of enforcing environmental procedural rights while assessing the environmental impacts of developmental projects. Hence, it is required for relevant departments to solicit comments from relevant entities, experts, and the general public while working out the plan to develop renewable energy resources. Since renewable energy sources have minimal adverse environmental impacts, this provision is in no way a hindrance but reiterates its eco-friendliness.

8.4.4 Support Framework

The REN Law 2005 is the principal framework that supports renewable energy development, including OGRE.[83]Several ingredients in it have precipitated the upsurge indevelopment of renewable energy, including OGRE in China. First, against the context and targets for the development of OGRE, it provides that 'the stateshall give priority to the development and utilisation ofrenewable energy in energy development'.[84] It has been argued that the latter provision has given OGREthe upper hand as per receiving support from the government to the extent that the disequilibrium between it and fossil fuel is being addressed.[85]The priority clause is supported by preferential bank loan packages for renewable energy, including OGRE.[86] Some authors report that this has helped OGRE developers address the initial capital cost's inaccessibility and/or lack of affordability.[87] The law also provides an industrial guide to channel private investment towards coordinated renewable energy development, including OGRE. Among other things, it establishes a Renewable Energy Development Fund (REDF), which disburses financial incentives to help investors overcome the problem of funding in the area.[88] The positive impact of the mentioned has been discussed extensively regarding how it has catalysed the development of OGRE.[89]

The provision of article 18 of REN Law makes for the inclusion of lower tiers of government in the adoption of targets, preparation of renewable energy development plans, and implementation measures that have played a significant role in garnering grassroot actions and support for OGRE development and deployment. This provision has been relevant for addressing the identified barriers, such as lack of awareness of the OGRE technology, particularly in rural communities. The requirement for the local governments to provide financial support for renewable energy utilisation projects in rural areas also contributes to addressing the barrier of non-affordability of OGRE by those in rural areas. Nonetheless, nothinghas been said about how REN Law interacts with current barriers in the sector.

For one, while it prioritises renewable energy development, including OGRE, over and above fossil fuel electricity options, it does not preclude continuing support for the latter. According to Evelyn Cheng, 'China is still increasing construction of coal-fired power plants. Analysis by US-based Global Energy Monitor indicates that last year, China built more than triple the amount of new coal power capacity as the rest of the world combined'.[90] The continuing support for fossil fuel electricity and the accumulated benefits of a decade of financial aid coupled with experience and usage, cumulatively gives it the upper hand against renewable, including OGRE. Thus, unsurprisingly, fossil fuel continues to dominate the sector despite measures adopted to the contrary.[91]

In addition, the preferential loan packages provided does not benefit household-based projects. The point has been made that OGRE is necessary to address the problem of a power blackout and replace fossil fuel electricity in line with the general mandate of combating climate change. Unfortunately, the banks' provision on preferential loan packages for renewable energy has been criticised for not being mandatory and incorporates some elements of competition to the extent that some OGRE projects will not benefit.[92] The banks are targeting costcompetitive renewable energy projects.[93]Small-scale OGRE, primarily used by households at the grassroots level are not benefitting from the funds.[94] Thus, its development is not as can be in comparison to competing fossil fuel electricity options.

The point has been made that the REDF disburses grants and loans to facilitate the development of renewable energy, including OGRE, by boosting the latter's awareness.Regardless of the REDF, the lack of awareness persists, especially at the grassroots level. In the words of the Global Environmental Facility (GEF), 'the concept of low-carbon development among rural and provincial government personnel is weak, and there is a lack of knowledge in the application of RE-based energy generation in rural areas, especially for electric power generation'.[95] It is apt to mention that the REDF operates at the central level. While it has benefitted some awareness projects at the grassroots level, its impact has not been widespread among the rural populace in China. The REN Law provides that:

> the administrative departments in charge of the energy work of the people's governments of all provinces, autonomous regions and municipalities directly under the Central Government shall, according to the national plan for the development and utilisation of renewable energy… work out plans for the development and utilisation of renewable energy resources for their respective administration regions …and implement them after they are approved.[96]

Regrettably, the REN Law does not expressly mandate them to adopt a fund such as the REDF to support awareness measures at their level. Thus, while some provinces and regions are giving financial incentives to support the development of renewable energy projects,[97] they are doing very little to address the problem of awareness at their respective units.

What is more, the REDF does not disburse financial aid or loans for storage facilities that will enable developers to overcome the problem of intermittency. The point was made that, in China, OGRE is perceived from a liberal perspective to the extent that it also encompasses the feeding of electricity from decentralised renewable electricity to a main grid.[98] For such OGRE, the developers must get a storage device to prevent the problem associated with the intermittency of most renewable energy.[99] Such devices are always expensive and most developers cannot easily afford them.[100]Regrettably, the REDF does not disburse funds that cover storage devices.[101] Thus, renewable energy developers, including OGRE, continue to be constrained amidst government affirmation of its commitment to addressing climate change and recovery from COVID-19.

Another issue worth mentioning is the withdrawal of grants for some renewable electricity projects, including OGRE photovoltaic projects.[102] The rationale for the withdrawal is that the same has become costcompetitive with fossil fuel electricity options.[103] Another reason is that the continuous payment of subsidies have become burdensome on the government, and they have been failing to pay it timeously.[104] While the withdrawal of subsidy is inevitable to avoid the downside of a false market, it must be done carefully to prevent a relapse in attaining the objectives behind it. While some OGRE projects have become cost competitive with fossil fuel, such a comparison is per unitcost. The initial capital cost of such OGRE projects remains relatively high compared to fossil fuel electricity projects.[105] The latter is further complicated by the accumulated support for fossil fuel electricity. Thus, such withdrawal will constrain renewable energy development, including OGRE, especially as it will continually allow the initial capital cost of fossil fuel to remain cheaper and more attractive to investors. Again, the argument that the subsidy is becoming onerous on the government is defeated by the amount of financial support for fossil fuel electricity projects. The latter can be diverted to support renewable energy, including OGRE.

8.5 Conclusion

Renewable energy is a crucial part of the global energy transition agenda. It has a germane role in the speedy recovery from the downsides of COVID-19. Increasingly, therefore, renewables have become the first choice for expanding, upgrading, and modernising power systems around the world. China has become one of the leading jurisdictions in the world for renewable energy development, including OGRE. Arguably, this has been made possible by recent years of policymaking and legislation on renewable energy. The journey to supporting renewable energy started with the Electricity Law of 1996. In 2005, a milestone was recorded with the adoption of the most significant regulatory support framework in the sector, i.e.,the Renewable Energy Law 2005.

The principal lesson extrapolated from the China experience is the use of an overarching supportive renewable energy law to drive the development of the sector by removing the barriers. Some authors have dismissed the use of one instrument for supporting fossil fuel and RE because it introduces a conflict of interest that negatively impacts its development.[106]The use of one overarching RE law with well-calculated measures that address existing barriers with mathematical accuracy, make it easier for coordination, consistency, simplicity, and clarity of support measures regarding every facet of OGRE.[107]The China RE Law 2005 set a useful precedent as to how one instrument can be used to address the many barriers to OGRE development.[108]Other countries should draw from their experience in light of the defects in the regulatory support framework for OGRE.

Regardless, this chapter has analysed the RE law and other mentioned laws to establish the extent of its support for the development of OGRE in the context of addressing climate change, attaining energy security, and recovery from the COVID-19 pandemic. It identified that some barriers still impede the development of OGRE in China. First, the persisting disequilibrium between the investment clime of fossil fuel and renewable energy, including OGRE, exacerbated by the continuous support for the latter, remains a barrier. Some residents, especially rural dwellers, are still in the dark as to the need to develop OGRE. As such, they are opposed to its development. Another challenge is that mini-grid developers are currently struggling to feed into the existing grid because of the intermittency of most renewable energy sources.[109] Against this background, it analyses the policy and regulatory environment for electricity development. It argues that the Renewable Energy Law 2005 paints the picture of the energy security and climate change rationale for developing law. However, the COVID-19 Government Work Report 2021 does not stipulate the role of renewable energy, including OGRE, in the recovery pathway from the COVID-19 pandemic. Against which it was argued that it could never fully reach its potential in the context of recovery. Additionally, while China's 14th five-year plan 2021–2025 recognises the role of renewable energy in developing the economy, there is no national target for it and, consequently, sectoral targets for OGRE. Thus, there is currently no benchmark for adopting and measuring the legislative interventions for the sector's development.

It analyses the licensing framework and environmental impact assessment for electricity ventures in China. It argues that the exemption of OGRE projects for personal consumption from licensing requirements is good law as it reduces the financial burden of initial capital project costs. However, other OGRE projects have to pay licensing fees, which adds to the initial capital cost faced by developers. It was argued that the Electricity Law 1996 on EIA provisions is a needed impetus for the increased development of OGRE. The specific requirement of the Renewable Energy Law 2005 on EIA was argued not to impede the sector's development.

The support framework for OGRE, i.e.,the Renewable Energy Law 2005, was analysed in line with the central research theme. It was argued that the Renewable Energy Law 2005 had helped address barriers to the sector's development, including financial and technological barriers. Regardless, the law does not preclude support for fossil fuel electricity projects. Such support coupled with the decade of accumulated years of experience with usage and subsidies has translated to fossil fuel still maintaining the upper hand in the electricity sector. The preferential loan packages that it creates does not benefit household OGRE projects. It establishes the REDF, disbursing grants and loans for renewable energy projects, including awareness measures. At the same time, there is a mandate for lower tiers of government to adopt measures to facilitate renewable energy development. There is no obligation to adopt the equivalent of the REDF at the grassroots level. There are no sufficient awareness measures to combat renewable energy illiteracy at the grassroots level. It is also argued that there is no financial support for enabling investors to overcome the problem of storage. Finally, the withdrawal of support for certain OGRE projects was argued to be bad law, given that it will undermine the sector's development.

The point has been made that China proposes attaining carbon neutrality by 2060. The latter entails a radical shift in the technological base from fossil fuel electricity strategies, especially coal, to renewable energy. The NDRC in the Action Plan for Carbon Dioxide Peaking before 2030 states, 'to build a legal system that supports the green and low-carbon development, we will promote the formulation and revision of … the Electric Power Law, the Coal Industry Law, the Renewable Energy Law…'.[110] Given the proposal to review the regulatory framework for renewable energy, including OGRE, it is expected that some of the flaws mentioned will be addressed. In the first instance, the NDRC should reinstate the disbursal of grants and loans to all forms of OGRE until the former is leading over fossil fuel electricity strategies in the sector. The issue of the China government struggling to pay such incentives can be addressed by directing the funding for fossil fuel electricity to renewable energy. Besides the withdrawal of the funds for fossil fuel, taxation may be introduced to level the disequilibrium between it and OGRE. The REDF should be stretched to accommodate support for storage facilities for minigrid renewables to enable its feeding into the national grid. The Renewable Energy Law should mandate the municipal government, province, and autonomous regions to adopt measures to educate the grassroots populace on the benefits of OGRE.

Notes

1 Ngozi Chinwa Ole, 'The Role of Renewable Energy Law 2005 in Supporting the Development of Off-Grid Renewable Electricity in China' (2019) 5 International Energy Law Review 132, 133.
2 Wang Zhong Ying, 'China's Achievement for Expanding Electricity Access for the Poor' (2006) 10 Energy for Sustainable Development 5.
3 Trading Economics, 'China Access to Electricity' (2021) <https://tradingeconomics.com/china/access-to-electricity-percent-of-population-wb-data.html> accessed 8th December 2021.

4 Ngozi Chinwa Ole, 'The Role of Renewable Energy Law 2005 in Supporting the Development of Off-Grid Renewable Electricity in China' (n 1) 133.

5 K.R. Ajao, 'Electric Energy Supply in Nigeria, Decentralized Energy Approach' (2009) 24 Cogeneration and Distributed Generation Journal 42.

6 Gopal K. Sarangi and others, 'Working Paper Decentralised Renewable Energy Systems in China, India and Thailand: Assessing the Role of Policies and Incentive Structures' (2017) 18 <www.prospernet.com>.

7 Ngozi Chinwa Ole, 'The Role of Renewable Energy Law 2005 in Supporting the Development of Off-Grid Renewable Electricity in China' (n 1) 133.

8 Peter Hoskins, 'China Power Cuts: What is Causing the Country's Blackouts?' (2021) <https://www.bbc.com/news/business-58733193> accessed 8th December 2021.

9 Ibid. See also Muyu Xu and Ryan Woo, 'China's State Grid Warns of Tight Supply in Winter, Despite Easing Power Crunch' (2021) <https://www.reuters.com/business/energy/chinas-state-grid-warns-tight-supply-winter-despite-easing-power-crunch-2021-11-07/> accessed 8th December 2021.

10 Ibid. See Aljazeera, "Unprecedented' Power Cuts in China Hits Homes, Factories' (2021) <https://www.aljazeera.com/news/2021/9/28/chinas-northeast-suffers-power-crunch>accessed 8th December 2021.

11 Ngozi Chinwa Ole, 'The Role of Renewable Energy Law 2005 in Supporting the Development of Off-Grid Renewable Electricity in China' (n 1) 133. See also Ngozi Chinwa Ole, 'The Nigerian Electricity Regulatory Framework: Hotspots and Challenges for Off-Grid Renewable Electricity Development' (2020) 38(4) Journal of Energy and Natural Resources Law 367.

12 Ibid.

13 Zhu Liu, 'China Carbon's Emissions Report 2015' (2015) <http://www.belfer-center.org/sites/default/files/legacy/files/China%20Carbon%20Emissions%202016%20final%20web.pdf> accessed 20th August 2021.

14 CSIC, 'China's Energy Footprint' (2021) <https://chinapower.csis.org/energy-footprint/#:~:text =Over%20the%20last%20half%20century,percent%20of%20China's%20energy%20use.> accessed 9th December 2021.

15 Yu Wang, 'A Review of Renewable Energy Legislation and Policy in China' in Espen Moe and Paul Midford (eds), *The Political Economy of Renewable Energy and Energy Security: Common Challenges and National Responses in Japan, China and Northern Europe* (Pengrave and Macmillian 2014) 207.

16 China's Government, 'China Signs Paris Agreement on Climate Change' (2016) <http://english.www.gov.cn/state_council/vice_premiers/2016/04/23/content_281475333331232.htm> accessed 9th December 2021.

17 China, 'Nationally Determined Contributions' (2016) 35<https://www4.unfccc.int/sites/ndcstaging/PublishedDocuments/China%20First/China%E2%80%99s%20Achievements,%20New%20Goals%20and%20New%20Measures%20for%20Nationally%20Determined%20Contributions.pdf>accessed 9th December 2021.

18 Ibid.

19 In the latest China's NDC document (Addendum version) in November 2021, China presented a number of success projects which is claimed to have been successfully increasing poverty alleviation and supporting green transformation. One of best example is the development of solar power in the Liangshan, Ganzi and Aba Prefectures in the western part of Sichuan Province which are relatively underdeveloped regions in China. Today, this region has become a best example of photovoltaic village. UNFCCC, 'China first Nationally Determined Contributions (Addendum)' (2021) 9 <https://www4.unfccc.int/sites/ndcstaging/PublishedDocuments/China%20First/China%E2%80%99s%20Achievements,%20New%20Goals%20and%20New%20Measures%20for%20Nationally%20Determined%20Contributions.pdf> accessed 15th February 2021.

20 Miao Hong and Peng Peng, 'Powering China with Clean Energy after COVID-19' (2020) <https://www.wri.org/insights/powering-china-clean-energy-after-covid-19> accessed 9th December 2021.
21 Ibid.
22 Qiang Tu and others, 'Using Green Finance to Counteract the Adverse Effects of COVID-19 Pandemic on Renewable Energy Investment-The Case of Offshore Wind Power in China' (2021) 158 Energy Policy 112542.
23 The 14th Five-Year Plan of the People's Republic of China—Fostering High-Quality Development 2021–2025, para 2.
24 National Development and Reform Commission, 'Action Plan for Carbon Dioxide Peaking before 2030' (2021) <https://en.ndrc.gov.cn/policies/202110/t20211027_1301020.html> accessed 10th December 2021.
25 Ngozi Chinwa Ole, 'The Role of Renewable Energy Law 2005 in Supporting the Development of Off-Grid Renewable Electricity in China' (n 1) 133.
26 Ibid.
27 Dilip Ahuja and others, 'Sustainable Energy for Developing Countries' (2009) 2(1) SAPIENS 2.
28 Wendy Annecke, 'Monitoring and Evaluation of Energy for Development: The Good, the Bad and the Questionable in M&E Practice' (2008) 36 Energy Policy 2839. See also Tuntivate Voravate, *Assessing Markets for Renewable Energy in Rural Areas of Northwestern China* (World Bank Publications 2000) Vol.23–492, p.16; Bhattacharyya and Ohiare, 'The Chinese Model of Rural Electrification and Electricity Access' in *Rural Electrification through Decentralised Off-Grid Systems in Developing Countries* (Springer 2012), p.119.
29 Craig A. Hart and Dominic Marcellino, 'Subsidies or Free Markets to Promote Renewables?' (2012) Renewable Energy Law and Policy Review 196.
30 Ngozi Chinwa Ole, 'The Role of Renewable Energy Law 2005 in Supporting the Development of Off-Grid Renewable Electricity in China' (n 1) 134.
31 Ibid.
32 IRENA, 'Majority of New Renewables Undercut Cheapest Fossil Fuel on CostTweet' (2021) <https://www.irena.org/newsroom/pressreleases/2021/Jun/Majority-of-New-Renewables-Undercut-Cheapest-Fossil-Fuel-on-Cost> accessed 16th December 2021.
33 Ngozi Chinwa Ole, 'The Role of Renewable Energy Law 2005 in Supporting the Development of Off-Grid Renewable Electricity in China' (n 1) 134.
34 Mikey Redding, 'Renewable Energy in China' (2021) <https://www.borgenmagazine.com/renewable-energy-in-china/> accessed 16th December 2021.
35 Dominic Dudley, 'China Is Set to Become the World's Renewable Energy Superpower, According to New Report' (2019) <https://www.forbes.com/sites/dominicdudley/2019/01/11/china-renewable-energy-superpower/> accessed 16th December 2021.
36 Charlie Campbell, 'China Is Bankrolling Green Energy Projects around the World' (2021) <https://time.com/5714267/china-green-energy/>accessed 16th December 2021.
37 CSIC, 'China's Energy Footprint' (n 14).
38 'China to Control, Phase Down Coal Consumption in Next Decade: Xi' <http://www.xinhuanet.com/english/2021-04/22/c_139899323.htm> accessed 15th February 2021.
39 Bloomberg News, 'China's Coal Support Continues with $31 Billion Finance Promise' *Bloomberg News* (2021) <https://www.bloomberg.com/news/articles/2021-11-18/china-s-coal-support-continues-with-31-billion-finance-promise>accessed 16th December 2021.
40 'China's Solar Desire Dims' *Financial Times*, 8th June 2018 <https://www.ft.com/content/985341f4-6a57-11e8-8cf3-0c230fa67aec> accessed 16th December 2021.

41 Charles Vast, 'China Turn to Energy Storage to Push Renewable Energy' (2021) <https://chinadialogue.net/en/energy/9635-china-turns-to-energy-storage-to-push-renewables/> accessed 16th December 2021.
42 Ibid.
43 Behrang Vand and others, 'Consumers' Attitudes to Support Green Energy: A Case Study in Shanghai' (2021)12 Energies 2379.
44 Sara Schuman and Alvin Lin, 'China's Renewable Energy Law and its impact on renewable power in China: Progress, challenges and recommendations for improving implementation' (2012) 51 Energy Policy 89.
45 The Renewable Energy Law of the People's Republic of China 2005 (as amended in 2009), Art 1.
46 Ibid.
47 Ngozi Chinwa Ole, 'The Role of Renewable Energy Law 2005 in Supporting the Development of Off-Grid Renewable Electricity in China' (n) 132.
48 See Jeffrey Heys, 'Electricity in China: Production, Grids, Inefficiencies and Shortages' <http://www .factsanddetails.com/china/cat13/sub85/item1726.html> accessed 20th December 2021.
49 Ngozi Chinwa Ole, 'The Nigerian Electricity Regulatory Framework: Hotspots and Challenges for Off-grid Renewable Electricity Development' (2020) 38(4) JENRL 367.
50 UN-Energy/Africa, *Energy for Sustainable Development: Policy Options for Africa 2011; UNEP, Financing Renewable Energy in Developing Countries* (UNEP 2012) 29.
51 The Chinese Government Work Report 2021 <https://www.cmbi.com/upload/202103/20210308682658.pdf> accessed 23rd December 2021.
52 Miao Hong and Peng Peng, 'Powering China with Clean Energy After COVID-19' (2020) <https://www.wri.org/insights/powering-china-clean-energy-after-covid-19> accessed 9th December 2021.
53 Ibid.
54 Jorrit Gosens and Frank Jotzo, 'China's Post-COVID-19 Stimulus: No Green New Deal in Sight' (2020) 36 Environmental Innovation and Societal Transitions 250.
55 The Renewable Energy Law of the People's Republic of China 2005 (as amended in 2009), Art 8. See also Richard J. Campbell, 'China and the United States—A Comparison of Green Energy Programs and Policies' (2014) 3 <http://www.digital-commons.ilr .cornell.edu/key_workplace/826> accessed 20th December 2021.
56 National Development and Reform Commission, People's Republic of China, 'Medium and Long-Term Development Plan for Renewable Energy in China' (September 2007) <http://www.chinaenvironmentallaw.com/wp-content/uploads/2008/04/medium-and-long-term- development- plan-forrenewable-energy.pdf> accessed 22nd December 2021.
57 Ngozi Chinwa Ole, 'The Role of Renewable Energy Law 2005 in Supporting the Development of Off-Grid Renewable Electricity in China' (n 1) 132
58 Ibid.
59 IRENA, *Renewable Energy Target Setting* (IRENA, 2015) 29. For further commentaries on the role of targets, See Karin Edvardsson and S. Hansson, 'When is a Goal Rational?' (2005) 24 Social Choice and Welfare 343.
60 The 14th Five-Year Plan of the People's Republic of China—Fostering High-Quality Development 2021-2025, para 2.
61 Ibid, para 9.
62 Ibid, para 3.
63 NDRC, 'Action Plan for Carbon Dioxide Peaking before 2030' (n 24).
64 Ibid.
65 Ibid.

66 Chung-min Tsai, 'Regulating China's Power Sector: Creating an Independent Regulator without Autonomy'(2014) 218 The China Quarterly 452, 462.
67 The Administrative License Law of the People's Republic of China 2004, CLI.1.49280.
68 State Council, 'National Energy Administration' <http://english.www.gov.cn/state_council/2014/10/01/content_281474991089761.htm> accessed 25th December 2021.
69 The Electricity Law of China 1996, Art 6.
70 Ibid.
71 Qing Zhang, 'Regulatory Framework for the Electricity Industry' (2015) <https://www.oecd.org/gov/regulatory-policy/41888874.pdf> accessed 26th December 2021.
72 Chung-min Tsai, 'Regulating China's Power Sector: Creating an Independent Regulator without Autonomy'(n 64).
73 Provisions on the Administration of Electric Power Business Licenses 2005 (as amended in 2015) (CH).
74 Provisions on the Administration of Electric Power Business Licenses 2005 (as amended in 2015) Art. 4.
75 Ibid, Art 17.
76 Ibid, Art 4.
77 Electric Power Sector Reform Act 2005 (NG), s.62(1).
78 *A and P Foods Ltd v Exusia Power and Gas Ltd* [2013] NERC/14/001.
79 The Electric Power Law 1996 (CH), Art 5.
80 Philippe Sands and others, *Principles of International Environmental Law* (Cambridge University Press 2012) 601.
81 The Electric Power Law 1996 (CH), Art 10.
82 Ibid.
83 Jingli Fan and others, 'The Development of China's Renewable Energy Policy and Implications to Africa' (2018) IOP Conference Series: Materials Science and Engineering 394.
84 The Renewable Energy Law 2005 (CH) Art 4. See also IBP Inc., *China Energy Policy Law, Regulation and Handbook*, Vol. 1 (Global Investment Center 2015) 83.
85 Ngozi Chinwa Ole, 'The Role of Renewable Energy Law 2005 in Supporting the Development of Off-Grid Renewable Electricity in China' (n 1) 133.
86 The Renewable Energy Law 2005 (CH), Art.25. See also Z. Ming, 'Review of Renewable Energy Investment and Financing in China: Status, Mode, Issues and Counter Measures' (2014) 31 Renewable and Sustainable Energy Review 23, 31.
87 Ngozi Chinwa Ole, 'The Role of Renewable Energy Law 2005 in Supporting the Development of Off-Grid Renewable Electricity in China' (n 1) 133.
88 The Renewable Energy Law 2005 (CH), Art24.
89 Ngozi Chinwa Ole, 'The Role of Renewable Energy Law 2005 in Supporting the Development of Off-Grid Renewable Electricity in China' (n 1) 133.
90 Evelyn Cheng, 'China Has No Other Choice but to Rely on Coal Power for Now' (2021) <https://www.cnbc.com/2021/04/29/climate-china-has-no-other-choice-but-to-rely-on-coal-power-for-now.html> accessed 28th December 2021.
91 CSIC, 'China's Energy Footprint' (n 14).
92 Charlie Zhu and Adam Rose, 'China Solar Expansion Needs Billions from Wary Investors' (2015) <http://www.reuters.com/article/us-china-solar-financing/china-solar-expansion-needs-billions-from-wary-investors-idUSKBN0N K2L120150430> accessed 31st December 2021.
93 Ngozi Chinwa Ole, 'The Role of Renewable Energy Law 2005 in Supporting the Development of Off-Grid Renewable Electricity in China' (n 1)132.
94 Ibid.
95 GEF, 'Enabling Zero Carbon Energy in Rural Towns and Villages in China (EZCERTV) Project' (2019)<https://www.thegef.org/project/enabling-zero-carbon

-energy-rural-towns-and-villages-china-ezcertv-project> accessed 31st December 2021.

96 The REN Law 2005, Art 9.

97 Z. Ming, 'Review of Renewable Energy Investment and Financing in China: Status, Mode, Issues and Counter Measures' (n 86) 32.

98 Gopal K. Sarangi, 'Working Paper Decentralised Renewable Energy Systems in China, India andThailand: Assessing the Role of Policies and Incentive Structures' (n 6).

99 Charles Vast, 'China Turn to Energy Storage to Push Renewable Energy' (2021) <https://chinadialogue.net/en/energy/9635-china-turns-to-energy-storage-to-push-renewables/> accessed 16th December 2021.

100 Ibid.

101 Ibid.

102 Ivy Yin and Eric Yep, 'China to Halt Subsidies for Some Types of Wind, Solar Projects: NDRC' (2021) <https://www.spglobal.com/platts/en/market-insights/latest-news/electric-power/061121-china-to-halt-subsidies-for-some-types-of-wind-solar-projects-ndrc> accessed 2nd January 2022.

103 Ibid.

104 Ibid.

105 Yan Zu and others, 'Concentrated Solar Power: Technology, Economy Analysis, and Policy Implications in China' (2022) 29 Environmental Science and Pollution Research1324.

106 Janet L. Sawin,*National Policy Instruments: Policy Lessons for the Advancement & Diffusion of Renewable Energy Technologies around the World*(World Bank 2004).

107 GIZ,*Legal Frameworks for Renewable Energy: Policy Analysis for 15 Developing and Emerging Countries* (Deutsche Gesellschaft für 2012) 13.

108 Ibid.

109 Semich Impram, 'Challenges of Renewable Energy Penetration on Power System Flexibility: A Survey' (2020) 31 Energy Strategy Reviews 100539.

110 National Development and Reform Commission, 'Action Plan for Carbon Dioxide Peaking before 2030' (n 24).

References

A and P Foods Ltd v Exusia Power and Gas Ltd (2013) NERC/14/001.

Aljazeera, '"Unprecedented" Power Cuts in China Hits Homes, Factories' (2021) <https://www.aljazeera.com/news/2021/9/28/chinas-northeast-suffers-power-crunch>

Behrang V and and others, '"Consumers" Attitudes to Support Green Energy: A Case Study in Shanghai' (2021) 12Energies2379.

Subhes Bhattacharyya and Sanusi Ohiare, 'The Chinese Model of Rural Electrification and Electricity Access' in Bhattacharyya S (ed), *Rural Electrification Through Decentralised Off-Grid Systems in Developing Countries* (Springer, London 2012).

Bloomberg News, 'China's Coal Support Continues with $31 Billion Finance Promise Bloomberg News' (2021) <https://www.bloomberg.com/news/articles/2021-11-18/china-s-coal-support-continues-with-31-billion-finance-promise>

Charles Vast, 'China Turn to Energy Storage to Push Renewable Energy' (2021) <https://chinadialogue.net/en/energy/9635-china-turns-to-energy-storage-to-push-renewables/>

Charlie Campbell, 'China is Bankrolling Green Energy Projects Around the World' (2021) <https://time.com/5714267/china-green-energy/>

Charlie Zhu and Adam Rose, 'China Solar Expansion Needs Billions from Wary Investors' (2015) <http://www.reuters.com/article/us-china-solar-financing/china-solar-expansion-needs-billions-from-wary-investors-idUSKBN0N K2L120150430>

China, 'Nationally Determined Contributions' (2016) <https://www4.unfccc.int/sites/ndcstaging/PublishedDocuments/China%20First/China%E2%80%99s%20Achievements,%20New%20Goals%20and%20New%20Measures%20for%20Nationally%20Determined%20Contributions.pdf>

China, 'Nationally Determined Contributions' Addendum' (2021) <https://www4.unfccc.int/sites/ndcstaging/PublishedDocuments/China%20First/China%E2%80%99s%20Achievements,%20New%20Goals%20and%20New%20Measures%20for%20Nationally%20Determined%20Contributions.pdf>

China's Government, 'China Signs Paris Agreement on Climate Change' (2016) <http://english.www.gov.cn/state_council/vice_premiers/2016/04/23/content_281475333331232.htm>

'China's Solar Desire Dims' *Financial Times*, 8 June 2018<https://www.ft.com/content/985341f4-6a57-11e8-8cf3-0c230fa67aec>

Chung-min Tsai, 'Regulating China's Power Sector: Creating an Independent Regulator Without Autonomy' (2014) 218The China Quarterly452.

Craig A. Hart and Dominic Marcellino, 'Subsidies or Free Markets to Promote Renewables?' (2012) Renewable Energy Law and Policy Review196.

CSIC, 'China's Energy Footprint' (2021) <https://chinapower.csis.org/energy-footprint/#:~:text=Over%20the%20last%20half%20century,percent%20of%20China's%20energy%20use.>

Dilip Ahuja and others, 'Sustainable Energy for Developing Countries' (2009) 2(1) SAPIENS2.

Dominic Dudley, 'China Is Set to Become the World's Renewable Energy Superpower, According to New Report' (2019) <https://www.forbes.com/sites/dominicdudley/2019/01/11/china-renewable-energy-superpower/>

EvelynCheng, 'China Has No Other Choice But to Rely on Coal Power for Now' (2021) <https://www.cnbc.com/2021/04/29/climate-china-has-no-other-choice-but-to-rely-on-coal-power-for-now.html>

GEF, 'Enabling Zero Carbon Energy in Rural Towns and Villages in China (EZCERTV) Project' (2019) <https://www.thegef.org/project/enabling-zero-carbon-energy-rural-towns-and-villages-china-ezcertv-project>

GIZ, *Legal Frameworks for Renewable Energy: Policy Analysis for 15 Developing and Emerging Countries* (Deutsche Gesellschaft für2012).

Gopal K. Sarangi and others, 'Working Paper Decentralised Renewable Energy Systems in China, India and Thailand: Assessing the Role of Policies and Incentive Structures' (2017) <www.prospernet.com>

IBP Inc., *China Energy Policy Law, Regulation and Handbook*, Vol. 1 (Global Investment Center2015).

International Renewable Energy Agency, 'Majority of New Renewables Undercut Cheapest Fossil Fuel on Cost Tweet' (2021) <https://www.irena.org/newsroom/pressreleases/2021/Jun/Majority-of-New-Renewables-Undercut-Cheapest-Fossil-Fuel-on-Cost>

International Renewable Energy Agency, *Renewable Energy Target Setting* (International Renewable Energy Agency2015).

Ivy Yin and Eric Yep, 'China to Halt Subsidies for Some Types of Wind, Solar Projects: NDRC' (2021) <https://www.spglobal.com/platts/en/market-insights/latest-news/electric-power/061121-china-to-halt-subsidies-for-some-types-of-wind-solar-projects-ndrc>

Janet L. Sawin, *National Policy Instruments: Policy Lessons for the Advancement & Diffusion of Renewable Energy Technologies Around the World* (World Bank2004).

JeffreyHeys, 'Electricity in China: Production, Grids, Inefficiencies and Shortages' <http://www.factsanddetails.com/china/cat13/sub85/item1726.html>

Jingli Fan and others, 'The Development of China's Renewable Energy Policy and Implications to Africa' (2018) IOP Conference Series: Materials Science and Engineering394.

Jorrit Gosens and Frank Jotzo, 'China's Post-COVID-19 Stimulus: No Green New Deal in Sight' (2020) 36Environmental Innovation and Societal Transitions250.

K R Ajao, 'Electric Energy Supply in Nigeria, Decentralized Energy Approach' (2009) 24Cogeneration and Distributed Generation Journal42.

Karin Edvardsson and S Hansson, 'When is a Goal Rational?' (2005) 24Social Choice and Welfare343.

Miao Hong and Peng Peng, 'Powering China with Clean Energy After COVID-19' (2020) <https://www.wri.org/insights/powering-china-clean-energy-after-covid-19>

Mikey Redding, 'Renewable Energy in China' (2021) <https://www.borgenmagazine.com/ renewable-energy-in-china/>

Muyu Xu and Ryan Woo, 'China's State Grid Warns of Tight Supply in Winter, Despite Easing Power Crunch' (2021) <https://www.reuters.com/business/energy/chinas-state-grid-warns-tight-supply-winter-despite-easing-power-crunch-2021-11-07/>

National Development and Reform Commission, 'Action Plan for Carbon Dioxide Peaking before 2030' (2021) <https://en.ndrc.gov.cn/policies/202110/t20211027_1301020.html>

National Development and Reform Commission, People's Republic of China, 'Medium and Long-Term Development Plan for Renewable Energy in China' (September 2007) <http://www.chinaenvironmentallaw.com/wp-content/uploads/2008/04/medium-and-long-term- development- plan-forrenewable-energy.pdf>

Nigeria, Electric Power Sector Reform Act 2005.

Ole Ngozi Chinwa, 'The Nigerian Electricity Regulatory Framework: Hotspots and Challenges for Off-grid Renewable Electricity Development' (2020) 38(4) Journal of Energy and Natural Resources Law367.

Ole Ngozi Chinwa, 'The Role of Renewable Energy Law 2005 in Supporting the Development of Off-grid Renewable Electricity in China' (2019) 5 International Energy Law Review 132, 133.

Peter Hoskins, 'China Power Cuts: What is Causing the Country's Blackouts?' (2021) <https://www.bbc.com/news/business-58733193>

Philippe Sands and others, *Principles of International Environmental Law* (Cambridge University Press2012).

Qiang Tu and others, 'Using Green Finance to Counteract the Adverse Effects of COVID-19 Pandemic on Renewable Energy Investment-The Case of Offshore Wind Power in China' (2021) 158Energy Policy112542.

Qing Zhang, 'Regulatory Framework for the Electricity Industry' (2015) <https://www.oecd.org/gov/regulatory-policy/41888874.pdf>

Richard J. Campbell, 'China and the United States—A Comparison of Green Energy Programs and Policies' (2014) <http://www.digitalcommons.ilr.cornell.edu/key_workplace/826>

Sara Schuman and Alvin Lin, 'China's Renewable Energy Law and Its Impact on Renewable Power in China: Progress, Challenges and Recommendations for Improving Implementation' (2012) 51Energy Policy89.

Semich Impram, 'Challenges of Renewable Energy Penetration on Power System Flexibility: A Survey' (2020) 31Energy Strategy Reviews100539.

State Council, 'National Energy Administration' <http://english.www.gov.cn/state_council/2014/10/01/content_281474991089761.htm>

The 14th Five-Year Plan of the People's Republic of China—Fostering High-Quality Development 2021–2025<https://www.adb.org/publications/14th-five-year-plan-high-quality-development-prc>

The Administration of Electric Power Business Licenses Law of the People's Republic of China 2005 (as amended in 2015).

The Administrative License Law of the People's Republic of China 2004, CLI.1.49280.

The Chinese Government Work Report 2021<https://www.cmbi.com/upload/202103/20210308682658.pdf>

The Electricity Law of China 1996.

The Renewable Energy Law of the People's Republic of China 2005 (as amended in 2009).

Trading Economics, 'China Access to Electricity' (2021) <https://tradingeconomics.com/china/access-to-electricity-percent-of-population-wb-data.html>

Tuntivate V.Oravate, *Assessing Markets for Renewable Energy in Rural Areas of North Western China* (World Bank Publications2000).

UN-Energy/Africa, *Energy for Sustainable Development: Policy Options for Africa 2011; UNEP, Financing Renewable Energy in Developing Countries* (UNEP2012).

Wang Yu, 'A Review of Renewable Energy Legislation and Policy in China' in EspenMoe and PaulMidford (eds), *The Political Economy of Renewable Energy and Energy Security: Common Challenges and National Responses in Japan, China and Northern Europe* (Palgrave Macmillan2014).

Wang Zhong Ying, 'China's Achievement for Expanding Electricity Access for the Poor' (2006) 10Energy for Sustainable Development5.

Wendy Annecke, 'Monitoring and Evaluation of Energy for Development: The Good, the Bad and the Questionable in M&E Practice' (2008) 36Energy Policy2839.

Yan Zu and others, 'Concentrated Solar Power: Technology, Economy Analysis, and Policy Implications in China' (2022) 29Environmental Science and Pollution Research1324.

Z. Ming, 'Review of Renewable Energy Investment and Financing in China: Status, Mode, Issues and Counter Measures' (2014) 31Renewable and Sustainable Energy Review23.

Zhu Liu, 'China Carbon's Emissions Report 2015' (2015) <http://www.belfercenter.org/sites/default/files/legacy/files/China%20Carbon%20Emissions%202016%20final%20web.pdf>

9 Philippines

Ngozi Chinwa Ole, Dickson E. Omukoro, and Ziyana Nazeemudeen Mohamed

9.1 Introduction

Off-grid renewable energy (OGRE) is growing in popularity as a preferred option of electrification for remote rural areas and archipelagos.[1] The archipelagic nature of the Philippines, coupled with concerns about the problem of climate change have made the development of renewable energy including OGRE, an exigency. Regardless, its development is constrained by the problem of affordability and accessibility of the initial capital costs, problem of financial sustainability of projects, and lack of technological capacity.

Against this background, the Renewable Energy Act 2008 was adopted as the supportive framework for the sector in the Philippines. This chapter analyses the Renewable Energy Act 2008 in order to determine its effectiveness in promoting OGRE through addressing the identified barriers to its development. It bolsters confidence in some of its features that has helped in the development of the sector like a clear context, implementation measures, delineation of institutional functions, and financial incentives. However, it finds that other factors constrain its effectiveness in this context. Some of the factors include the lack of priority for OGRE projects, the delay in the set-up of the Renewable Energy Trust Fund (RETF), and the absence of sectoral targets.

9.2 Imperatives and Barriers for OGRE

The Philippines is an archipelago with over 7,107 islands. This makes off-grid electricity systems a viable option for the electrification of some parts of the country.[2] It is therefore not surprising that, as far back as the late 1980s, off-grid electricity was already developed and mostly dominated by diesel electricity technologies.[3] The use of diesel electricity technologies, however, came with a major drawback because it made the Philippines' energy supply insecure since the country was not an oil-producing state.[4] As such, the country depended on the importation of Middle East–sourced oil to meet its electricity needs.[5] Consequently, the electricity supply was threatened by oil supply disruptions and volatility in price.[6] Energy security is considered to be of strategic importance to countries due to the role of energy in both war and peace, particularly seeing that no nation can

DOI: 10.4324/9781003178088-12

develop without energy. The latter made the development of OGRE exigent in the context of achieving energy security in such an archipelago.

Beyond energy security, the use of off-grid diesel electricity technologies exacerbates the global problem of climate change.[7] The Philippines is one of the most vulnerable countries to climate change.[8] It is therefore not surprising that the country did not hesitate to sign and ratify the Paris Climate Change Agreement 2015.[9] The use of diesel generators in the Philippines was resulting in high emissions of GHGs and other pollutants, which were harmful to the environment.[10] Mitigating climate change in the Philippines entails a transition from such off-grid diesel to OGRE.

The concerns resulted in the initiation of programmes for the development of renewable energy (RE), including OGRE, as far back as the 1980s.[11] It is important to mention at this point that both the government and donor agencies funded these policies and programmes.[12] Even with the governmental assisted participation in the sector, there was an exponential increase in population,[13] which exerted pressure on the limited electricity supply, resulting in frequent blackouts and shortages.[14] The government saw privatisation as an option that would end the electricity shortages.[15] This is because it would bring in private players and culminate in greater electricity generation and supply that would meet the current electricity demand.[16] As a result, in 2001, the Philippines electricity market including OGRE was privatised.[17]

While the OGRE sector has recorded some progress, the off-grid electricity sector is still dominated by fossil fuel.[18] In comparison to on-grid renewable electricity, the development of OGRE has been constrained.[19] The latter is attributable to the problem of affordability and accessibility of OGRE. The initial capital cost of OGRE technologies and projects are remarkably higher than that of fossil fuel, making it difficult for investors and consumers to afford them without recourse to financial assistance from banks.[20] Regrettably, such assistance is not readily available because banks struggle to finance long-term projects, including OGRE and they also perceive the sector as risky.[21] Even when investors manage to overcome the latter barrier, they are also faced with the problem of the sustainability of such projects because most rural dwellers cannot afford a cost-reflective tariff.[22] The availability of relevant technological capacity to manufacture and maintain renewable electricity technologies is also a major inhibitor of development in the OGRE sector.[23]

9.3 Policy Direction for OGRE

Unlike some countries,[24] the Philippines does not have a separate policy document for renewable electricity. The policy directives are contained in the Electric Power Industry Reform Act of 2001. It provides that the policy of the state is 'to ensure and accelerate the total electrification of the country'.[25] Another objective is 'to assure socially and environmentally compatible energy sources and infrastructure'.[26] Renewable energy options including OGRE are more environmentally friendly in comparison to those based on fossil fuel.[27] The point has already been made that OGRE is also exigent in the context of electrifying islands in the Philippines. Thus, arguably, the two abovementioned objectives create the needed impetus for the optimal development of OGRE.

However, the above argument is whittled down by the absence of a definition of 'socially and environmentally compatible energy sources'. A liberal definition of environmentally compatible energy incorporates fossil fuel energy especially when an efficient technology is used to produce electricity from the latter.[28] Thus, the abovementioned policy objectives are also a driver for the development of such fossil fuel electricity as it is for OGRE. However, such fossil fuel electricity options, irrespective of their efficiency, contributes more to climate change than the dirtiest renewable electricity option.[29] Consequently, OGRE is still preferable in the context of addressing climate change. Regrettably, the two policy objectives are not strong enough to garner the needed regulatory support for the optimal development of OGRE to the extent that it is also affording fossil fuel energy the same support.

Notwithstanding, the Electric Power Act 2001 provides, as one of their policy objectives, the promotion of the utilisation of renewable energy resources in power generation to reduce dependency on imported energy.[30] The point has already been made that dependency on imported fossil fuel energy cause oil supply disruptions and volatility in price, making renewable energy a preferable option.[31] Thus, it can be argued that the mentioned objective is implicitly nudging the state to commit to the promotion of OGRE over and above fossil fuels in the context where it is needed. However, the fact that OGRE was not mentioned means that lawmakers may not be presented with the latter logic when adopting measures for the promotion of investments in the electricity sector.

Finally, another policy goal provided for in the Electric Sector Reform Act 2001 is a commitment to reducing government investment in the sector by gathering private capital.[32] The implication of the latter is to remove the barriers to the influx of such private capital in the electricity sector including OGRE. The point was made that the development of OGRE is constrained by financial and capacity barriers. Thus, gathering the needed private capital to develop the OGRE sector, entails addressing the mentioned barriers to the influx of such capital. It was against the latter background that the Philippines Renewable Energy Act 2008 was enacted.

9.4 Regulatory Support Framework

9.4.1 Introduction

The Philippines Renewable Energy Act 2008 was aimed at affirming the government's commitment to accelerate the utilisation of renewable energy (RE) resources in the country and as such is the supportive framework for OGRE.[33] It creates support for capacity building and financing for OGRE, which will be discussed *in seriatim*.[34] Additionally, it provides a clear context for OGRE development. Regardless, such clear context was not reinforced by a priority clause. Also, there are no sectoral targets for OGRE. There are other provisions geared towards addressing financing and capacity building in the sector. These include a clear delineation of institutional functions. This reduces the costs of OGRE projects by eliminating the requirement for a multiplicity of permits. A Renewable Energy Trust Fund (RETF) is provided for which should disburse initial capital grants to OGRE investors including for capacity building. Additional financial incentives are provided for to help private investors to overcome the problem of financial sustainability in remote rural areas. There are also some provisions for financial loan packages for private investors. These provisions will be analysed *in seriatim*.

9.4.2 Clear Context

The Philippine Renewable Energy Act 2008 recognises energy security and environmental protection as the social objectives for the development, utilisation, and commercialisation of RE. It provides that it will achieve 'energy self-reliance, through the adoption of sustainable energy development strategies to reduce the country's dependence on fossil fuels and thereby minimize the country's exposure to price fluctuations in the international markets'.[35] Additionally, it intends to 'encourage the development and utilization of RE resources to prevent or reduce harmful emissions and thereby balance the goals of economic growth and development with the protection of health and the environment'.[36] Consequently, it proposes to adopt fiscal and non-fiscal incentives to attract private sector capital to commercialise the sector.[37]

The Philippines Government commits to promoting the development of OGRE in areas not covered by their grid and in remote rural areas.[38] The Act defines off-grid areas as areas that are not connected to the national grid and remote rural areas.[39] The point has been made that the development of OGRE is constrained by the problem of financing,[40] sustainability of such projects,[41] and absence of technological capacity.[42] The Philippines had a local manufacturing base for RE technologies even before the Renewable Energy Act 2008.[43] While some local companies are holders of the patent of some renewable electricity technologies including OGRE,[44] the government aspires to innovate and commercialise more RE technologies.[45] The Philippines Renewable Energy Act 2008 can be commended for providing the context for the development of OGRE. On this contextual basis, the government can remove the barriers impeding private sector participation in the development of OGRE.[46]

However, the absence of a provision that prioritises renewable energy over fossil fuel whittles down the effectiveness of the law in driving the development of OGRE. The Philippines Act (2008) does not have any provision that prioritises renewable energy including OGRE. This has been identified by the IRENA as a lapse because some government bodies such as the Small Power Utilities Group of the National Power Corporation (SPUG)[47] are still undertaking off-grid diesel electricity projects to a greater degree than OGRE projects in such areas.[48] This will not have been the case if there is a provision that prioritises renewable energy including OGRE in the context of where it is needed. Given the noted lapses, off-grid fossil fuel electricity has dominated the sector. A study conducted in 2016 reports that the fossil fuel electricity (including off-grid) capacity installed by the government and private actors between 2008 and 2016 is double that of RE.[49] What is more, another study in 2019 confirms that while the OGRE sector has recorded some growth, fossil fuel electricity options continue to grow faster than the latter.[50]

9.4.3 National Targets and Implementation Measures

The Philippines Renewable Energy Act 2008 creates the legal basis for the adoption of goals and targets for the development of the OGRE sector. The Department of Energy (DOE) is 'the dominant regulating authority over the energy sector, the ministry is responsible for energy issues, plans, laws, and programmes'.[51] It has the power under the Act to formulate 'long-term policy … which identifies among others, the goals, and targets for the development and utilization of renewable energy in the country'.[52] This provision does not expressly create an obligation on the DOE to make sector-specific targets for OGRE. However, given the archipelagic nature of the Philippines that render the imperatives for OGRE more pronounced, it would be expected that the DOE would exercise its discretion to make a specific sectoral target for it.[53]

Regrettably, the DOE National Renewable Energy Plan 2011 does not do so. There is also no deadline for the setting of such targets by the DOE, and as such, the National Renewable Energy Plan was not produced until 2011. The Plan sets a general target of increasing the on-grid RE capacity in the Philippines from 5,438 MW/h capacity to 15,304 MW/h.[54] However, it does not contain any sectoral targets for individual OGRE technologies.[55] Commenting on this, the IRENA opines that 'the National Renewable Energy Plan issued by the DOE focuses only on targets for renewable energy on the main grid'.[56] One best practice for a support framework for the financing of OGRE projects is the adoption of a 'SMART'[57] target.[58] The IRENA has added that the characteristics of a SMART target include it being specific and measurable.[59] This will make it possible to 'know and measure to what extent it has been achieved; this entails that it must be possible to monitor and check compliance'.[60] Flowing from this, it is based on such a SMART target that the government would effectively adopt legal measures to support the financing of such targeted OGRE.[61] It is also on that basis that progress can be measured and monitored.[62] Given the absence of an express target for OGRE, there is no benchmark for the evaluation and monitoring of the success of the measures for capacity building and financing of OGRE in the 2011 Plan.[63]

The Philippines Government consists of autonomous local government units including autonomous regions, provinces, municipalities, and barangays.[64] Even though the central government has control of the development of the electricity sector, these local governments still reserve powers for the issuance of any licence or permit and may undertake OGRE projects.[65] Regrettably, the Philippines Renewable Energy Act 2008 does not provide a basis for the inclusion of the lower units of government in the design or implementation of measures for the promotion of OGRE.[66] Marquardt reports that this lack of inclusion of local government authorities has stalled progress in the development of the RE sector, including OGRE.[67] This is because OGRE projects are increasingly facing opposition from local governments, who feel side-lined by the central government in the formulation of relevant laws for the sector.[68] In addition, the Philippines OGRE sector does not benefit from the pronounced participation of local governments in financing and capacity building. This lack of inclusion in planning also makes it harder to coordinate efforts to achieve the overall development of the sector.

9.4.4 A Clear Delineation of Institutional Functions

The Electric Power Industry Reform Act 2001 created the Energy Regulatory Commission (ERC) as the quasi-regulatory body in the electricity industry.[69] As such, it is charged with the responsibility of 'encouraging market development' through the adoption of rules and regulations that will remove the barriers to private sector participation.[70] Consequently, it has the power to develop regulations and policies for the governance of the OGRE sector including matters such as licencing, financing, and capacity building.[71] These functions coincide with the DOE's regulatory function. The Department of Energy Act 1992 confers the DOE with powers to 'regulate private sector activities' to increase investment in the energy sector.[72] In addition, the Electric Power Industry Reform Act 2001 confers on them the power to make regulations to 'encourage private sector investments in the electricity sector and promote the development of indigenous and renewable energy sources'.[73] Consequently, it also has the power to regulate OGRE development including financing and capacity building. This conflict of interest affected the development of the sector by resulting in the duplication of actions in the OGRE sector.[74]

The Renewable Energy Act 2008 lays this issue to rest by providing that 'DOE shall be the lead agency mandated to implement the provisions of this Act'.[75] Consequently, the DOE[76] is now the lead regulator for the support of OGRE.[77] Flowing from this, it was mandated to formulate the Implementation Rules and Regulations (IRR) 2008 within six months after the commencement of the Act.[78] In addition to the existing ERC, the Act created the National Renewable Energy Board (NREB) as the information hub and policy house for the Philippines RE sector, including OGRE.[79] In the IRR 2008, the DOE clarifies the functions of all the institutions involved in the administration of the sector to avoid problems of overlap.[80]

9.4.5 Renewable Energy Trust Fund (RETF)

The Philippines Act 2008 establishes the Renewable Energy Trust Fund (RETF), which provides financial support for the development of every sphere of the OGRE sector. It is aimed at financing 'the research, development, demonstration, and promotion of the widespread and productive use, of RE Systems for power and non-power applications, as well as to provide funding for research and development institutions'.[81]

In light of the need to innovate and commercialise more OGRE technologies, financing the development of advanced capacity through advanced studies and R & D would be beneficial for the development of the sector. Given that the RETF is committed to the widespread and productive use of RE systems, it contributes to addressing the problem of financing which hinders the development of the OGRE sector. In addition, the fund aspires to 'propagate RE knowledge by accrediting … training, and providing benefits to institutions, entities and organizations, which can extend the promotion and dissemination of RE benefits to the national and local levels'.[82] As such, it addresses the problem of capacity to operate and maintain RE technologies in remote rural areas.

The Act provides for the sources of the RETF. These include, '(a) proceeds from the emission fees collected from all generating facilities consistent with Republic Act No. 8749 or the Philippines Clean *Air* Act'[83] and 'One and a half percent (1.5%) of the net annual dividends remitted to the National Treasury of the Philippines National Oil Company and its subsidiaries'.[84] The Philippines Clean Air Act 1999 imposes emission fees on electricity generating plants, including off-grid (but not thermal) plants.[85] The inclusion of the Clean Air Fund and a specified percentage of the proceeds from the Philippines National Oil Companies as a source of the RETF is commendable because it contributes to addressing the level playing field problem that contributes to the financial barriers in the OGRE sector.[86] The RETF is funded from other sources, such as 1.5% of the proceeds from the Philippines Charity Sweepstakes Office and the Philippines Amusement and Gaming Corporation.[87] Sources of funding also come from contributions, grants, and donations.[88] The last provision is not surprising, because even before the Act, several international bodies like USAID, through its Alliance for Mindanao Off-Grid Renewable Energy (AMORE) Project had been funding development and capacity building in the OGRE sector.[89] As such, it is expected that international bodies and donor agencies will also contribute to the RETF. All in all, the RETF benefits from a variety of sources; this has led DOE officers to opine that the availability of funding for the RETF will not be an issue.[90]

Notably, the Act provides that the DOE, who use a wide range of funding vehicles to achieve the objectives above, shall administer the RETF.[91] It provides that the use of the RETF 'may be through grants, loans, equity investments, loan guarantees, insurance, counterpart funds or such other financial arrangements necessary for the attainment of the objectives of this Act'.[92] Just like China, the Philippines OGRE sector is open to small, medium, and large industrial actors with different financial needs.[93] The flexibility accorded to the DOE to use various tools for funding enables them to meet the differing needs of various actors allowing them to invest in the sector. In the same vein, the DOE can use any financial tools to address the problem of financial sustainability, which is peculiar to OGRE projects in remote rural areas in the Philippines.[94]

Regrettably, the RETF is yet to be set up by the DOE partly because of the absence of a deadline for its establishment on the one hand and a lack of political will to do so on the other. The Act provides that the RETF 'shall be administered by the DOE as a special account'.[95] It adds that the NREB, which is the policy house of the sector, shall oversee and monitor the utilisation of the RETF, which is administered by the DOE.[96] In other words, the DOE will trigger the actual establishment of the RETF by formulating rules that govern its operations.[97] On that basis, the NREB would initiate the collection of funds from various sources.[98] However, the Act is silent on the deadline for the actual establishment of the fund by the DOE. Thus, there is no needed push, including political will, for its set-up. As a result, the RETF is yet to be set up as of 2017, nine years after the Renewable Energy Act was enacted. Commenting on this delay, the OECD remarked in 2017 that 'the Act stipulates financial support for renewable energy through the Renewable Energy Trust Fund … however, this is yet to materialise'.[99] According to NREB officials, the reason for the delay is attributed to the DOE's delay in producing the governing rules for the fund, which would trigger them to start collecting the funds for the RETF from various sources.[100] This delay ordinarily would not have happened if the Act provided for a deadline for the set-up of the fund.[101]

9.4.6 Financial Incentives and Financial Assistance Programme for the Banks

Aside from the RETF, the Act provides for subsidies of up to 50% for OGRE generated in areas where consumers cannot afford a cost-reflective tariff.[102] As previously stated, there are remote rural areas where some of the dwellers are so poor that they cannot afford to pay for OGRE.[103] These are called 'missionary areas' under the Act.[104] Consequently, OGRE investors are deterred from developing projects in such areas because they cannot sustain them financially.[105] This provision addresses the barrier by paying 50% operational subsidies for electricity generated and supplied to these rural dwellers.

The DOE, in the IRR, provides that the ERC shall set up the mechanism for the disbursal of the fund and formulate rules for the set-up.[106] In 2011, the ERC produced the rules for the set-up.[107] It provides that an OGRE developer:

> Shall be entitled to the cash incentive under the Renewable Energy Act from the time it commences its commercial operation in the missionary (non-viable remote) area as an NPP under a PSA approved by the ERC: Provided that the said RE Developer is established after the effectivity of the Renewable Energy Act.[108]

The said cash incentive is defined under the resolution as 'fifty percent (50%) of the universal charge for the power needed to service missionary areas where the renewable developer operates the same'.[109] Undoubtedly, since 2011, this subsidy has been helping to remove the problem of financial sustainability for OGRE projects in such remote rural areas.[110]

However, no phase-out or gradual reduction programme is regulating how the subsidy would give way to a competitive market. The Philippines Electric Power

Industry Reform Act 2001 provides that one of the ingredients of the privatisation of the electricity sector is the realisation of a competitive electricity market.[111] Regrettably, the ERC Resolution that provides for the 50% subsidy does not have any phase-out plan or graduation scheme.[112] This undermines the actualisation of the competitive market as stipulated in the Electric Power Industry Reform Act 2001.[113]

Furthermore, the Renewable Energy Act 2008 incorporates a financial assistance programme that addresses the problem of access to bank facilities for renewable energy projects. The Act provides that

> Government financial institutions such as the Development Bank of the Philippines (DBP), Land Bank of the Philippines (DBP), Phil-Exim Bank and other government financial institutions shall … provide preferential financial packages for the development, utilization, and commercialization of renewable energy projects.[114]

As such, the IRR requires them to formulate financial packages and to submit them for DOE approval within six months after it takes effect.[115] This provision has resulted in financial packages for renewable energy projects in various forms. Before the Renewable Energy Act 2008, access to financial facilities from banks was a problem for some RE projects because of their higher risk profiles.[116] Because of the Renewable Energy Act's provision, some banks have established financial packages for renewable energy projects, including mini-grid projects undertaken by end-user associations such as electricity cooperative groups.[117] For instance, the Development Bank of the Philippines has launched the LGU Guarantee Scheme.[118] This provides cheaper credit facilities to end-user electricity cooperative societies to enable them to undertake mini-grid renewable electricity projects.[119] Similarly, the Banks of the Philippine Islands, in their sustainable energy finance programmes, provide loans and other financial facilities for renewable electricity projects of varying scales.[120] Additionally, the sustainable energy finance programmes provide funding for manufacturers of RE technologies, including OGRE projects.[121]

On the one hand, there is no express provision mandating the financial institutions to create such packages specifically for standalone renewable electricity projects. For most archipelagic areas, such standalone OGRE is exigent for the electrification of individual households especially those that do not form part of any cooperatives.[122] In the absence of any mandatory obligation to design specific programmes for such standalone projects, banks are rarely providing financial packages to address the problem of funding in the sector.[123] As such, investors and consumers are struggling to afford such initial capital costs needed for the standalone projects.[124]

9.5 Conclusion

Undoubtedly, the Philippines Renewable Energy Act 2008 has spurred the development of renewable electricity including OGRE beyond what it used to be before

2008. However, the electricity sector is dominated by fossil fuel energy sources which account for 69% of the available capacity: while renewable electricity including OGRE accounts for 31%.[125] As such, if energy security will be attained and the climate change objectives of the country achieved, renewable electricity including OGRE should be developed beyond its current capacity. This work has shown that while the Renewable Energy Act 2008 has some features that is suited to support the development of OGRE, it still has some way to go in the context of attaining the optimal development of the sector. While the Act provides for a clear context for renewable energy including OGRE, it does not prioritise its development over that of fossil fuel electricity options. As such, investments in fossil fuel electricity options continue to increase at the expense of the development of OGRE.

It was argued that the absence of sectoral targets for OGRE is a clog in the wheel of progress of the Renewable Energy Act because there is no basis for the development of the sector or a benchmark for measuring progress made. It was argued that the Act did not attempt to garner local support for the development of the sector by the inclusion of the lower tiers of government in the implementation of its provisions. As such, there is a dichotomy between the activities of the central government and autonomous units which impacts negatively on the development of the sector. Another milestone recorded by the Act is that it lays to rest the overlap in institutional functions in the sector by making the DOE the lead agency for the implementation of its provisions. It provides for the establishment of a RETF designed to disburse various grants to address the identified financial barriers in the sector. Regrettably, it is yet to be set up. The 2008 Act creates an obligation on banks to create various loan packages for different renewable energy projects. It was noted that this provision has crystallised into different financial packages for OGRE projects, specifically mini-grid ones undertaken by cooperative societies. Nonetheless, this provision has not benefited standalone projects because there is no obligation on banks to specifically design loan packages for them.

It is currently forecast that the development of fossil fuel electricity options will continue to increase, thereby occasioning a diminishing effect on the growth of renewable energy including OGRE.[126] To avert this, the Philippine Renewable Energy Act 2008 should be fine-tuned to best optimise the development of the sector. The latter can be achieved by amending the Act in order to trim out the defects noted in this paper. Government had already recognised the superior benefits of renewable energy over fossil fuel electricity options. The policy statements should be backed by the political commitment to expedite the implementation of the Renewable Energy Act 2008, especially the setting up of the RETF.

Notes

1 See International Energy Agency, *Renewables Information 2010: With 2009 Data* (IEA 2010) 41. A Yadoo and others, 'Low-Carbon Off-Grid Electrification for Rural Areas in the United Kingdom: Lessons from the Developing World' (2011) 39(10) EP 6400e7. See also Zbigniew Chmiel and Subhes C Bhattacharyya, 'Analysis of Off-

Grid Electricity System at Isle of Eigg (Scotland): Lessons for Developing Countries' (2015) 81 RE 578.

2 Henry S Bensurto, *Archipelagic Philippines: A Question of Policy and Law* (Maritime Border Diplomacy 2012) 323.

3 Jens Marquardt, *How Power Shapes Energy Transitions in Southeast Asia: A Complex Governance Challenge* (Taylor and Francis 2016) 108.

4 Sahara Piang Brahim, 'Renewable Energy and Energy Security in the Philippines' (2014) 52 Energy Procedia 480.

5 Ibid.

6 Felix Chang, 'Running Out of Gas: Phillipine Energy Security and the South China Sea' (2019) <https://www.fpri.org/article/2019/09/running-out-of-gas-philippine-energy-security-and-the-south-china-sea/> accessed 15th August 2021. Most of its oil and gas supplies were imported from the Middle East countries like Qatar, Oman *etc*.

7 Ibid. See also, Sara Jane Ahmed and Jose D Logarta, 'Electricity-Sector Opportunity in the Philippine the Case for Wind- and Solar-Powered Small Island Grids' (2017) <http://ieefa.org/wpcontent/uploads/2017/05/Electricity-Sector-Opportunity-in-the-Philippines_May-2017.pdf> accessed 2nd August 2021.

8 D Eckstein and others, *Global Climate Risk Index 2018, Who Suffers Most from Extreme Weather Events? Weather-Related Loss Events in 2016 and 1997 to 2016* (Germanwatch 2017).

9 Han Chen-Alum, 'Philippines Joins the Paris Agreement on Climate Change' (2017) <https://www.nrdc.org/experts/han-chen/philippines-joins-paris-agreement-climate-change> accessed 15th August 2021.

10 Sahara Piang Brahim, 'Renewable Energy and Energy Security in the Philippines' (n 4).

11 Mario C Marasigan, 'Renewable Energy Development in the Philippines' (2012) (copy in file). Also cited by Sahara P Brahim, 'Renewable Energy and Energy Security in the Philippines' (2014) 52 Energy Procedia 480.

12 Ibid.

13 Ibid. See also World Bank, *Philippines: Meeting Infrastructure Challenges* (World Bank 2005) 17.

14 Ibid.

15 Fernando Roxas and Andrea Santiago, 'Broken Dreams: Unmet Expectations of Investors in the Philippines Electricity Restructuring and Privatization' (2010) 38(11) Energy Policy 7269.

16 World Bank, *Philippines - Power Sector Study: Structural Framework for the Power Sector* (World Bank 1994).

17 'The Electric Sector Reform Act 2001 (PH). See Power Engineering Industry, 'Privatization Programmes: The Philippines Reforms' (2001) <https://www.powerengineeringint.com/world-regions/asia/privitization-progra mmes-the-philippines-reforms/> accessed 16th August 2021.

18 Jose Barroco and Maria Herrera, 'Clearing Barriers to Project Finance for Renewable Energy in Developing Countries: A Philippines Case Study' (2019) 135 Energy Policy 111008, 111012.

19 Ibid, 111013.

20 Hooman Peimani, 'Financial Barriers to Development of Renewable and Green Energy Projects in Asia' (2018) Working Paper No 862, Asian Development Bank Institute <https://www.adb.org/publications/ financial-barriers-development-renewable-green-energy-projects-asia > accessed 17th August 2021.

21 Ibid.

22 Ibid.

23 Rabindra Nepal and others, 'Green Technological Development and Deployment in the Association of Southeast Asian Economies (ASEAN)—At Crossroads or Roundabout?' (2021) 13 Sustainability 758, 761.

24 For instance, Nigeria has separate policies in the electricity sector including the National Energy Policy (NEP) 2003 and, the National Renewable Energy and Energy Efficiency Policy (NREEEP) 2015. See Ngozi Chinwa Ole, 'The Nigerian Electricity Regulatory Framework: Hotspots and Challenges for Off-Grid Renewable Electricity Development' (2020) 38(4) Journal of Energy and Natural Resources Law 3.
25 The Electric Power Reform Act, Republic Act No 9136, s. 2(a).
26 Ibid, s. 2(g).
27 Xu Jianzhong and others, 'Renewable Energy and Sustainable Development in a Resource-Abundant Country: Challenges of Wind Power Generation in Kazakhstan' (2018) 10 Sustainability 3315.
28 K R Ajao and others, 'Global Warming and Environmental Change Problems: Renewable Energy and Cleaner Fossil Fuels Technology as a Solution' (2010) 10th International Conference on Clean Energy (ICCE-2010) 1.
29 Ibid.
30 The Electric Power Reform Republic Act No 9136, 2001, s.2 (h).
31 Felix Chang, 'Running Out of Gas: Philippine Energy Security and the South China Sea' (n 6).
32 The Electric Power Reform Republic Act (n 31) s 2(d).
33 M A D Rosellon, 'The Renewable Energy Policy Debate in the Philippines' (2017) Discussion Paper Series 2017-17 <https://pidswebs.pids.gov.ph/CDN/PUBLICATIONS/pidsdps1717.pdf> accessed 20th August 2021.
34 Ibid. For commentaries on the effect of the Renewable Energy Act on the Philippines, See Connie Tan Hui Ann, 'The Philippines' Renewable Energy Sector is Booming (and it Could Get Bigger)' (9 August 2016) <https://www.cnbc.com/2016/08/09/the-philippines-renewable-energy-sector-is-booming-and-it-could-get-bigger.html> accessed 21st August 2021.
35 The Philippines Renewable Energy Act 2008 (PH) s 2(a).
36 Ibid, s 2(b).
37 Ibid, s 2.
38 Ibid.
39 Ibid, s 4 (jj). For a list of the off-grid areas in the Philippines, See Philippines Energy Plan 2007–2014, 115 <https://www.doe.gov.ph/sites/default/files/pdf/pep/2007-pep_update.pdf> accessed 3rd August 2021. Note Philippines has since released the Philippines Energy Plan 2018–2040 embodying a clear set of objectives. <https://www.doe.gov.ph/sites/default/files/pdf/pep/pep-2018-2040_20210323.pdf> accessed 21st August 2021.
40 Allan Joseph Mesina, 'Rethinking Off-Grid Rural Electrification in the Philippines' (2016) 11(9) Energy Sources, Part B: Economics, Planning and Policy 815.
41 Alliance for Mindanao, *Off-Grid Renewable Energy Project Final Performance Report* (Winrock International 2005) 9.
42 United Nations Development Programme (UNDP) (2011), *Final Evaluation of the GEF/UNDP/DOE Philippines: Capacity Building to Remove Barriers to Renewable Energy Development (CBRED) Project* (Manila Centre for Environmental Studies and Management 2006) 1.
43 The development of the local manufacturing capacity for different renewable electricity technologies including off-grid was achieved because of the 'Pole Vaulting' program for the energy sector, which was initiated to boost technology transfer of newy renewable electricity technologies into the Philippines. See Timothy Forsyth, *International Investment and Climate Change: Energy Technologies for Developing Countries* (Routledge 2014) 79; Copenhagen Economics and the IPR Company, 'Are IPR a Barrier to the Transfer of Climate Change Technology?' (19 January 2009) <https://www.copenhageneconomics.com/dyn/resources/Publication/publication-

PDF/7/27/0/Are_IPR_a_barrier_to_the_transfer_of_climate_change_technology.pdf> accessed 15th August 2021.

44 Ibid.

45 DOE, 'National Renewable Energy Programme' (2008) <www.doe.gov.ph/national-renewable-energy-program> accessed 10th August 2021.

46 Ahmad Murtaza Ershad, 'Institutional and Policy Assessment of Renewable Energy Sector in Afghanistan' (2017) JRE 1.

47 The SPUG is a unit of the National Power Corporation that is responsible for overseeing the provision of electricity to OGRE areas.

48 IRENA, *Accelerating Renewable Mini-Grid Deployment: A Study on the Philippines* (IRENA 2017) 70.

49 Energiewende Team, 'A Struggle between Coal and Renewable Energy in the Philippines' (2016) <https://energytransition.org/2016/07/a-struggle-between-coal-and-renewable-energy-in-the-philippines/#:~:text=The%20struggle%20between%20coal%2Dfired,in%20the%20Philippines%20is%20heated.&text=New%20coal%20and%20natural%20gas,share%20in%20the%20power%20mix.> accessed 11th August 2021.

50 Jose Barroco and Maria Herrera, 'Clearing Barriers to Project Finance for Renewable Energy in Developing Countries: A Philippines Case Study' (n 18) 111012.

51 Marquardt Jens, 'How Power Affects Policy Implementation: Lessons from the Philippines' (2017) 36(1) JCSAA 3, 12.

52 The Philippines Renewable Energy Act 2008 (NG) s 4(rr).

53 IRENA, *Accelerating Renewable Mini-Grid Deployment: A Study on the Philippines* (n 48) 11.

54 DOE, 'National Renewable Energy Program 2011'.

55 Ibid.

56 IRENA, *Accelerating Renewable Mini-Grid Deployment: A Study on the Philippines* (n 48) 70.

57 SMART target mean that the target is 'Specific, Measurable, Achievable, Realistic and Time-bound'. See Karin Edvardsson and S Hansson, 'When is a Goal Rational?' (2005) 24 SCW 343.

58 Sadie Cox and others, 'Solar Power Policy Overview and Good Practices' (2015) NREL/TP-6A20-64178, 4.

UNEP, *Financing Renewable Energy in Developing Countries* (UNEP 2012) 48.

59 IRENA, *Renewable Energy Target Setting* (IRENA 2015) 29.

60 Ibid.

61 Sadie Cox (n 59).

62 Ibid.

63 N Holvoet and R Renard, 'Monitoring and Evaluation under the Poverty Reduction Strategy Papers (PRSP): Solid Rock or Quicksand?' (2007) 30 EPP 66–81.

64 The Local Government Code of the Philippines 1991 (PH).

65 S Yilmaz and V Venugopal, 'Local Government Discretion and Accountability in the Philippines' (2013) 25(2) JID 227–250.

66 The Philippines Renewable Energy Act 2008.

67 Marquardt Jens, 'How Power Affects Policy Implementation: Lessons from the Philippines' (n 51).

68 Ibid.

69 The Electric Power Sector Reform Act 2001 (PH), s 38.

70 Ibid, s 48.

71 Energy Regulatory Commission, 'Objectives' <www.erc.gov.ph/ContentPage/12> accessed 22nd August 2021.

72 The Department of Energy Act 1992, s 5(j) and (k).

73 The Electric Sector Reform Act 2001 (PH), s 37(p) and (q).

74 Ibid.
75 The Philippines Renewable Energy Act 2008 (PH), s 5 and 33.
76 Ibid.
77 Department of Energy, 'Off-Grid' <www.doe.gov.ph/search/node/off-grid> accessed 21st August 2021.
78 The Philippines Renewable Energy Act 2008 (PH) s 33.
79 Ibid, s 27.
80 The Rules and Regulations Implementing Republic Act 2008 (PH), ss 22–31.
81 The Philippines Renewable Energy Act 2008 (PH), s 28(a).
82 Ibid, s 28(d).
83 Ibid, s 28.
84 Ibid.
85 The Clean Air Act 1999 (PH) s 19.
86 C A Hart and D Marcellino argued that the elimination of subsidies for fossil fuel subsidies without more will not be enough to address the problem of the absence of a level playground that disfavours renewable energy. As such actions such as taxation of emission from off-grid renewable electricity technologies level the playground indirectly. See Craig A Hart and Dominic Marcellino, 'Subsidies or Free Markets to Promote Renewables?' (2012) 3(3) Renewable Energy Law and Policy Review 196, 196.
87 The Philippines Renewable Energy Act 2008 (PH), s 28.
88 Ibid.
89 WINROCK INTERNATIONAL & USAID, 'Alliance for Mindanao Off-Grid Renewable Energy (AMORE) Project' (July 15 2005) <https://pdf.usaid.gov/pdf_docs/PDACF122.pdf> accessed 22nd August 2021.
90 Amy Remo, 'DOE Readies Renewable Energy Trust Fund' (2012) <https://business.inquirer.net/64443/doe-readies-renewable-energy-trust-fund> accessed 18th July 2021.
91 The Philippines Renewable Energy Act 2008 (PH), s 28.
92 Ibid, s 28.
93 SGV, 'Doing Business in the Philippines' (21 March 2018) <https://www.sgv.ph/publications/search?s=Doing+Business+in+the+Philippines> accessed 10th August 2021.
94 World Bank, 'Philippines: Poor Families in Remote Areas to Benefit from Renewable Energy Grants' (2016) <https://www.worldbank.org/en/news/press-release/2016/08/23/philippines-poor-families-in-remote-areas-to-benefit-from-renewable-energy-grants> accessed 11th August 2021.
95 The Philippines Renewable Energy Act 2008 (PH), s 28.
96 Ibid.
97 Ibid, s 34.
98 Ibid, s 27.
99 OECD, *OECD Green Growth Studies Green Growth in Cebu* (Philippines 2017) 85.
100 Lenie Lectura, 'NREB Urges DOE to Finally Approve RE Trust Fund Rules' *Business Mirror*, 25 July 2017, <https://businessmirror.com.ph/2017/07/25/nreb-urges-doe-to-finally-approve-re-trust-fund-rules/> accessed 28th August 2021.
101 The Philippines case laws provide a legal basis for an order of mandatory injunction to compel an administrative body such as DOE to perform a function stipulated by law. See *Bel-Air Village Association Inc v Magsino* [1981] 11803 (SC PH) Experiences in the Philippines power sector have shown that individuals and bodies often use the mandatory injunction to coerce government bodies in the sector to perform their legal functions as stipulated by the law. See *National Power Corporation v Hon Abraham Vera* [1989] G.R. No. 83558, 170 SCRA 721. However, where there is no express provision in law like the deadline in this context, the courts cannot intervene to award

such mandatory injunction against DOE. See *Francisco Garcia v John C Sweeney* [1904] G.R. No. 1693 (SC PH).
102 The Philippines Renewable Energy Act 2008 (PH), s 15(h).
103 See USAID, 'Alliance for Mindanao Off-Grid Renewable Energy (AMORE) Project' (n 89).
104 The Philippines Renewable Energy Act 2008 (PH), s 2(cc).
105 Allan Joseph Mesina (n 40) 820.
106 The Rules and Regulations Implementing Republic Act 2008 (PH), s 17(d).
107 ERC Resolution No 21 Series 2011, Art 4.
108 Ibid.
109 The IRR 2008 (PH), s 2(b).
110 Janet Sawin, 'National Policy Instruments: Policy Lessons for the Advancement and Diffusion of RET around the World' in Dirk Assmann (ed), *Renewable Energy: A Global Review of Technologies, Policies and Markets* (Routledge 2012) 105.
111 The Electric Power Industry Reform Act 2001 (PH), s 3.
112 IRENA, *Accelerating Renewable Mini-Grid Deployment: A Study on the Philippines* (n 48) 23.
113 The Electric Sector Power Reform Act 2001 (PH), s 3.
114 The Philippines Renewable Energy Act 2008 (PH), s 29.
115 Ibid, s 17(b).
116 Hong Mark Tat Soon and others, *Asia's Energy Trends and Developments* (World Scientific 2013) 114.
117 Ibid.
118 World Bank, *LGU Guarantee Corporation for a Philippines Renewable Energy Development Project* (World Bank 2016) 5, 35.
119 Ibid.
120 IFC, 'Sustainable Energy Finance Program' <https://unfccc.int/climate-action/momentum-for-change/financing-for-climate-friendly/sustainable-energy-finance-program-in-the-philippines> accessed 27 August 2021.
121 Ibid.
122 Charmaine Samala Guno and others, 'Optimal Investment Strategy for Solar PV Integration in Residential Buildings: A Case Study in The Philippines' (2021) 10(1) Guno 79–83.
123 Ibid.
124 Ibid. See Ana PaulaFarias-Rocha and others, 'Solar Photovoltaic Policy Review and Economic Analysis for On-Grid Residential Installations in the Philippines' (2019) 223 Journal of Cleaner Energy 45–46.
125 Ronald Dime and Edward Eviota, 'The Renewable Energy Law Review: Philippines' (2021) <https://thelawreviews.co.uk/title/the-renewable-energy-law-review/philippines> accessed 21st August 2021.
126 Ibid.

Bibliography

Articles and Journals

Ana Paula F-R, Hassan K, Milimata J and Sanchez-Cubedo, 'Solar Photovoltaic Policy Review and Economic Analysis for On-grid Residential Installations in the Philippines' (2019) 223 Journal of Cleaner Energy 45–46.

Barroco J and Herrera M, 'Clearing Barriers to Project Finance for Renewable Energy in Developing Countries: A Philippines Case Study' (2019) 135 Energy Policy 111008, 111012.

Brahim S P, 'Renewable Energy and Energy Security in the Philippines' (2014) 52 Energy Procedia 480.
Chmiel Z and Bhattacharyya S C, 'Analysis of Off-grid Electricity System at Isle of Eigg (Scotland): Lessons for Developing Countries' (2015) 81 RE 578.
Cox S, Walters T and Esterly S, 'Solar Power Policy Overview and Good Practices' (2015) NREL/TP-6A20- 64178, 4.
Edvardsson K and Hansson S, 'When is a Goal Rational?' (2005) 24 SCW 343.
Ershad A M, 'Institutional and Policy Assessment of Renewable Energy Sector in Afghanistan' (2017) JRE 1.
Guno S C, Agaton C B, Villanueva R O and Villanueva R O, 'Optimal Investment Strategy for Solar PV Integration in Residential Buildings: A Case Study in The Philippines' (2021) 10(1) Guno 79–83.
Hart C A and Marcellino D, 'Subsidies or Free Markets to Promote Renewables?' (2012) 3(3) Renewable Energy Law and Policy Review 196.
Holvoet N and Renard R, 'Monitoring and Evaluation Under the Poverty Reduction Strategy Papers (PRSP): Solid Rock or Quicksand?' (2007) 30 EPP 66–81.
IRENA, *Accelerating Renewable Mini-Grid Deployment: A Study on the Philippines* (IRENA 2017) 70.
Jens M, 'How Power Affects Policy Implementation: Lessons from the Philippines' (2017) 36(1) JCSAA 3.
Jianzhong X Assenova A and Erokhin V, 'Renewable Energy and Sustainable Development in a Resource-Abundant Country: Challenges of Wind Power Generation in Kazakhstan' (2018) 10 Sustainability 3315.
Lectura L, 'NREB Urges DOE To Finally Approve RE Trust Fund Rules' *Business Mirror* (25 July 2017) <https://businessmirror.com.ph/2017/07/25/nreb-urges-doe-to-finally-approve-re-trust-fund-rules/> accessed 28 August 2021.
Mesina J A, 'Rethinking Off-grid Rural Electrification in the Philippines' (2016) 11(9) Energy Sources, Part B: Economics, Planning and Policy 815.
Nepal R, Phoumin H and Khatri A, 'Green Technological Development and Deployment in the Association of Southeast Asian Economies (ASEAN)—At Crossroads or Roundabout?' (2021) 13 Sustainability 758, 761.
Ole N C, 'The Nigerian Electricity Regulatory Framework: Hotspots and Challenges for Off-grid Renewable Electricity Development' (2020) 38(4) Journal of Energy and Natural Resources Law 3.
Roxas F and Santiago A, 'Broken Dreams: Unmet Expectations of Investors in the Philippines Electricity Restructuring and Privatization' (2010) 38(11) Energy Policy 7269.
Yadoo A, Gormally A and Cruickshank H, 'Low-Carbon Off-grid Electrification for Rural Areas in the United Kingdom: Lessons from the Developing World' (2011) 39(10) EP 6400e7.
Yilmaz S and Venugopal V, 'Local Government Discretion and Accountability in the Philippines' (2013) 25(2) JID 227–250.

Book Chapters/Books

Bensurto, H S, 'Archipelagic Philippines: A Question of Policy and Law' in *Maritime Border Diplomacy* (Brill, Leiden, The Netherlands 2012).
Forsyth T, *International Investment and Climate Change: Energy Technologies for Developing Countries* (Routledge 2014).

Hong, Mark Tat Soon and Amy V R Lugg, *Asia's Energy Trends and Developments* (World Scientific 2013).

Marquardt J, *How Power Shapes Energy Transitions in Southeast Asia: A Complex Governance Challenge* (Taylor and Francis 2016).

Sawin J, 'National Policy Instruments: Policy Lessons for the Advancement and Diffusion of RET around the World' in Dirk Assmann (ed.), *Renewable Energy: A Global Review of Technologies, Policies and Markets* (Routledge 2012).

UNEP, *Financing Renewable Energy in Developing Countries* (UNEP 2012).

Cases

Bel-Air Village Association Inc v Magsino [1981] 11803 (SC PH).

Francisco Garcia v John C Sweeney [1904] G.R. No. 1693 (SC PH).

National Power Corporation v Hon Abraham Vera [1989] G.R. No. 83558, 170 SCRA 721.

Papers, Projects and Seminars

Ajao K R, Ajimotokan H A and Raji W O, 'Global Warming and Environmental Change Problems: Renewable Energy and Cleaner Fossil Fuels Technology as a Solution' (2010) 10th International Conference on Clean Energy (ICCE-2010) 1.

Alliance for Mindanao, *Off-Grid Renewable Energy Project Final Performance Report* (Winrock International 2005) 9.

Eckstein D, Künzel V and Schäfer L, *Global Climate Risk Index 2018, Who Suffers Most from Extreme Weather Events? Weather-Related Loss Events in 2016 and 1997 to 2016* (German Watch 2017).

Peimani H, 'Financial Barriers to Development of Renewable and Green Energy Projects in Asia' (2018) Working Paper No 862, Asian Development Bank Institute <https://www.adb.org/publications/financial-barriers-development-renewable-green-energy-projects-asia > accessed 17 August 2021.

Rosellon M A D, 'The Renewable Energy Policy Debate in the Philippines' (2017) Discussion Paper Series 2017–17 <https://pidswebs.pids.gov.ph/CDN/PUBLICATIONS/pidsdps1717.pdf> accessed 20 August 2021.

United Nations Development Programme (UNDP) (2011), *Final Evaluation of the GEF/UNDP/DOE Philippines: Capacity Building to Remove Barriers to Renewable Energy Development (CBRED) Project* (Manila Centre for Environmental Studies and Management 2006) 1.

World Bank, *Philippines - Power Sector Study: Structural Framework for the Power Sector* (World Bank 1994).

World Bank, *LGU Guarantee Corporation for a Philippines Renewable Energy Development Project* (World Bank 2016) .

Online Materials/Sources

Ahmed S J and Logarta J D, 'Electricity-Sector Opportunity in the Philippine the Case for Wind- and Solar-Powered Small Island Grids' (2017) <http://ieefa.org/wpcontent/uploads/2017/05/Electri city-Sector-Opportunity-in-the-Philippines_May-2017.pdf> accessed 2 August 2021.

Amy Remo, 'DOE Readies Renewable Energy Trust Fund' (2012) <https://business.inquirer.net/64443/doe-readies-renewable-energy-trust-fund> accessed 18 July 2021.

Chang F, 'Running Out of Gas: Phillipine Energy Security and the South China Sea' (2019) <https://www.fpri.org/article/2019/09/running-out-of-gas-philippine-energy-security-and-the-south-china-sea/> accessed 15 August 2021.

Chen-Alum H, 'Philippines Joins the Paris Agreement on Climate Change' (2017) <https://www.nrdc.org/experts/han-chen/philippines-joins-paris-agreement-climate-change > accessed 15 August 2021.

Connie Tan Hui Ann, 'The Philippines' Renewable Energy Sector is Booming (And It Could Get Bigger)' (9 August 2016) <https://www.cnbc.com/2016/08/09/the-philippines-renewable-energy-sector-is-booming-and-it-could-get-bigger.html> accessed 21 August 2021.

Copenhagen Economics and the IPR Company, 'Are IPR a Barrier to the Transfer of Climate Change Technology?' (19 January 2009) <https://www.copenhageneconomics.com/dyn/resources/Publication/public ationPDF/7/27/0/Are_IPR_a_barrier_to_the_transfer_of_climate_change_technology.pdf> accessed 15 August 2021.

Department of Energy, 'Off-grid' <www.doe.gov.ph/search/node/off-grid> accessed 21st August 2021.

Dime R and Eviota E, 'The Renewable Energy Law Review: Philippines' (2021) <https://thelawreviews.co.uk/title/the-renewable-energy-law-review/philippines> accessed 21st August 2021.

DOE, 'National Renewable Energy Programme' (2008) <www.doe.gov.ph/national-renewable-energy-program> accessed 10th August 2021.

Energiewende Team, 'A Struggle Between Coal and Renewable Energy in the Philippines' (2016) < https://energytransition.org/2016/07/a-struggle-between-coal-and-renewable-energy-in-the-philippines/#:~ :text= The%20struggle%20between%20coal%2Dfired,in%20the%20Philippines%20is%20heated.&text=New%20coal%20and%20natural%20gas,share%20in%20the%20power%20mix.> accessed 11 August 2021.

Energy Regulatory Commission, 'Objectives' <www.erc.gov.ph/ContentPage/12> accessed 22 August 2021.

IFC, 'Sustainable Energy Finance Program' <https://unfccc.int/climate-action/momentum-for-change/financing-for-climate-friendly/sustainable-energy-finance-program-in-the-philippines> accessed 27 August 2021.

Philippines Energy Plan 2007–2014, 115 <https://www.doe.gov.ph/sites/default/files/pdf/pep/2007-pep_update.pdf> accessed 3 August 2021.

Power Engineering Industry, 'Privatization Programmes: The Philippines Reforms' (2001) <https://www.powerengineeringint.com/world-regions/asia/privitization-progra mmes-the-philippines-reforms/> accessed 16 August 2021.

SGV, 'Doing Business in the Philippines' (21 March 2018) <https://www.sgv.ph/publications/search? s=Doing+Business+in+the+Philippines> accessed 10 August 2021.

The Philippines Energy Plan 2018–2040 embodying a clear set of objectives. <https://www.doe.gov.ph/sites/default/files/pdf/pep/pep-2018-2040_20210323.pdf> accessed 21 August 2021.

WINROCK INTERNATIONAL & USAID, 'Alliance for Mindanao Off-Grid Renewable Energy (AMORE) Project' (July 15 2005) <https://pdf.usaid.gov/pdf_docs/PDACF122.pdf> accessed 22 August 2021.

World Bank, 'Philippines: Poor Families in Remote Areas to Benefit from Renewable Energy Grants' (2016) <https://www.worldbank.org/en/news/press-release/2016/08/23

/philippines-poor-families-in-remote-areas-to-benefit-from-renewable-energy-grants>
accessed 11 August 2021.

Statute

ERC Resolution No 21 Series 2011.
Republic Act No 9136.
The Clean Air Act 1999 (PH).
The Electric Power Sector Reform Act 2001 (PH).
The Electric Sector Reform Act 2001 (PH).
The Department of Energy Act 1992.
The Local Government Code of the Philippines 1991 (PH).
The Rules and Regulations Implementing Republic Act 2008 (PH).
The Philippines Renewable Energy Act 2008 (PH).

10 Indonesia

Anton Latief and Randy Hendrika

10.1 Introduction

Indonesia is an archipelago with 16,056 islands, spanning around 1,919,906 square kilometres.[1] At the end of 2020, the country had a population of more than 269 million people, who predominantly reside in Indonesia's western islands (more than 85% of the population reside in Java, Sumatra, and Bali).[2]

The Ministry of Energy and Mineral Resources (Ministry of EMR) reports that in January 2021, the electrification ratio in Indonesia is at 99.45%.[3] However, there is a stark disparity between the electrification ratio in Indonesia's main islands in western part and that in the rural eastern islands. The western parts of Indonesia, where the areas are demographically dense and consist of major islands with high gross domestic product (such as Java, Sumatra, and Bali), are oversupplied with electricity. On the other hand, the eastern parts of Indonesia, which are characterised by low demographic densities spread through vast mountainous or islandic areas, are often deprived of a basic supply of electricity. This lack of electricity in the rural eastern parts of Indonesia is mainly because it is still prohibitively expensive to connect these areas to the on-grid electricity system or to build and maintain an off-grid electricity system.

Indonesia's archipelagic nature and geographic location on the equator means that it has an abundant potential for renewable electricity sources, such as hydropower, geothermal, solar power, and even the ocean. Based on an official report from the Government of Indonesia, Indonesia has a potential renewable electricity of 441.7 GW.[4] Amidst a growing concern over climate change in Indonesia, renewable electricity sources have seen a surge of interest as a preferred option for electrification.

However, at the end of 2018, the total installed capacity of renewables was reported to be only 9.4 GW – 17% of which was off-grid.[5] By the end of 2019, total renewables installed capacity (both on-grid and off-grid) was reported to only be 10.17 GW.[6] Hydropower dominated with 5.4 GW, followed by geothermal at 2.13 GW, bioenergy at 1.9 GW, mini/micro hydro at 464.7 MW, wind at 148.5 MW, solar photovoltaic (PV) at 152.4 MW, and waste power plant at 15.7 MW. Renewables investment was noted to have reached USD1.7 billion by September 2019.[7] Due to the COVID-19 outbreak, and the resulting energy

DOI: 10.4324/9781003178088-13

development budget cuts and halted investment realisations, Indonesia only saw a minor increase in the total renewable electricity installed capacity at 10.50 GW (a 1.8% year-on-year increase) by the end of 2020.[8] At the end of 2021, the Ministry of EMR reported that the total installed capacity is at 11.15 GW and that the share of renewable electricity is 13.5% of Indonesia's total electricity mix.[9]

As shown by the energy mix percentage above, renewable energy development and investments in Indonesia generally remain low. Like many other developing countries, Indonesia still relies heavily on fossil fuels to support its economy, and its electrification projects are still mainly driven by coal power plants. These are caused mainly by constraints such as affordability and accessibility of initial capital costs, financial sustainability of projects, and lack of technological capacity. Improvements in key regulations are needed to help accelerate clean energy development in the country, in order to ensure energy security, economic competitiveness, and successful transition towards sustainable energy systems.

Further, with interest in renewable energy as a preferred electrification system still in its infancy, developments specific to off-grid renewable electricity are still very limited in Indonesia. Although there is ample opportunity to electrify rural areas by using off-grid renewable systems, Indonesia does not have many regulations that specifically opt for, or incentivise, the development of off-grid renewable electricity systems.[10] In addition, there is very little demand for electricity in Indonesia's geographically challenged, rural areas—which reduces profitability and bankability of off-grid renewable electricity projects.[11] It is also notable that PT Perusahaan Listrik Negara (PERSERO) (PLN) (which is a state-owned enterprise and the main provider of electricity and its infrastructure in Indonesia with work areas covering around 90% of Indonesian territory[12]) continues to resort to grid extension as its primary approach to electrify rural areas.[13] As a result, there are very limited developments on off-grid electricity systems. There is also little information available on aggregate, nationwide data for off-grid electricity systems due to the systems' stand-alone nature.

Given the lack of developments and information specific to off-grid renewable electricity in Indonesia, this chapter analyses the drivers and barriers to the development of the country's renewable electricity systems (as a whole – both on-grid and off-grid), national targets, and role of the current regulatory framework in supporting the development of renewable electricity in Indonesia.

10.2 Drivers and Barriers for OGRE

Despite the Government of Indonesia's efforts to increase the role of renewable energy, dating as early as 2004, Indonesia has remained mainly reliant on fossil fuel energy to support its economy and society functions. Only in recent years has the Government of Indonesia given more attention and effort to make palpable growth in the utilisation of renewable energy including in the utilisation of power generation based on renewable energy.

Given its archipelagic nature, Indonesia is among the most vulnerable countries to climate change. In addition, as a developing country, it is generally considered more vulnerable to the damaging effects of a natural hazard but has lower coping capacity.[14] The above observations are reflected in the Global Climate Risk Index 2021 which noted Indonesia as the 14th most affected country by climate change.[15]

Through Law No. 16 of 2016 concerning the Ratification of the Paris Agreement to the United Nations Framework Convention on Climate Change, Indonesia has ratified the legally binding global climate change target under the Paris Agreement. Indonesia was among the top 10 largest greenhouse gas emitters in 2018 and, therefore, its contribution in cutting the amount of greenhouse gas is crucial. The Directorate General of Climate Change Control (*Direktorat Jenderal Pengendalian Perubahan Iklim* – DGCCC) reports that the effects of climate change are apparent and affecting Indonesian society and its economy. Among others, the DGCCC reports a number of crop failures in various locations and several cases of tide change and increase in acidity of sea waters which cause disturbances to the fisheries industry.[16] These phenomena are seen as a serious threat to food security in Indonesia by the DGCCC. Meanwhile, several global models from various institutions concluded that reaching the targets in the Paris Agreement means that Indonesia's share of renewable electricity must be increased so that the share is around 50%–85%. For context, in 2021, the share of renewable electricity was only 13.5%.[17] Against the above background, the Government of Indonesia has slowly but steadily issued or amended regulations in order to foster more development in the renewable energy and electricity sector.

On the other hand, it is notable that from the perspective of energy usage, Indonesia's total electricity consumption has increased by more than 50% between 2010 and 2021. In 2020, more than 60% of Indonesian power plants were coal powered.[18] However, coal production plummeted from 616 million tons in 2019 to only around 230 million tons in 2020. Investment in coal power plants is expected to continue plummeting as regulations around carbon emission cause the long-run marginal cost (LRMC) of the coal power plant to continue to skyrocket. While coal is becoming less ideal and more costly for power generation, the demand for electricity consumption is expected to only increase and Indonesia's current energy sufficiency is deemed to be in a less than ideal position.

Based on the Presidential Regulation No. 18 of 2020 concerning the Mid Term National Development Plan for 2020–2024 (*Rencana Pembangunan Jangka Menengah Nasional Tahun 2020–2024*), Indonesia is aiming to have an economic growth of 5.7% to 6% per annum. Along with the agenda for economic growth that is aimed at the non-consumer sector, energy consumption is also expected to significantly increase in order to support such economic growth. At the end of 2021, the total installed capacity was reported to be at 71 GW.[19] However, in order to reach 1,000 watt/capita, an additional 200 GW of installed capacity (or around three times the currently installed capacity) is needed.[20]

Given the foregoing, securing Indonesia's energy security by developing renewable energy and electricity is not only a normative obligation, but also represents a real imperative for Indonesia to maintain its economic growth.

Notwithstanding the foregoing, developing renewable electricity in Indonesia requires higher capital costs when compared with fossil fuel energy sources electricity. This translates to less installed capacity per initial dollar invested. In addition, as capital intensive projects generally require higher financing amounts, these generally result in higher lending rates for the same capacity developed with conventional fossil fuel. Further, sustainability of the renewable energy projects often presents another economic challenge. This is especially true for off-grid renewable electricity or other renewable electricity power plants which are built in areas where there are very low demographic densities and gross domestic product. These raise concerns over whether rural dwellers can afford the cost reflective tariff in order to maintain the facilities.

Another barrier to renewable energy development in Indonesia is subsidies provided for the non-renewable electricity sector. Pursuant to the Government Regulation No. 45 of 2013 on Procedure to Execute the State Revenue and Budget (*Anggaran Pendapatan dan Belanja Negara* – APBN) as amended by Government Regulation No. 50 of 2018, a portion of the APBN is allocated for subsidy, including subsidy in energy.[21] Thereunder, the Minister of Finance is given the authority to further establish the procedure to implement such subsidy.[22] Electricity subsidy has been provided in the APBN as a way to maintain price stability to maintain the buying power of the society, to help the poor, and to maintain the available energy supply.[23] The electricity subsidy is given to the customers through PLN.[24]

With Indonesia's mission to reach net zero emission, the subsidy that the state has to provide is expected to increase. Pursuant to the 2021 until 2030 Electricity Supply Business Plan of PT Perusahaan Listrik Negara (PERSERO) (*Rencana Umum Penyediaan Tenaga Listrik* –RUPTL),[25] the decrease of coal and the increase of co-firing biomass and gas/liquified natural gas in the energy mixture will cause an increase to the Basic Electricity Generation Cost (*Biaya Pokok Penyediaan Pembangkitan* – BPP). Given the foregoing, the BPP is estimated to increase up to Rp 183/kWh, or to require an additional subsidy and compensation up to IDR 72,800,000,000,000 in year 2030[26] to maintain society's buying power.

The Government of Indonesia realises the potential increase of BPP and subsidy/compensation caused by the utilisation of renewable energy and would provide the required subsidy/compensation.[27] The government hopes that by regulating the purchase price of renewable energy generated electricity to PLN, the increase of BPP due to the price of electricity generated by renewable energy may be suppressed.[28] However, as the subsidy increases the financial impact to the government, its development is naturally hampered due to the provision of such subsidy.

Separately, although the continuous efforts of the Government of Indonesia to ease the development of renewable power plant facilities are to be applauded, there are some regulations that still hinder the pace of such development. For example, the Minister of Industry Regulation No. 54 of 2012 concerning the Guideline on the Domestic Product Utilization for Electricity Infrastructure as amended by the Minister Regulation No. 5 of 2017 (MOI Regulation 54/2012) stipulates minimum domestic content requirements for geothermal, hydro, and solar PV projects that will generate power for public consumption and funded by the state budget, regional government budget, grant, or foreign loan. This requirement applies broadly to projects conducted by state-owned enterprises, regional government-owned enterprises, cooperatives, and private companies.

For example, in the solar PV power generation sector, MOI Regulation 54/2012 set out the minimum local content of solar modules at 40% in 2017 and increased to 50% in January 2018 and 60% in January 2019. The high local content requirements for solar modules are deemed to be one of the main barriers to rapid deployment of solar PV in Indonesia because local modules are more expensive than imported modules from China. Imported module prices are reported to range between USD0.25–USD0.37/Wp compared to local modules at USD0.47/Wp on average. In addition, in some cases the use of local modules would also lower the bankability of solar projects as lenders regard local modules less durable than those imported.

Although a decade has passed since the enactment of MOI Regulation 54/2012, the Ministry of EMR acknowledged that domestic industry is not yet ready to meet these targets. The application of local content regulation for renewable electricity, while the market is still dependent on foreign technologies,[29] is seen to create a chicken-and-egg conundrum. Investment for domestic production of renewable electricity equipment would naturally be attractive if the market is sufficiently large. However, the market growth is constrained by the high cost of locally produced equipment, which is not yet able to reach economies of scale.

10.3 Policy Direction and National Targets for OGRE

First of all, it is notable that Article 33 paragraph (3) of the 1945 Constitution of the Republic of Indonesia provides the following mandate: 'The land and waters and the natural wealth contained in it shall be controlled by the state and utilised for the optimal welfare of the people'. Given the foregoing, in general, renewable energy that is generated from land, waters, and natural resources of the Republic of Indonesia shall be under the control of the Government of Indonesia.

In that respect, renewable energy finds its legal basis under Law No. 30 of 2007 concerning Energy (Energy Law). The Energy Law provides that, in principle, renewable energy shall be managed by the Government of Indonesia and utilised for the optimal welfare of society. With the approval from the House of Representatives (*Dewan Perwakilan Rakyat*), the Government of Indonesia shall issue a national-level energy management policy document called the National Energy Policy (*Kebijakan Energi Nasional*).

Currently, the applicable National Energy Policy is provided for under Government Regulation No. 79 of 2014 concerning the National Energy Policy (GR 79/2014) and shall remain effective from 2014 until 2050. GR 79/2014 stipulates that the National Energy Policy is provided as a guideline for national energy management in order to reach energy independency and endurance to support sustainable national development. The National Energy Policy provides for broad policies regarding energy development and its utilisation among others on: (i) the availability of energy for national energy needs, (ii) energy development priority, (iii) utilisation of national energy resources, and (iv) national energy reserves. As part of energy development priority, it is notable that one of the principles of energy development is to 'maximize the utilization of Renewable Energy with due regard to its economic levels'.

In addition, GR 79/2014 also provides concrete national energy utilisation goals. With respect to the electricity sector, Article 8 of GR 79/2014 provides that the Government of Indonesia aims to reach (i) power generation capacity of 115 GW by 2025 and 430 GW by 2050 and (ii) power utilisation per capita of 2,500 kWh by 2025 and 7,000 kWh by 2050.

Through the Presidential Regulation of the Republic of Indonesia No. 22 Year 2017 regarding the National Energy General Plan (*Rencana Umum Energi Nasional* – RUEN), the Government of Indonesia sets out the implementation of the multi-sectoral energy policies to achieve the National Energy Policy.[30,31] The current RUEN is applicable for the period of 2017 until 2050.[32]

The RUEN sets out the Government of Indonesia's target to increase the utilisation of renewable energy to at least 23% of the total primary energy mix in 2025, and at least 31% of the total primary energy mix in 2050.[33] Aside from being used as a primary energy for power plants, renewable energy shall also be developed to:[34]

(i) fulfil the need of electricity with the target to use as much as 2,500 kWh/capita of electricity in 2025, and 7,000 kWh/capita in 2050, and implement the plan to develop non-electricity renewable energy based on the current production realisation and optimal usage of biofuel in the consumer sector; and
(ii) achieve the balance between the demand and supply of energy for the purpose of achieving at least 23% of renewable energy mix in 2025 and at least 31% in 2050.

In order to achieve the aforementioned targeted renewable energy development, the Indonesian government will take, among others, the following measures:[35]

(i) establish a separate renewable energy business entity, which is mandated by the Indonesian government to develop, use, and/or buy renewable energy;
(ii) apply and improve the feed in tariff of renewable energy power plants to the applicable electricity power plant as long as the price of electricity produced by renewable energy is higher than the price of electricity produced from other primary energy sources;
(iii) draft a guideline for the local government to provide energy subsidy, which budget is allocated in the Regional Revenue and Expenditure Budget (*Anggaran Pendapatan Belanja Daerah*);
(iv) budget for continuous development of renewable energy infrastructure for villages that will not be provided with electricity in a long period of time;
(v) mandate the national infrastructure financing institution to finance renewable energy construction projects; and

(vi) develop a small renewable energy-based electricity system to provide electricity to areas not reached by the expansion of electricity grid.

It is indicated that the following principles are followed in the 2016–2025 development plan of renewable energy other than the hydro power plant and geothermal power plant:[36]

(i) prioritisation of building renewable energy power plant is performed based on the amount of consumed electricity per capita per province/area and the available potential renewable energy per province;
(ii) the province/area with the least amount of electricity consumption per capita will be prioritised in the development of renewable energy; and
(iii) the province/area with the largest renewable energy potential will be prioritised in the development of renewable energy.

To achieve the amount of renewable energy mix as targeted under the RUEN, renewable energy power plants should have the capacity to provide around 45.2 GW in 2025 and 167.6 GW in 2050. Based on committed and potential projects, RUEN also provides the following guidelines for the development of each renewable energy power plant:

1. *Geothermal Power Plant*
   Development in producing electricity through geothermal is projected to result in 7.2 GW by 2025 and 17.6 GW by 2050 which is around 59% of the estimated available 29.5 GW geothermal potential.[37]
2. *Hydro Power Plant*
   Development in producing electricity through hydropower is projected to result in 18 GW in 2025 and 38 GW in 2050 which is around 51% of the estimated available 75 GW hydropower potential.[38]
3. *Mini Hydro and Micro Hydro Power Plant*
   Development in producing electricity through mini hydro power and micro hydro power is projected to result in 3 GW in 2025 and 7 GW in 2050 which is 37% of the estimated available 19 GW mini hydro power and micro hydro potential.[39]
4. *Biofuel Power Plant*
   Development in producing electricity through biofuel is projected to result in 5.5 GW in 2025 and 26 GW in 2050 which is 80% of the estimated available 32.7 GW biofuel potential.[40]
5. *Solar Power Plant*
   Development in producing electricity through solar power is projected to result in 6.5 GW in 2025 and 45 GW in 2050 which is 22% of the estimated available 207.9 GW solar power potential. The projection of the solar power plant is optimistic considering the investment trend and electricity price of the global solar power plant, which becomes cheaper with the passage of time and the advancement of technology.[41]

6. *Wind Power Plant*
   Development in producing electricity through wind power is projected to result in 1.8 GW in 2025 and 28 GW in 2050 which is 46% of the estimated available 60.6 GW wind power potential.[42]
7. *Other Renewable Energy Power Plant*
   Development in producing electricity through other renewable energy sources is projected to result in 3.1 GW in 2025 and 6.1 GW in 2050. Other renewable energy power plant includes diesel power plant with mixture of biofuel, wave power station, ocean current power station, and ocean thermal energy.[43] RUEN also provides certain measures required to achieve the abovementioned targets.

Finally, it is notable that the National Energy Policy and RUEN do not provide any specific target or programme to increase the utilisation of off-grid renewable electricity systems.

10.4 Regulatory Framework for OGRE

The regulatory framework of renewable electricity is provided for under the Energy Law and Law No. 30 of 2009 concerning Electricity as amended by Law No. 11 of 2020 (Electricity Law). In general, provisions on power generation (e.g., the required licences) under the Electricity Law also apply to renewable energy power generation.

The Ministry of EMR is the main regulator that governs the Indonesian power market and consists of several directorates with separate areas of responsibility. Its Directorate General of New Energy, Renewable Energy, and Conservation of Energy (DGRE) has the authority to regulate and supervise the renewable energy sector.[44] In performing its task, the DGRE's functions include the preparation and implementation of policies concerning the management, control, and supervision of the renewable energy sector.[45] Separately, its Directorate General of Electricity is responsible for general electricity supply, including formulating and implementing policies among others relating to the economics, technical, and safety aspects of the electricity supply business.[46]

Other than the Ministry of EMR, several other ministries and institutions also have roles in setting and implementing policies related to renewable energy electricity. Some of the notable ministries and institutions and of their roles in relation to renewable energy electricity are, among others, as follows:

- the Ministry of Finance is in charge of the formulation of budget allocations and subsidies;
- the National Energy Council (*Dewan Energi Nasional*) has the authority to formulate the National Energy Policy with the approval from the House of Representative of the Republic of Indonesia, issue the National Energy Policy, and determine steps to recover from energy crisis and emergencies;[47] and

- the Ministry of Environment and Forestry is responsible for formulating, issuing and implementing policies relating to environmental protection, issuing environmental permits, and formulating environmental standards.[48]

Pursuant to Article 4 of the Electricity Law, the power generation business activities of the Government of Indonesia shall be conducted by a State Owned Entity (*Badan Usaha Milik Negara* or BUMN) or a Regional Government Owned Entity (*Badan Usaha Milik Daerah* or BUMD). However, power generation business activities are not limited only to a BUMN and a BUMD, but also open for investment by private entities, cooperatives, and society organisations (*lembaga swadaya masyarakat*).

Although foreign investment limitations in the renewable electricity sector have been substantially reduced in recent years, there are still some foreign investment limitations that may be applicable depending on the size of the power plant. Pursuant to President Regulation No. 10 of 2021 concerning Business Activities for Investment as amended by President Regulation No. 49 of 2021, power generation of less than 1 MW is reserved for cooperatives and Micro, Small and Medium Enterprises (*Usaha Mikro, Kecil dan Menengah* or known as UMKM). Further, pursuant to Law No. 25 of 2007 concerning Investment, foreign investment may only be made to large scale business. Therefore, although Indonesia's foreign investment limitation in the power sector has seen progress,[49] currently no foreign investments can be made to power plant business activities with power generation less than 1 MW.

Finally, as electricity generation may be a stand-alone business activity in Indonesia, it is notable that an independent power producer (IPP) may build a renewable energy-based power plant and sell the electricity that it produced to PLN who then transmit, distribute, and/or sell such electricity to the general public. However, based on the EMR Minister Regulation 50 of 2017 concerning Utilization of Renewable Energy Sources for Power Generation (EMR Minister Regulation 50/2017), all renewable electricity generation projects that sell electricity to PLN (with the exception of waste-to-energy) had to be structured as a build-own-operate-transfer (BOOT). In that regard, IPPs had to transfer the project to PLN at the end of their power purchase agreement. This requirement was heavily criticised as one of the main barriers for investment in renewable electricity and significantly reduces the bankability of the projects. In response, the EMR Minister has removed such transfer requirement with the issuance of the EMR Minister Regulation 4 of 2020 concerning the Second Amendment to EMR Minister Regulation 50/2017 (EMR Minister Regulation 4/2020) on February 26, 2020. Currently the renewable electricity generation may be structured as a build-own-operate (BOO). Therefore, IPPs may now continue to own its power plants even after their power purchase agreements with PLN have expired.

1. Licensing

Pursuant to the Electricity Law and the EMR Minister Regulation No. 11 of 2021 regarding Conducting Electricity Business, electricity supply business activity (which includes power generation) is categorised into electricity supply for public interests and electricity supply for own interests.

Electricity Supply for Public Interests

Electricity supply for public interests refers to one or more of the following activities:

- electricity generation;
- electricity transmission;
- electricity distribution; and
- sale of electricity.

Further, electricity supply for public interests is conducted based on work area (*wilayah usaha*), whereby one business entity shall be responsible for the integrated electricity supply of one work area (an area determined by the Central Government as the area where a business entity conducts the distribution and/or sale of electricity power).[50] Pursuant to Article 11 of Electricity Law, State Owned Entity is given first priority to conduct electricity supply for public interests. Regional Government Owned Entity, private entities, cooperatives, and society organisations may be provided with an opportunity to conduct integrated electricity supply for public interests in areas which have not had electricity. For information, as at 2019, PLN's work area covers around 90% of Indonesian territory and the remainder is open for development by the aforementioned entities.[51] Any party who conducts electricity supply for public interests must obtain an Electricity Supply for Public Interest Business License (*Izin Usaha Penyediaan Tenaga Listrik untuk Kepentingan Umum* – IUPTLU).

Pursuant to Article 10 of Government Regulation No. 14 of 2012 concerning Electricity Supply Business Activities as amended by Government Regulation No. 23 of 2014 (GR 14/2012), the authority to grant IUPTLU shall vest in (i) the EMR Minister for business entity whose work area is interprovincial, is a BUMN, and sells electricity and/or lease electricity network to an electricity supply business licence holder whose licence is granted by the EMR Minister; (ii) the governor for business entity whose work area is intercity/interdistrict and sells electricity and/or lease electricity network to an electricity supply business licence holder whose licence is granted by a governor; and (iii) a regent/mayor for business entity whose work area is within a district/city and sells electricity and/or lease electricity network to an electricity supply business licence holder whose licence is granted by the regent/mayor.

In order to obtain an IUPTLU for power plant, a business entity is required to submit to the OSS system:[52] (i) electricity supply business feasibility study (which covers financial feasibility analysis, operational feasibility analysis, network interconnection study, installation location, single line diagram, business type and capacity that will be conducted, construction schedule, and operational schedule) that is prepared by a certified institution and (ii) power purchase agreement between the applicant and prospective electricity purchaser. Please note that the power purchase price in the power purchase agreement must obtain prior approval from the EMR Minister or a governor as relevant.

2. *Electricity Supply for Own Interests*

On the other hand, electricity supply for own interests refers to the following activities: electricity generation, electricity generation and distribution, or electricity generation, distribution, and transmission.

The electricity supply for own interests' business activities are conducted according to the nature of its utilisation which covers:

- main utilisation to operate the power plant continuously, in order to fulfil a part or all of its own electricity needs;
- back-up utilisation to operate the power plant only from time to time to guarantee the continuity and reliability of electricity supply for its own interests;
- emergency utilisation to operate the power plant only when there is a power disruption to the local IUPTLU holder; and
- temporary utilisation to operate the power plant only for temporary activities, including mobile and portable power plants.

The Electricity Supply Business License for Own Interest (*Izin Usaha Penyediaan Tenaga Listrik untuk Kepentingan Sendiri* – IUPTLS) shall be required to conduct any electricity supply for own interest business activities. Pursuant to Article 28 of GR 14/2012, the authority to grant IUPTLS shall vest in: (i) the EMR Minister for facilities whose installation covers multiple provinces, (ii) governor for facilities whose installation covers district/city, (iii) regent/mayor for facilities whose installation scope is within a district/city.

To obtain the IUPTLS, the business entity must fulfil administrative, technical, and environmental requirements. The technical requirements that need to be fulfilled and provided along with the application for IUPTLS are details on the location of installation, single line diagram, type and capacity of the power generator installation, construction schedule, and operational schedule. The administrative requirements that need to be provided concern the entity's profile (i.e., company profile, board of directors, board of commissioners, and shares composition).

Both IUPTLU and IUPTLS can be obtained via the OSS system. These business licences are also applicable as a business licence for electricity supply using renewable energy. Please note that other than the foregoing main business licences, other licences, and permits (e.g., environmental licences and permits) may also be required to build and operate a power plant.

3. *Environmental Regulations*

Based on Government Regulation No. 22 of 2021 regarding Provision of Protection and Management of the Environment, any business and/or activity must have an Environmental Approval (*Persetujuan Lingkungan*) – which is a prerequisite to obtain business licences or other governmental approvals. Environmental Approval is provided in the following 3 (three) forms:

- Analysis on Impacts to the Environment (*Analisis Mengenai Dampak Lingkungan Hidup* – AMDAL). AMDAL must be obtained by business and/or activity that is classified as having an Important Impact to the environment. In turn, Important Impact is defined as fundamental change to the environment, caused by a business and/or activity. The criteria for an Important Impact to the environment are business or activities that:[53]
  (i) alters the shape of land and landscape;
  (ii) exploits the natural resources, either renewable or non-renewable;
  (iii) potentially causes environmental pollution and/or environmental damage and dissipation and deterioration of natural resources in its utilisation;
  (iv) have products which may affect the environment, man-made environment, and the social and cultural environment;
  (v) have products which affects a conservation and/or a protected cultural heritage area;
  (vi) introduces types of vegetation, animal, and micro-organism;
  (vii) creates and utilises living substance and non-living substance;
  (viii) has high risk and/or influence on state defence; and/or
  (ix) implements technology that is envisaged to have a large-scale potential to influence the environment.
- Measures to Manage the Environment and Measures to Supervise the Environment (UKL-UPL), for businesses/activities:[54]
  (i) that do not have an Important Impact;
  (ii) whose location is outside of and/or not directly neighbouring with conservation areas; and
  (iii) that is included as a type of business and/or activity plan that are excluded from those which are mandatorily required to obtain AMDAL.
- Statement Letter on Capability to Manage and Supervise the Environment (*Surat Pernyataan Kesanggupan Pengelolaan dan Pemantauan Lingkungan Hidup* – SPPL) for activities/business that:
  (i) do not have an Important Impact and are not obligated to have a UKL-UPL;
  (ii) are classified as micro and small business and/or activity that does not have Important Impact to the environment; and/or
  (iii) are included as a business and/or activity plan that is excluded from being obligated to have a UKL-UPL.

Pursuant to the Minister of the Environment and Forestry Regulation No. 4 of 2021 regarding the List of Businesses and/or Activities Obligated to have an Analysis on the Impact to the Environment, Measures to Manage the Environment and Measures to Supervise the Environment or a Statement Letter on Capability to Manage and Supervise the Environment, with regard to power plants, the

Table 10.1 Required Environmental Approval in relation to Power Plant Type and Power Capacity

Type of Power Plant	*Scale That Needs AMDAL*	*Scale That Needs UKL-UPL*	*Scale That Needs SPPL*
Wind Power Plant Construction (*Pembangunan PLT Bayu*)	≥ 50 MW	≥ 1 MW to < 50 MW	< 1 MW
Solar Power Plant Construction (*Pembanguan PLT Surya*)	≥ 50 MW	≥ 1 MW to < 50 MW	< 1 MW
Mini Hydro until Mid- Hydro Power Plant Construction (*Pembangunan PLT Air (Skala Kecil (Mini) s/d Skala Menengah*)	–	≥ 1 MW to < 50 MW	–
Pico Hydro until Micro Hydro Power Plant Construction (*Pembangunan PLT Air (Skala Kecil (Piko Hidro) s/d Mikro Hidro)*	–	–	< 1 MW
Biomass Power Plant Construction (*Pembangunan PLT Biomassa*)	≥ 50 MW	< 50 MW	–
Biogas Power Plant Construction (*Pembangunan PLT Biogas*)	≥ 50 MW	< 50 MW	–
Biofuel Power Plant Construction (*Pembangunan PLT Bahan Bakar Nabati*)	≥ 100 MW	< 100 MW	–

required Environmental Approval depends on the power capacity that will be generated. Some of the notable requirements are as provided in Table 10.1 (Required Environmental Approval in relation to Power Plant Type and Power Capacity).

In general, the Government of Indonesia's main approaches to provide electricity include expanding existing grid and building off-grid power plants and infrastructures.[55] In order to raise the electrification ratio, the Government of Indonesia provides a special programme for the provision of small-scale electricity (with total capacity of up to 50 MW) that is generated by renewable energy and designated for: (i) undeveloped villages, (ii) remote villages, (iii) villages located near border areas, and (iv) small inhabited islands (Small Scale Off-Grid Power Plant Program). Such a programme is provided for under the EMR Regulation No. 38 of 2016 concerning Acceleration of Electrification in Undeveloped, Remote, Border Area Villages, and Inhabited Small Islands Through Conducting Small Scale Electricity Supply Business (EMR Minister Regulation 38/2016).

Under EMR Minister Regulation 38/2016, a BUMD, private party, or cooperative may manage a work area that is currently unreached by other work area holders. Pursuant to EMR Minister Regulation 38/2016, in general, the framework for the determination of the work area and business entity to conduct the programme is as follows:[56]

- the EMR Minister has the authority to determine the work area for the programme based on recommendation from a governor after coordination with PLN or based on the EMR Minister's own discretion;

- the relevant governor shall offer the work area (through auction) to qualified business entities (i.e., a BUMD, private entity, or cooperative). If there are no interested investor/business entities, then the governor may instruct a relevant BUMD to conduct the provision of small-scale renewable energy electricity generation;
- the business entity that is selected to carry out such programme will be provided with an IUPTLU by the governor;
- thereafter, the governor will propose such business entity to the EMR Minister and, in turn, the EMR Minister shall give an assignment to such business entity to conduct the provision of small-scale renewable energy electricity generation.

Notwithstanding the foregoing, the Small Scale Off-Grid Power Plant Program is still deemed to be unattractive for the private sector and BUMD due to perceived poor project economics.[57] The Government of Indonesia needs to evaluate this policy so that more private parties and BUMDs may participate in building off-grid power plant projects for remote areas where grid expansion makes no commercial sense. Further, the programme should also include sustainable services so that at the time the power plant is handed over to the relevant community and operated by the appointed institution, its operations may be sustained for the continuous benefit of the community.[58]

4. Pricing Mechanisms

In general, renewable energy electricity prices are negotiated between power supplier and purchaser (e.g., PLN and IPP), and price of renewable energy electricity to be purchased shall subsequently need to be approved by the EMR Minister or the regional government as relevant.[59] Nevertheless, it is notable that the price of renewable energy electricity to be purchased by PLN is regulated. Pursuant to Presidential Regulation No. 112 of 2022 concerning Acceleration of Renewable Energy Development for Power Generation ("**PR 112/2022**"), in general the price of renewable energy electricity to be purchased by PLN from power plants which are not developed (wholly or partially) by the Government of Indonesia shall be either (i) the price which may be negotiated and agreed up to the applicable Highest Reference Price (*Harga Patokan Tertinggi*) or (ii) for certain types of renewable energy power plant projects, the price agreed by the power producer and PLN.[60]

With regard to the applicable Highest Reference Price, it shall be determined depending on a number of factors including the type of renewable energy used for power generation, the location of the power plant project, and the capacity of the power plant. Further, PR 112/2022 also provides the following principles relating to the Highest Reference Price and the purchase price of renewable energy electricity:

- the purchase price shall be applicable from the commercial operation date and generally such purchase price shall be fixed (cannot be increased) for the term of the power purchase agreement;[61]

- the purchase price shall be the price at the meeting point between the electrical equipment at the distribution installation (plant busbar).[62] This excludes transmission electricity price, which shall be set based on agreement between the IPP and PLN for a maximum of 30% of electricity purchase price.[63] An exception to such transmission electricity price cap is possible by obtaining a separate approval from the EMR Minister;[64]
- the formula for the Highest Reference Price is provided for in the attachment of PR 112/2022 and a purchase price that is agreed based on the Highest Reference Price formula serves as the price approval from EMR Minister[65]. In such case, the absence of a separate EMR Minister price approval is expected to expedite the procurement process;
- the Highest Reference Price is divided into two (2) stages (i.e., the first stage from year one (1) to year ten (10), and the second stage from year 11 (eleven) and onwards);[66]
- the Highest Reference Price is calculated by taking into account the location of the project as one of the factors. Essentially, the Government of Indonesia is providing incentive for the development of power plants in frontier areas of Indonesia; and
- the Highest Reference Prices are subject to annual review by the Ministry of EMR.

Notwithstanding the foregoing, for certain types of renewable energy power plant projects (i.e., (i) biofuel and ocean energy renewable power plants developed by an IPP or the Government of Indonesia or (ii) hydro peaker power plant developed by an IPP), the price is not capped by the Highest Reference Price and may be agreed by the IPP and PLN. However, in such cases the price must obtain approval from the EMR Minister.

Some issues remain unclear under PR 112/2022 including the following:

- PR 112/2022 provides different total duration Highest Reference Prices depending on the type of the power plant (e.g., biomass has 25 years period, biogas has 20 years period, and others with 30 years period)[67]. Currently, based on EMR Minister Regulation No. 10 of 2017 concerning the Principles of Power Purchase Agreement as amended by the EMR Minister Regulation No. 10 of 2018 (EMR Minister Regulation 10/2017), the term of a power purchase agreement may be provided up to 30 years. It is unclear whether there will be a future regulation that will limit the term of the power purchase agreement depending on the type of the renewable energy utilized for power generation; and
- as the purchase price is applicable from the commercial operation date, it remains unclear whether PLN will not be allowed to pay for any electricity produced prior to the commercial operation (e.g., during commissioning) or whether other prices may be agreed for such electricity.

PR 112/2022 is enacted in September 2022 and is still quite nascent as at the writing of this chapter. Further implementing regulations from various ministries are to be expected and should provide more clarity on the implementation of the provisions under PR 112/2022.

For off-grid renewable energy electricity, in general, there is no express regulation that specifically regulates the applicable electricity price, with the exception of electricity generated under the scope of the EMR Minister Regulation 38/2016. Under the EMR Minister Regulation 38/2016, tariff is set depending on whether the business entity under the Small Scale Off-Grid Power Plant Program utilises subsidy from APBN:

- if subsidy is utilised, PLN's tariff for household consumer with 450 VA connected capacity shall apply;[68]
- if no subsidy is utilised, the EMR Minister or the relevant governor shall determine the tariff in accordance with their respective authority. If the regional government is unable to determine the applicable tariff, then Central Government shall determine the tariff based on the applicable PLN's tariff.[69]

The current pricing mechanism has raised some concerns for the development of renewable energy:

1. The Highest Reference Price does not reflect and takes into account any non-financial benefits of renewable energy generation (e.g., reduction of air pollution as a result of substituting coal-fired power generation); and
2. The existing price cap is still deemed insufficient to make renewable projects attractive to some investors as compared to Feed-In Tariff mechanism—especially for hydro power and solar power.[70]

5. Support Schemes

In line with the National Energy Policy and RUEN, the Energy Law encourages the utilisation of renewable energy for power generation by providing certain incentives from the government and/or regional government.

Tax and Duty Allowances

There is a suite of government regulations and Minister of Finance regulations, such as the Government Regulation No. 78 of 2019 concerning Income Tax Facility for Investment in Certain Business Fields and/or Certain Areas, and the Minister of Finance Regulation No. 11/PMK.010/2020 of 2020 concerning the Implementation of Government Regulation No. 78 of 2019 concerning Income Tax Facility for Investment in Certain Business Fields and/or Certain Areas as lastly amended by the Minister of Finance Regulation No. 96/PMK.010/2020 of 2020 that provide tax and duty allowances for greenfield investment or expansion of business activities in micro hydro and mini hydro power plants:

- reduction of net income in the amount equivalent to 30% of investment values in the form of tangible fixed assets such as lands which are utilised for the main business activity, which shall be charged over 6 years at 5% per annum;
- accelerated depreciation and amortisation;
- reduction of tax on dividends remitted to foreign tax payers which are not permanent establishment in Indonesia to become 10% or a lesser rate as applicable based on the prevailing tax treaty; and
- extension of tax loss carry-forward period to become longer than 5 years but not more than 10 years, with an additional 1 year also being provided for investment in the field of renewable energy.

In addition, pursuant to the Minister of Finance Regulation No. 66/PMK.010/2015 of 2015 concerning Exemption of Import Duty on Investment Goods for the Construction or Development of Power Plant Industry for Public Interest, business actors that hold an IUPTL, or has a power purchase agreement with PLN under which all of the electricity it produces shall be purchased by PLN, or has a finance lease agreement with PLN, or has a sale and purchase of electricity agreement with an IUPTL holder that has a work area and all generated electricity shall be purchased by such IUPTL holder, shall be entitled to apply to the Head of the Investment Coordinating Board (*Badan Koordinasi Penanaman Modal*) for exemption of import duty over investment goods. Such goods must be utilised in the power plant industry and fulfil any one of the following: (i) has yet to be produced domestically, (ii) has been produced domestically but has yet to fulfil the required specifications, or (iii) has been produced domestically but the quantity has yet to fulfil the industry's need.

Tax Holiday

The Minister of Finance Regulation No. 130/PMK.010/2020 on Provision of Corporate Income Tax Reduction Facility (MOF Regulation 130/2020) provides deduction of corporate income tax (known as tax holidays) for investments in 'pioneer industries'. Renewable energy power plants are part of the 'economic infrastructure' business which is included in the list of pioneer industries stipulated in the Investment Coordinating Board Regulation No. 7 of 2020 on Details of Business Fields and Production Types of Pioneering Industry.

The facilities being provided are the deductions of the following taxes:[71]

a. 100% of outstanding corporate income tax for greenfield investment of at least IDR 500,000,000,000 (approx. USD 35,000,000) and shall be applicable for:
   - 5 (five) fiscal years for greenfield investment of at least IDR 500,000,000,000 (approx. USD 35,000,000) but less than IDR 1,000,000,000,000 (approx. USD 70,000,000). Thereafter, a deduction of 50% income tax shall be applicable for the ensuing 2 (two) fiscal years;

- 7 (seven) fiscal years for greenfield investment of at least IDR 1,000,000,000,000 (approx. USD 70,000,000) but less than IDR 5,000,000,000,000 (approx. USD 350,000,000) Thereafter, a deduction of 25% income tax shall be applicable for the ensuing 2 (two) fiscal years;
- 10 (ten) fiscal years for greenfield investment of at least IDR 5,000,000,000,000 (approx. USD 350,000,000) but less than IDR 15,000,000,000,000 (approx. USD 1,050,000,000);
- 15 (fifteen) fiscal years for greenfield investment of at least IDR 15,000,000,000,000 (approx. USD 1,050,000,000) but less than IDR 30,000,000,000,000 (approx. USD 2,100,000,000); and
- 20 (twenty) fiscal years for greenfield investment of at least IDR 30,000,000,000,000 (approx. USD 2,100,000,000); and

b. 50% of outstanding corporate income tax for greenfield investment of at least IDR 100,000,000,000 (approx. USD 7,000,000) but less than IDR 500,000,000,000 (approx. USD 35,000,000) and shall be applicable for 5 (five) fiscal years.

Specific Incentives and Support for Development of Geothermal Energy Electricity

Under PR 112/2022, the supports provided specific for purchase of electricity by PLN from geothermal energy power plants are as follows:

- the appointment of a public service agency (*badan layanan umum*) or a state-owned company to acquire additional geothermal data during the exploration phase;[72]
- the appointment of a prospective IPP to conduct preliminary survey and exploration;[73]
- for Geothermal Business License (Izin Panas Bumi), power of attorney, and contract holder, a 'de-risking' scheme (where the Government of Indonesia and the IPP will share exploration risks) and certain financing facilities may be applicable;[74]
- the possibility to escalate the electricity purchase price (which is not permitted for any other type of renewable energy electricity) during the term of the power purchase agreement with PLN;[75] and
- the direct appointment of IPP by PLN as further elaborated below.[76]

Accelerated Termination of Coal Fired Power Plant

Pursuant to PR 112/2022, the Government of Indonesia has formally prohibited the development of new coal-fired power plants with the following exceptions:[77]

- coal-fired power plants which development has been stipulated prior to the enactment of PR 112/2022; or
- coal-fired power plants which:

- are integrated with industries which are oriented to the increase of added value of natural resources or included in National Strategic Project which has major contribution to the creation of employment opportunities and/or the growth of national economy;
- are committed to reduce green gas emission at least 35% (thirty five percent) within a period of 10 (ten) years since the power plant conduct commercial operation as compared to the average coal-fired power plants' emission in 2021 through the development of technology, carbon offset, and/or mix utilization of renewable energy; and
- will operate up to 2050 at the latest.

PR 112/2022 also requires the Ministry of EMR (in coordination with the Ministry of Finance and Ministry of State Owned Enterprises) to prepare a roadmap for the early termination of existing coal-fired power plants operated by PLN and IPPs.[78] Further, PR 112/2022 also provides that the Government of Indonesia may issue fiscal contribution through financing framework (including in the form of blended financing) from the State Budget or other sources to the relevant IPPs to support the accelerated termination.[79] While it is common for power purchase agreements to include government policy as one of force majeure events, where PLN would normally be required to pay a termination fee to the relevant IPP, currently it is unclear whether the financing will be used to pay such termination fees or for other purposes. PR 112/2022 also mandates the Ministry of Finance to issue an implementing regulation to further detail the form of such fiscal contribution[80]; however, such implementing regulation has not been issued yet.

Other Incentives under PR 112/2022

Other than the foregoing, PR 112/2022 provides that the Government of Indonesia will provide the following incentives:[81]

- fiscal incentives in the form of income tax facility, import duty, land and building tax facilities, geothermal development support, financing or project guarantees by Indonesian state-owned companies;
- non-fiscal incentives from central and regional governments.

PR 112/2022 also requires the EMR Minister, various ministers, head of institutions or regional government relevant to the development of power plants to provide support as required in the development of renewable energy power plants. PR 112/2022 mandates the following forms of support from the relevant ministers:

- the EMR Minister is required to provide support in the form of the formulation of the development plan for renewable energy power plant;[82]
- the Minister of Finance is required to provide support in the form of the provision of fiscal incentives;[83]

- the Minister of Agrarian Affairs and Spatial Planning is required to provide support in the form of provision of priority to the development of renewable energy power plants in the national spatial planning, as well as ease of licensing in the agrarian/land affairs and spatial planning;[84]
- the Minister of Environment and Forestry is required to provide support in the form of ease of licensing for land use in the forest areas and reduced costs for the development of renewable energy[85];
- the Minister of Public Works and Public Housing is required to provide support in the form of ease of licensing and reduced costs for the development of renewable energy;[86]
- the Minister of Home Affairs is required to provide support in the form of formulation of policies to support the development of renewable energy power plant within the scope of regional government;[87]
- the Minister of State Owned Enterprises is required to provide support in the form of the stipulation of a certain target for the implementation of renewable energy in the performance indicator of PLN;[88]
- the Minister of Industry is required to provide support to business entities by prioritizing the use of domestic products through creation of supply capability including in the aspects of quality, costs, reasonable delivery and increase depth of industry structure, determining import quota for renewable energy power plant components, domestic content utilization verification for renewable energy power plant components, and providing a roadmap for the development of supporting industry for power generation;[89]
- the Indonesian Investment Coordinating Board is required to provide support in the form of ensuring the ease of licensing and reduced costs for the development of renewable energy;[90]
- regional government is required to provide support in the form of ease of licensing, incentive and guarantee of availability of land in accordance with its designated utilization for the development of renewable energy power plant.[91]

Currently, the implementation of these incentives and supports is still unclear at the moment and further details are to be expected after the enactment of the implementing regulations of PR 112/2022.

6. *Purchasing Arrangements*

In general, the purchase of electricity shall be conducted through a general auction. Nevertheless, the purchase of renewable energy electricity by IUPTL holders (other than PLN) may be conducted through direct appointment.[92] On the other hand, PLN may purchase renewable energy electricity be through direct selection or direct appointment depending on the type of renewable energy source utilized for power generation as follows:[93]

- direct appointment may be conducted for electricity purchase from (i) hydro power plant (which utilises dam/water reservoir or multipurpose irrigation

stream constructed by the Government of Indonesia), (ii) geothermal power plant (from certain geothermal business license/power of attorney holders stipulated under PR 112/2022), (iii) additional capacity from the expansion of renewable energy power plant (except for those from ocean energy and biofuel power plants), and (iv) excess power of geothermal, hydro, biomass and biogas power plants; and
- direct selection may be conducted for electricity purchase from (i) solar or wind power plants equipped or not equipped with battery facilities or electricity storage facilities which land is provided for by the Government of Indonesia or is procured by itself, (ii) biofuel, ocean energy, hydro, biomass, biogas power plants, and (iii) a hydro power plant that functions as peaker.

With regard to power purchase agreements, in principle, they are subject to negotiations between the power producer and purchaser. There is no regulation for power purchase agreements between IPPs except for when PLN is the purchaser. Based on the EMR Minister Regulation 10/2017, the EMR Minister sets out principles that need to be covered in the power purchase agreement where PLN is a purchaser. A few notable points renewable energy power purchase agreements may cover include:

- transaction can be structured as build-own-operate (BOO) instead of build-own-operate-transfer (BOOT) scheme;[94]
- sale and purchase of electricity can be made for a term of 30 (thirty) years from the commercial operation date;[95] and
- payment must be in Indonesian Rupiah, unless there is an exception made by Bank of Indonesia and if the payment is made in United States Dollars, the parties shall utilise the Jakarta Interbank Spot Dollar Rate (JISDOR) as the applicable USD-IDR exchange rate.[96]

There is no specific regulation that expressly provides for purchasing arrangements of off-grid renewable electricity.

10.5 Conclusion

As laid out in the National Energy Policy, the Government of Indonesia recognises the importance of renewable energy, clean energy production, and maximisation of energy efficiency as key priorities to drive sustainable economic development. However, although concrete actions have been taken by the Government of Indonesia for the development of renewable energy in in recent years, this development and its implementation are notably still in its infancy stage.

Despite the high electrification ratio claimed by the government, in 2020 a report mentioned that there are still more than 1 million households experiencing no access to electricity in remote regions of Indonesia. The geographical and economic conditions of these areas and the costs and funding scheme issues are key hindrances to universal electrification. Despite the potential to provide electricity to these rural areas, there is also very limited utilisation of off-grid renewable electricity generation. Given the foregoing, in order to foster the development of renewable energy and off-grid renewable electricity, the following factors are among those that need to be addressed:

First of all, the Government of Indonesia may need to prioritise the review and revision of the legal and regulatory framework of support for the development of off-grid renewable electricity. Currently the legal bases for renewable electricity are still scattered in various regulations which signifies a lack of focus in this sector and there are no specific regulations issued for the development of off-grid renewable electricity. In that regard, it will be imperative to ensure that there are laws and regulations that are tailored to the renewable energy sector and off-grid renewable electricity to aim for a more streamlined and efficient legal framework to support development. Such review and revision of the legal and regulatory framework holds the potential to reduce overhead costs in the development of renewable electricity. Further, currently there is a bill on renewable energy law that is listed in the National Legislation Priority Program (*Program Legislasi Nasional Prioritas*) and still under the promulgation process. Nevertheless, there is no news yet on when the bill will be passed. The enactment of this new renewable energy law is expected to create a more focused legal and regulatory framework for the development and management of renewable energy and electricity that will further foster its development.

Second, renewable electricity generation projects' economics need to be better facilitated with a more conducive investment climate. Although foreign limitation has seen relaxation in early 2021, investment in renewable electricity in general has not seen much progress given investors' concerns on sustainability of projects' economics. After all, renewable electricity generation projects are still considered capital intensive projects and need higher financing amounts than investment in fossil fuel power plants. In that regard, a more conducive investment climate is necessary for the renewable electricity generation industry's economics to be able to compete with that of the conventional electricity generation industry. To increase the economics of renewable electricity generation projects, among others the Government of Indonesia may consider setting more attractive tariffs on renewable electricity sold to PLN. Although the updated pricing mechanism for the sale of renewable electricity to PLN introduced under PR 112/2022 is much appreciated by the power industry, the price is essentially still capped at the Highest Reference Price for most renewable energy types.[97] The industry players hope for the implementation of Feed-In Tariff instead which would improve the profitability and economics of renewable electricity generation projects, and creates unattractive tariffs for renewables.

As noted above, RUEN's target is to have 23% renewable energy in the energy mix by 2025. At the end of 2021, the total renewable energy in the energy mix is at 13.5% (11.15 GW). Further, at current average increase of renewable electricity development of around 1 GW per annum, Indonesia is set to miss the current RUEN's target. It is therefore an imperative for the Government of Indonesia to facilitate a more conducive investment climate and appropriate policies to accelerate and safeguard the transition from fossil fuel energy to renewable energy.

Notes

1 Central Statistics Agency (*Badan Pusat Statistik*) <https://www.bps.go.id/indikator/indikator/view_data_pub/0000/api_pub/UFpWMmJZOVZlZTJnc1pXaHhDV1hPQT09/da_01/1> accessed 22 January 2022.
2 Central Statistics Agency (*Badan Pusat Statistik*) <https://www.bps.go.id/indicator/12/1886/1/jumlah-penduduk-hasil-proyeksi-menurut-provinsi-dan-jenis-kelamin.html> accessed 22 January 2022.
3 The Ministry of Energy and Mineral Resources, '*Capaian kinerja sektor ESDM tahun 2021 & rencana 2022*', 19.
4 Academic Script (*Naskah Akademik*) of bill for Renewable Energy Law, 4.
5 Institute for Essential Services Reform, 'Indonesia Clean Energy Outlook: Reviewing 2018, Outlooking 2019', 8.
6 Institute for Essential Services Reform, 'Indonesia Clean Energy Outlook: Tracking Progress and Review of Clean Energy Development in Indonesia', 7.
7 *Ibid.*
8 Institute for Essential Services Reform, 'Indonesia Energy Transition Outlook 2021: Tracking Progress of Energy Transition in Indonesia', 10.
9 The Ministry of Energy and Mineral Resources, '*Capaian kinerja sektor ESDM tahun 2021 & rencana 2022*', 17.
10 Institute for Essential Services Reform, 'Indonesia Clean Energy Outlook: Reviewing 2018, Outlooking 2019', 9.
11 Institute for Essential Services Reform, 'Igniting a Rapid Deployment of Renewable Energy in Indonesia: Lessons Learned from Three Countries', 16.
12 Institute for Essential Services Reform, '*Akses Energi yang Berkelanjutan untuk Masyarakat Desa: Status, Tantangan, dan Peluang*', 12.
13 *Ibid.*
14 German Watch, 'Global Climate Risk Index 2021', 5.
15 German Watch, 'Global Climate Risk Index 2021', 41.
16 Directorate General of Climate Change Control (*Direktorat Jenderal Pengendalian Perubahan Iklim*), '*Dampak & Fenomena Perubahan Iklim*' <http://ditjenppi.menlhk.go.id/kcpi/index.php/info-iklim/dampak-fenomena-perubahan-iklim≥ accessed 22 January 2022.
17 The Ministry of Energy and Mineral Resources, '*Capaian kinerja sektor ESDM tahun 2021 & rencana 2022*', 17.
18 Institute for Essential Services Reform, 'Indonesia Energy Transition Outlook 2021: Tracking Progress of Energy Transition in Indonesia', 16.
19 The Ministry of Energy and Mineral Resources, Pers Conference No. 286.Pers/04/SJI/2020 <https://www.esdm.go.id/id/media-center/arsip-berita/hingga-juni-2020-kapasitas-pembangkit-di-indonesia-71-gw#:~:text=Menteri%20Energi%20dan%20Sumber%20Daya,lalu%20sebesar%2069%2C7%20GW> accessed 22 January 2022.
20 Attachment I of Presidential Regulation No. 22 of 2017 concerning National Energy General Plan (*Rencana Umum Energi Nasional*), 8.
21 Government Regulation No. 45 of 2013 regarding Procedure to Execute the State Revenue and Budget as amended by the Government Regulation No. 50 of 2018 (collectively, the "**GR 45/2013**"), Article 94.
22 GR 45/2013, Article 98.
23 The Minister of Finance Regulation No. 174/PMK.02/2019 on Procedure to the Provision, Calculation, Payment, and Responsibility of Electricity Subsidy, as lastly amended by the Minister of Finance Regulation No. 178/PMK.02/2021 ("**PMK 174/2019**"), Article 2(1).
24 PMK 174/2019, Article 5(1) jo. Article 1(1).
25 The Minister of Energy and Mineral Resources ("**EMR Minister**") Decree No. 188.K/HK.02/MEM.L/2021 on Ratification of the 2021 until 2030 Electricity Supply

Business Plan of PT Perusahaan Listrik Negara (PERSERO) ("**RUPTL EMR Minister Decree**").

26 RUPTL EMR Minister Decree, V-101.

27 RUPTL EMR Minister Decree, V-101.

28 RUPTL EMR Minister Decree, VI-6.

29 Institute for Essential Services Reform, 'Igniting a Rapid Deployment of Renewable Energy In Indonesia: Lessons Learned from Three Countries', 13.

30 Presidential Regulation of the Republic of Indonesia No. 22 Year 2017 regarding the National General Energy Plan, Article 1(1).

31 RUEN, Article 1(3).

32 RUEN, Article 2.

33 RUEN Schedule 1, 61.

34 RUEN Schedule 1, 62.

35 RUEN Schedule 1, 66.

36 RUEN Schedule 1, 70–71.

37 RUEN Schedule 1, 63.

38 *Ibid.*

39 *Ibid.*

40 RUEN Schedule 1, 64.

41 *Ibid.*

42 *Ibid.*

43 *Ibid.*

44 Presidential Regulation of the Republic of Indonesia No. 97 Year 2021 regarding the Ministry of Energy and Mineral Resources ("**PR 97/2021**"), Article 20.

45 *Ibid.*

46 PR 97/2021, Article 15.

47 Presidential Regulation No. 26 of 2008 concerning the Establishment of National Energy Council and Procedure for the Selection of National Energy Council Member.

48 Minister of Environment and Forestry Regulation No. 15 of 2021 concerning the Organization and Framework of the Ministry of Environment and Forestry, Article 5.

49 The renewable electricity generation sector is relatively more capital intensive than conventional power generation. However, despite National Energy Policy's establishment in 2014, prior to March 2021 the following foreign investment limitation in the electricity sector was in place. Pursuant to President Regulation No. 44 of 2016 concerning List of Business Activities Closed for Investment and Business Activities Open with Conditions (also known as the "negative investment list"), for electricity generation projects (including renewable electricity): (i) no foreign investment shall be allowed for power generation less than 1MW;

(ii) foreign ownership is capped at 49% for power generation between 1MW and 10MW; and

(iii) foreign ownership is allowed up to 95% for power generation exceeding 10MW (except for (i) public-private partnership where 100% foreign ownership is allowed or (ii) geothermal power projects equal to or less than 10MW where 67% foreign ownership is allowed).

After much advocation, it was only in March 2021 that the foreign investment in the electricity generation sector (including renewable electricity) was relaxed.

50 Government Regulation No. 25 of 2021 concerning the Conduct of Energy and Mineral Resources Sector ("**GR 25/2021**"), Article 1 para 29.

51 Institute for Essential Services Reform, '*Akses Energi yang Berkelanjutan untuk Masyarakat Desa: Status, Tantangan, dan Peluang*', 12.

52 The Ministry of EMR Regulation 5 of 2021 concerning the Conduct of Risk Based Business License in the Energy and Mineral Resources Sector, Attachment II.A.I. (KBLI 35111 – Power Plant).

53 Government Regulation No. 22 of 2021 regarding Provision of Protection and Management of the Environment, Article 8.
54 Government Regulation No. 22 of 2021 regarding Provision of Protection and Management of the Environment, Article 6(2).
55 Institute for Essential Services Reform, '*Akses Energi yang Berkelanjutan untuk Masyarakat Desa: Status, Tantangan, dan Peluang*', 3.
56 Minstry of Energy and Mineral Resources, *'Permen ESDM Nomor 38 Tahun 2016: Upaya Pemenuhan Kelistrikan Daerah Terpencil'* <https://www.esdm.go.id/id/media-center/arsip-berita/permen-esdm-nomor-38-tahun-2016-upaya-pemenuhan-kelistrikan-daerah-terpencil> accessed 22 January 2022.
57 Institute for Essential Services Reform, '*Akses Energi yang Berkelanjutan untuk Masyarakat Desa: Status, Tantangan, dan Peluang*', 3.
58 Institute for Essential Services Reform, '*Akses Energi yang Berkelanjutan untuk Masyarakat Desa: Status, Tantangan, dan Peluang*', 4.
59 Electricity Law, Article 33 jo. GR 14/2012, Article 39 and Article 40.
60 PR 112/2022,Article 5(1)
61 PR 112/2022, Article 5(3)
62 PR 112/2022, Article 7
63 PR 112/2022, Article 8(1)
64 PR 112/2022, Article 8(3)
65 PR 112/0222, Article 6
66 PR 112/2022, Attachment I
67 *Ibid.*
68 EMR Minister Regulation 38/2016, Article 20.
69 EMR Minister Regulation 38/2016, Article 21.
70 Kontan, '*Perpres Harga Listrik EBT Terbit, Begini Tanggapan Asosiasi PLTMH*' <https://industri.kontan.co.id/news/perpres-harga-listrik-ebt-terbit-begini-tanggapan-asosiasi-pltmh> accessed 11 October 2022
71 MOF Regulation 130/2020, Article 2.
72 PR 112/2022, Article 27(2)
73 PR 112/2022, Article 27(3)
74 PR 112/2022, Article 27(4)
75 PR 112/2022, Article 6(2)(c)
76 PR 112/2022, Article 14(2)(b)
77 PR 112/2022, Article 3(4)
78 PR 112/2022, Article 3(1)
79 PR 112/2022, Article 3(9)
80 PR 112/2022, Article 3(10)
81 PR 112/2022, Article 22(2)
82 PR 112/2022, Article 23(2)
83 PR 112/2022, Article 23(3)
84 PR 112/2022, Article 23(4)
85 PR 112/2022, Article 23(5)
86 PR 112/2022, Article 23(6)
87 PR 112/2022, Article 23(7)
88 PR 112/2022, Article 23(8)
89 PR 112/2022, Article 23(9)
90 PR 112/2022, Article 23(11)
91 PR 112/2022, Article 23(12)
92 GR 14/2012, Article 25(4).
93 PR 112/2022, Article 14(1).
94 EMR Minister Regulation 50/2017 as amended by EMR Minister Regulation 4/2020, Article 26A.

95 EMR Minister Regulation 10/2017, Article 4 jo. Article 1(6).
96 EMR Minister Regulation 10/2017, Article 15.
97 PR 112/2022, Article 5 jo. Article 6.

Bibliography

Academic Script (Naskah Akademik) of bill for Renewable Energy Law, 4.

Attachment I of Presidential Regulation No. 22 of 2017 concerning National Energy General Plan (*Rencana Umum Energi Nasional*), 8.

Central Statistics Agency (Badan Pusat Statistik), 'Jumlah Penduduk Hasil Proyeksi Menurut Provinsi dan Jenis kelamin (Ribu Jiwa), 2018–2020' <https://www.bps.go.id/indicator/12/1886/1/jumlah-penduduk-hasil-proyeksi-menurut-provinsi-dan-jenis-kelamin.html> accessed 22 January 2022.

Central Statistics Agency (Badan Pusat Statistik), 'Luas Daerah dan Jumlah Pulau Menurut Provinsi, 2021' <https://www.bps.go.id/indikator/indikator/view_data_pub/0000/api_pub/UFpWMmJZOVZlZTJnc1pXaHhDV1hPQT09/da_01/1> accessed 22 January 2022.

Directorate General of Climate Change Control (Direktorat Jenderal Pengendalian Perubahan Iklim), 'Dampak & Fenomena Perubahan Iklim' <http://ditjenppi.menlhk.go.id/kcpi/index.php/info-iklim/dampak-fenomena-perubahan-iklim> accessed 22 January 2022.

Electricity Law, Article 33 jo. GR 14/2012, Article 39 and Article 40.

EMR Minister Regulation 10/2017, Article 15.

EMR Minister Regulation 10/2017, Article 4 jo. Article 1(6).

EMR Minister Regulation 38/2016, Article 20.

EMR Minister Regulation 38/2016, Article 21.

EMR Minister Regulation 50/2017 as amended by EMR Minister Regulation 4/2020, Article 4(1) GR 14/2012, Article 25(4).

EMR Minister Regulation 50/2017 as amended by EMR Minister Regulation 4/2020, Article 26A.

EMR Minister Regulation 50/2017 as amended by EMR Minister Regulation 4/2020, Article 5, 6, 8, 9, and 12.

German Watch, 'Global Climate Risk Index 2021', 41.

German Watch, 'Global Climate Risk Index 2021', 5.

Government Regulation No. 22 of 2021 regarding Provision of Protection and Management of the Environment, Article 8.

Government Regulation No. 22 of 2021 regarding Provision of Protection and Management of the Environment, Article 6(2).

Government Regulation No. 25 of 2021 concerning the Conduct of Energy and Mineral Resources Sector ("GR 25/2021"), Article 1 para 29 the Ministry of EMR Regulation 5 of 2021 concerning the Conduct of Risk Based Business License in the Energy and Mineral Resources Sector, Attachment II.A.I. (KBLI 35111 – Power Plant).

Government Regulation No. 45 of 2013 regarding Procedure to Execute the State Revenue and Budget as amended by the Government Regulation No. 50 of 2018 (collectively, the "GR 45/2013"), Article 94.

GR 45/2013, Article 98 the Minister of Finance Regulation No. 174/PMK.02/2019 on Procedure to the Provision, Calculation, Payment, and Responsibility of Electricity Subsidy, as lastly amended by the Minister of Finance Regulation No. 178/PMK.02/2021 ("PMK 174/2019"), Article 2(1).

Institute for Essential Services Reform 'Indonesia Clean Energy Outlook: Reviewing 2018, Outlooking 2019', 8.

Institute for Essential Services Reform, 'Akses Energi yang Berkelanjutan untuk Masyarakat Desa: Status, Tantangan, dan Peluang', 12.

Institute for Essential Services Reform, 'Akses Energi yang Berkelanjutan untuk Masyarakat Desa: Status, Tantangan, dan Peluang', 3.

Institute for Essential Services Reform, 'Akses Energi yang Berkelanjutan untuk Masyarakat Desa: Status, Tantangan, dan Peluang', 4.

Institute for Essential Services Reform, 'Igniting a Rapid Deployment of Renewable Energy In Indonesia: Lessons Learned from Three Countries', 16.

Institute for Essential Services Reform, 'Igniting a Rapid Deployment of Renewable Energy In Indonesia: Lessons Learned from Three Countries', 13.

Institute for Essential Services Reform, 'Indonesia Clean Energy Outlook: Tracking Progress and Review of Clean Energy Development in Indonesia', 7.

Institute for Essential Services Reform, 'Indonesia Clean Energy Outlook: Reviewing 2018, Outlooking 2019', 9.

Institute for Essential Services Reform, 'Indonesia Clean Energy Outlook 2022: Tracking Progress of Energy Transition in Indonesia: Aiming for Net-Zero Emissions by 2050', 38.

Institute for Essential Services Reform, 'Indonesia Energy Transition Outlook 2021: Tracking Progress of Energy Transition in Indonesia', 10.

Institute for Essential Services Reform, 'Indonesia Energy Transition Outlook 2021: Tracking Progress of Energy Transition in Indonesia', 16.

Katadata, 'Rancangan Perpres EBT Rampung, Begini Rincian Harga Jual Listrik EBT' <https://katadata.co.id/happyfajrian/ekonomi-hijau/61554a9bd37ab/rancangan-perpres-ebt-rampung-begini-rincian-harga-jual-listrik-ebt> accessed 22 January 2022.

Minister of Environment and Forestry Regulation No. 15 of 2021 concerning the Organization and Framework of the Ministry of Environment and Forestry, Article 5.

Minstry of Energy and Mienral Resources, 'Permen ESDM Nomor 38 Tahun 2016: Upaya Pemenuhan Kelistrikan Daerah Terpencil' <https://www.esdm.go.id/id/media-center/arsip-berita/permen-esdm-nomor-38-tahun-2016-upaya-pemenuhan-kelistrikan-daerah-terpencil> accessed 22 January 2022.

MOF Regulation 130/2020, Article 2.

PMK 174/2019, Article 5(1) jo. Article 1(1).

PR 97/2021, Article 15.

Presidential Regulation No. 26 of 2008 concerning the Establishment of National Energy Council and Procedure for the Selection of National Energy Council Member.

Presidential Regulation of the Republic of Indonesia No. 22 Year 2017 regarding the National General Energy Plan, Article 1(1).

Presidential Regulation of the Republic of Indonesia No. 97 Year 2021 regarding the Ministry of Energy and Mineral Resources ("PR 97/2021"), Article 20.

RUEN Schedule 1, 61.

RUEN Schedule 1, 62.

RUEN Schedule 1, 63.

RUEN Schedule 1, 64.

RUEN Schedule 1, 66.

RUEN Schedule 1, 70–71.

RUEN, Article 1(3).

RUEN, Article 2.

RUPTL EMR Minister Decree, V–101.

RUPTL EMR Minister Decree, V–101.

RUPTL EMR Minister Decree, VI-6.
The Minister of Energy and Mineral Resources ("EMR Minister") Decree No. 188.K/HK.02/MEM.L/2021 on Ratification of the 2021 until 2030 Electricity Supply Business Plan of PT Perusahaan Listrik Negara (PERSERO) ("RUPTL EMR Minister Decree").
The Ministry of Energy and Mineral Resources, 'Capaian kinerja sektor ESDM tahun 2021 & rencana 2022', 19.
The Ministry of Energy and Mineral Resources, 'Capaian kinerja sektor ESDM tahun 2021 & rencana 2022', 17.
The Ministry of Energy and Mineral Resources, Pers Conference No. 286.Pers/04/SJI/2020 <https://www.esdm.go.id/id/media-center/arsip-berita/hingga-juni-2020-kapasitas-pembangkit-di-indonesia-71-gw#:~:text=Menteri%20Energi%20dan%20Sumber%20Daya,lalu%20sebesar%2069%2C7%20GW> accessed 22 January 2022.

Part 4

North and South America

11 Brazil

Clarissa Emanuela Leão Lima, Luiz Gustavo Kaercher Loureiro, and Eduardo G. Pereira

11.1 Introduction

Despite the continental dimension of the Federative Republic of Brazil (Brazil), its electrical system was thought to be interconnected; the so-called National Interconnected System (in Portuguese *Sistema Interligado Nacional* – SIN). The history of the development of this sector demonstrates that the government wanted to implement a system that would meet local supply deficiencies and, at the same time, rationalise the production of a given region, considering the centralisation of production systems.[1] Thus, as a rule, the Brazilian electrical system was structured in a nationally interconnected arrangement, composed of four subsystems (South, Southeast/Midwest, Northeast, and North).

Considering that the rule of the electric sector is the interconnection of the system, represented by the SIN, which means approximately 99.4% of generation and consumption, the isolation of loads and generation is an exception, which, in the Brazilian case, is called the isolated system (in Portuguese *Sistema Isolado* – SISOL). In other words, even though it may make sense to think about off-grid solutions in specific situations,[2] those are not regarded as an end in themselves but as ways to make the supply viable until a future connection occurs.

It should be emphasised that low demographic densities characterise the regions where the isolated systems are located. Therefore, even though they represent a significant extension of the Brazilian territory, their non-connection to the system does not mean a failure in the universalisation of electric energy services.

The expansion of the transmission system allowed the country to benefit from better weather conditions spread across the subsystems. However, these benefits do not reach isolated systems, where most of the generation comes from fossil fuels, such as diesel oil. It was only in 2009, seeking to improve the procurement competition in Isolated Systems, that specific legislation was created to provide better supply in those areas.

As will be evidenced in this section, what can be observed in the Brazilian case is that security of supply and universalisation has been the electricity sector's priority. The greenhouse gas emissions' (GGEs) subsidiary importance can be explained because the energy sector is responsible for only 19% of the country's emissions, and only 12.9% of this amount comes from electricity generation.[3,4]

DOI: 10.4324/9781003178088-15

In addition, and as anticipated, off-grid consumption represents only 0.4% of the system, so even though it is eminently made up of polluting sources, it represents little of the system. However, it is worth pointing out that despite this low representation, the mitigation of greenhouse gases in this system also needs to be addressed.

Therefore, considering the two pillars of the off-grid renewable electricity (OGRE) – energy security and climate change[5] – there is no extensive discussion in the country on this topic. Ole (2020)[6] explains that OGRE refers to renewable energy generation close to the load, which is distributed directly, without connection to a national grid. Thus, it can be seen that discussions about this form of supply – necessarily environmentally sustainable – in the country are very recent. It is relevant to mention this issue to justify why there has never been the structuring of any national plan or defined goal for OGRE since the electric sector is considered 'clean' as a whole.

Notwithstanding the above, what has been observed in recent years is that the country has been trying to improve SISOL's supply conditions, seeking to create incentives for the implementation of renewable sources. The inclusion of renewable sources in isolated systems could mean a two-way gain for consumers living in these locations: (i) generation sources with lower greenhouse gas emissions; (ii) lower generation costs since petroleum diesel generation is considerably expensive.

11.2 Policy Targets and Framework for Off-Grid Renewables

11.2.1 Brief Context of the Brazilian Electric Sector

The Brazilian National Interconnected System Development: the Sectorial Rule

The implementation of the system interconnection dates to the 1950–1960 decade.[7] Through Decree No. 41,019 of 1957, mechanisms were created to direct the planning of the system towards its integration. This initiative consisted in: (i) the attribution of competence to the National Council of Water and Electric Energy (in Portuguese *Conselho Nacional de Águas e Energia Elétrica* – CNAEE) to promote the interconnection and rationalisation of the system; (ii) the creation of a specific chapter to discipline how the centralisation of production would take place, emphasising an effort to unify the electric system.

In the 'Centralisation of Production' chapter, it was explained that the centralisation of the generating systems could occur through the following processes: (i) 'by the simple interconnection of two or more generating systems of different entities, maintaining each one its criteria of operation and freely applying its energy availability'; (ii) 'by integration under a single ownership and centralized operation, both of generation and transmission, as well as of the distribution of energy to consumers'; and (iii) 'by coordinating generation and transmission operations for the wholesale supply of Distribution System Operators (DSO) for its consumers' (Article 149, I, II, and III).

The interconnecting effort of the electricity system increased over the years. In 1996, a structuring reform resulted from the hiring of a consulting firm to promote sector competition[8] that led to creating the New Model of the Electric Sector (in Portuguese *Novo Modelo do Setor Elétrico* – NMSE). The reform proposed creating the National System Operator (ONS) ,[9] an institution responsible for coordinating and controlling the generation and transmission operation in the interconnected system (article 13) to face these challenges and enable the process of the NMSE. Thus, ONS, a private law entity formed by stakeholders in the electricity sector and authorised by the National Agency of Electric Energy ('ANEEL' or just 'Energy Agency')[10] – assumed National Interconnected System operation.[11]

As anticipated, the result of this interconnection effort is that, by 2019, the SIN represented about 99.4% of Brazilian generation and consumption.[12]

11.2.2 The Isolated System – the Off-Grid Part of the Electric Sector: the Sectorial Exception

As briefly mentioned in the previous section, the Brazilian electrical system was structured to be interconnected. On the other hand, the isolated systems are the exception to the sectorial rule that results from the absence of technical or economic conditions of connecting these localities to the SIN.[13] Recent data from the Energy Research Office (in Portuguese *Empresa de Pesquisa Energética* – EPE) indicates that there are currently 271 isolated systems spread throughout the country, representing about 40% of the national territory. According to the last 'Isolated Systems Planning Report, 2024 Horizon – Cycle 2019' published by EPE, of the 271 existing systems, 45 are expected to become interconnect until 2024 and 13 after this period. These systems can be found in the North (where they are concentrated) and Midwest regions, and Fernando de Noronha Island.

Although the sizeable territorial extension of isolated systems brings the idea that the Brazilian electricity sector has failed in its universalisation, it is essential to point out that these regions are characterised by low demographic density, representing about 760,000 consumers, only 0.6% of the consumers (EPE, 2021).

Unlike the National Interconnected System,[14] in isolated systems, thermoelectric generation with diesel oil predominates, representing 96% of the installed capacity, followed by 2.2% natural gas, 1.1% biomass, and 0.7% of hydric power plants (EPE, 2020). On this matter, it is critical to point out that, despite the environmental impact caused by the emissions of these thermal power plants, the choice for this type of enterprise stems from the logistics involved in its implementation. As the northern Brazilian region is characterised by logistical difficulty, being the primary mode of transport, often the waterway, there are high costs related to the acquisition and displacement of fuels from the producing centres – often near the coast – to the generation plants. Even so, as diesel oil generation is considered simple to implement, DSO usually prefers to celebrate power purchase agreements (PPAs) with these suppliers (Souza, Calili, and Faria, 2017).[15] Additionally, this type of enterprise has 'flexibility to keep up with variations in energy consumption and can operate for long periods continuously',[16] and its operation is already known mainly in isolated systems with nearby spare parts suppliers. Thus, the resulting situation is that, despite the ease and low cost of maintaining these generation facilities, the fuels used increase operating costs, making this system highly dependent on sector subsidies (Ponte, 2019). Hence, renewable off-grid electricity solutions in these regions would be an appropriate alternative for climate change mitigation and also for the promotion of affordable energy.

Considering all the above, and that the existence of SISOL is incidental and not purposeful, the Ministry of Mines and Energy (MME),[17] supported by the Energy Research Office,[18] is responsible for isolated systems planning and integration. The Energy Research Office is responsible for evaluating the proposals for service of the isolated systems presented by the DSO accountable for supply in these locations, which must contain, among others, information about the prediction or unfeasibility of interconnections with other isolated systems or with the National Interconnected System.[19] The EPE also supports MME in sector planning, elaborating studies of the expansion of the transmission system, which may state if a specific isolated system connection to SIN[20] will be possible or not.

11.3 Regulatory Environment

As already stated, the isolated system represents the off-grid part of the Brazilian electric system and mainly consists of generation from fossil sources. As this off-grid generation is expressed very little in the country, only lately has there been an attempt to increase the participation of renewable sources in these locations. Until recently, the focus of public policy was on the population's access to electricity and universalisation. The search for electricity quality and efficiency has begun to be addressed in recent years, as discussed further below.

An example of this focus on universalisation was the creation of the National Program for the Universalization of Access and Use of Electric Energy (in Portuguese *Luz para Todos*) in 2003, through Decree No. 4.873. This public policy tried to address the accessibility to electricity at isolated systems even before implementing a legal framework to make it more competitive.

This programme, which will be implemented until 2022, aims to provide electric energy services to the Brazilian rural population that still does not have access to this public service. This programme is destined to be an existing subsidy in the electric sector (the so-called energy development account, in Portuguese *Conta de Desenvolvimento Energético* – CDE[21]) to fund universalisation and created a governance structure to manage these resources. The *Luz para Todos* is in force until 2022. It is estimated (data from 2018) that about 16.2 million residents of rural areas of the country had access to electricity services through the programme.

First Brazilian OGRE Programme: More Light for the Amazon

It is important to briefly mention the creation of a complementary new programme to *Luz para Todos*, called the National Program for Universalization of Access and Use of Electric Energy in the Legal Amazon – More Light for the Amazon (in Portuguese *Mais Luz para a Amazônia*), established by Decree No. 10.221/2020. Different from its predecessor, this programme is aimed explicitly at SISOL in the Amazon region and intends to increase the participation of renewable sources in these locations by using CDE resources, among others, to finance the implementation of renewable sources in isolated systems and to promote the substitution of fossil sources in these systems.

The participation of the DSOs in the Amazon region is mandatory, and recently, through Normative Resolutions 940/2021 and 2891/2021, ANEEL established the programme's targets to be met by the relevant public service providers. Because it is very recent, there is still no consolidated information about the advances of this public policy.

Returning to the specific regulation of isolated systems, Law No. 12,111/2009, the first particular norm to deal with SISOL resulted from the conversion of Provisional Measure No. 466/2009 (PM 466).[22] The Statement of Justification No. 00033 (SJ) of the Ministry of Mines and Energy and the Ministry of Finance, which followed the PM 466, pointed out that there was a legal gap in the treatment of these systems, which until that date had not received specific legal treatment.[23]

The SJ that followed PM 466 presented the Executive Branch's reasons for applying a particular law to rule SISOL. Between those objects, they emphasised: (i) the high costs of electricity in the isolated system as a whole and (ii) the risk of tariff increase by consumers that could follow every isolated system interconnection.

The SJ stressed a relevant economic and financial disparity in electrical activity in isolated systems, pointing out that the tariff structure would not cover the operating costs in this system. One of the factors responsible for the high costs of operation was mainly diesel oil to produce electricity. To solve the challenges identified, the SJ, and consequently the PM, proposed two solutions:

i) that the contracting of isolated system supply sources should follow a competitive procedure or regulated auctions. If the bidding procedure does not have participants, a regulation should be defined to ensure publicity and transparency for the DSO to celebrate PPAs directly.
ii) The Fuel Consumption Account should subsidise the energy parity costs of the isolated systems with the SIN.[24]

i. Isolated System Competitive Procurement

Until the enactment of Law 12.111/09, Distribution System Operators were responsible for contracting the supply of their consumers. Thus, until the implementation of the contracting rules of the SISOL legal framework, the DSO's preference for the security of supply was seen in fossil fuel power plants, which became predominant. As stated by Ole (2019), this is not exclusive to Brazil, as in other countries, such as China, fossil fuel generation 'have accumulated "benefits of experience" with usage, which translates into a cheaper initial capital project cost'.[25]

However, it is relevant to mention that, even with the new explicit rules, it is still possible for DSO's to directly celebrate PPA in the face of specific scenarios, embracing transparent procedures to do so. This aspect is relevant to understand why, even after the competition promotion, oil projects were still the majority of PPA due to DSO's preferences, such as their familiarity, as mentioned above. The result of this predominance of fossil sources is clear: it is estimated that greenhouse gas emissions (GGEs) by SISOL are about 7.4 times higher than National Interconnected System emissions (Ponte, 2019).

In 2010, Decree No. 7,246/2010 was issued to regulate the provisions of Law No. 12,111/2009. ANEEL gained the responsibility to (i) define the quality requirements of supply for isolated systems; and (ii) seek economic and energy efficiency for isolated system agents to mitigate the impact on the environment and achieve the economic sustainability of the generation (articles 3 and 4). In addition, it would be up to the sector regulator to also carry out – directly or indirectly – the bids for the contracting of all SISOL suppliers by the Distribution System Operators (articles 7 and 8), following the Ministry of Mines and Energy guidelines.

Decree No. 7,246/2010 highlights the role of distribution agents in the isolated systems, who must submit annually to the Ministry of Mines and Energy the plan of service to their markets for a five-year horizon. In line with PM 466's SM regarding the reduction of tariffs for isolated systems agents, article 8 of the decree demands that (i) competitive procedures should consider the lowest total cost of generation over the contractual horizon, including the costs of investment, operation, and maintenance of supply solutions (§ 4); (ii) the bidding should seek to reduce the total cost of generation in SISOL and the need for reimbursement by CCC (§ 8º), and (iii) Ministry of Mines and Energy would define the maximum bid price for each bid (§ 7). Article 9 details the procedures that MME will implement if the auction is unfeasible or has no participants. If those hypotheses occur, Ministry of Mines and Energy should authorise: (i) the DSO's direct procurement, limited to the maximum value proposed by the Energy Research Office and approved by the MME; (ii) addition of the supply PPA already signed (until July 30, 2009) for the increase of quantity and deadline; and (iii) emergency contracting of energy and electric power of a selling agent or electric power generation units' rental for operation directly by the DSO, through a public call to be made by an agent appointed by the Ministry of Mines and Energy.

It is also relevant to highlight some points in which Decree No. 7,246/2010 brought provisions that possibly hindered the hiring of renewable sources in isolated system auctions, which have only recently been corrected. First, until 2017, the Decree had a provision that predicted that Distribution System Operators would create 'Reference Projects', which were proposals for an electricity supply solution to guide the bid. The Reference Projects ended up directing the recommendations to choose thermoelectric plants,[26]especially diesel oil. The offer 'targeting' happened because:

a. the Distribution System Operators proposed this sort of project given that it was the type of solution they were familiar with (in terms of safety of supply, ease of implementing, and acquisition of replacement parts, as already remarked). Here is essential to highlight that in the last instance, the DSO could be responsible for the implementation of the solution itself, being an extra incentive for Reference Projects to be of a source that DSO's dominate the technical operation;
b. the bid participants had incentives to propose thermoelectric diesel oil projects to be closer to the Reference Project. This oil preference would happen because it would increase the project's probability of winning the tender.

Decree No. 9,047, 2017 revoked the existence of the Reference Project, and the effect of this change was felt in Auction No. 001/2019, where 14% of the hired capacity in the bid was diesel oil, and in Auction No. 003/2021, with 34% of oil diesel solutions contracted.

The second aspect from Decree No. 7,246/2010 that allowed the prevalence of fossil fuel power plants at SISOL was the express authorisation for DSO's to directly celebrate PPA if the auction remains desert or supply is unfeasible. The proposal to allow distributors to contract directly, even with transparent procedures, is an exceptional measure that occasionally happened. When it occurred, these agents have opted for the solution that they considered safer and more familiar, which were diesel oil thermoelectric plants (Ponte, 2019).

In addition, in 2010, the Ministry of Mines and Energy issued Ordinance No. 600 to define the guidelines that the Energy Agency should follow for the SISOL auctions, updated by MME Ordinance No. 67/2018. Those Ministry of Mines and Energy ordinances established the information that Distribution System Operators must present to the Energy Research Office to guide the off-grid system planning for the subsequent five years (article 4 of MME Ordinance No. 600/2010[27] and article 3 of Ordinance MME No. 67/2018[28]). Based on this planning, the MME defines whether it will be necessary to conduct auctions for the SISOL supply. With that information in hand, Ministry of Mines and Energy allows Granting Power to more appropriately structure the next steps that should be followed in these locations.

As the auctions for isolated system supply became mandatory – through Resolution No. 1,733/2014 – ANEEL approved the public bidding rules for SISOL auctions and its annexes. In this rule, the Regulatory Agency delegated its realisation to the DSO, and Resolution No. 2,172/2016 later updated those rules.

Two highlights should be made about this regulatory act. The first focus concerns the proposed methodology for calculating the reference price at the first version of the norm. This calculation methodology should be emphasised because it considered the remuneration of diesel oil thermoelectric generation projects.

Besides the Fixed Annual Revenue (in Portuguese *Receita Anual Fixa* – RAF), the other components are typically from oil diesel power plants (Ponte, 2019).[29] Thus, agents interested in investing in alternative sources, other than diesel oil, would not have a formula that meets the specificities of their generation form by that time, which evidences a regulatory asymmetry in the treatment of different sources at the beginning.[30]

The second highlight is about the Bidding Document rule's improvement. In the SISOL Bidding Document, there is a provision that allowed renewable generation solution employment after the bid. This provision seems to be an ANEEL attempt to implement its competence in mitigating the impact on the environment and the pursuit of the economic sustainability of electricity generation (Article 4, Decree No. 7,246/2010). That said, the resolutions approved those requirements that contained a willingness to encourage hybrid solutions, even after the competitive process. Take the example of this provision in Bidding Document No. 001/2014:

> 1.1.2 The generation of electricity and the availability of power provided for in the REFERENCE PROJECT(S) or the ALTERNATIVE(S) PROJECT(S) will be done through the construction, operation, and maintenance of generating power plant(s). They may include the use of renewable, fossil, or mixed-use of sources and technologies.
>
> 1.1.2.1 After the AUCTION and upon prior approval of the respective project(s) by the Granting Authority, SELLER may add to the central(s) generator(s) equipment(s) a renewable energy source generation, as well as the use of other fuels. The minimum amounts of power and energy established in item 1.1 of this Notice are guaranteed.
>
> 1.1.2.2 The use of the faculty referred to sub-item 1.1.2.1 may imply a partial reduction in the REFERENCE PRICE. (highlighted).

Subsequent public bidding rules – Auction No. 010/2015, No. 002/2016, No. 01/2019, and 003/2021 – also brought these provisions. In the latter ones, there was an improvement in the rule. This modification predicted that, in cases of change in technical characteristics to include renewable source generation equipment, it would be possible to reduce the expected operating costs, obeying the sharing of the gains in the 70%–30% ratio between seller and buyer, respectively.

Thus, to regulate this proposition that would enable the increase of renewable sources in isolated systems, even after the auction, in 2020, the National Agency of Electric Energy launched the Public Consultation No. 67 (CP 67). CP 67 aimed 'to obtain subsidies to the Regulatory Impact Analysis Report - AIR and Regulatory Act Draft regarding the regulation of criteria for adding a renewable source in diesel power plants in isolated systems'.

In the Regulatory Impact Analysis Report (AIR) that followed the CP 67, the Energy Agency explained that, of the 148 locations served by the auctions that took place between 2014 and 2018, there was the implementation of only one 'Alternative Project'. This project was the Oiapoque Thermal Power Plant, sold at Auction No. 001/2014, where the Oiapoque Photovoltaic Plant (U.F.V. Oiapoque) was installed. The implementation of U.F.V. Oiapoque occurred after the entrepreneur consulted ANEEL in 2016, obtaining the support of the Superintendence of Regulation of Generation Services to pursue the project. In this case, through National Agency of Electric Energy Order No. 428/2017, the Regulatory Agency allowed the photovoltaic generating unit with 4.3 MWp, which would be accounted for at the same contractual delivery point of the thermic power plant with the energy price unchanged.

Thus, through CP 67, ANEEL proposed a standard to regulate 'minimum criteria for adding renewable sources to diesel plants in Isolated Systems to support the evaluation by all interested selling agents'. The idea was that, based on clear rules, it would be possible to encourage such agents to implement these extra renewable solutions. The AIR analysed four methodology alternatives to account for the adding of renewable sources on the reference price of the enterprise, adopting alternative 2 in the regulatory Act draft presented for discussion. In this alternative, it was proposed that in cases of addition of renewable solution for more than five years, the reduction of the reference price until the end of the contract in R$/MWh would follow this formula:

$$\text{Reference price reduction} = 30\% \times \text{Estimated Benefit}$$

$$\text{Estimated Benefit} = \frac{\text{Avoided fuel cost - Photovoltaic energy} \times \text{Renewable Energy Produced}}{\text{Power Purchase Agreement required energy}}.$$

If the addition of the renewable solution happens closer to the end of the PPA (less than five years), there would be no reduction in the reference price of the contract.

This public consultation ended in January 2021. Until this date, the new regulation did not come into force.

ii. Fuel Consumption Account

The Fuel Consumption Account was a subsidy created to share fuel costs among the SIN agents in 1973, by Law 5.899. In 2009, with Law 12.111, the CCC became a subsidy for DSO to adjust SISOL's high operating costs.

The CCC had already been assigned for isolated systems under Law 8631/1993, which established that the apportionment of the cost of fuel consumption for the generation of electricity in isolated systems would be 'extended to all distributors' (art. 8). The innovation of the new law was that the amount to be reimbursed by the distribution agents would be the difference between the total cost of generation in isolated systems and the amount corresponding to the average cost of energy and energy commercialisation in the SIN (art. 3).

Therefore, it demonstrates that Brazilian policymakers realised that it was necessary to subsidise off-grid solutions of supply, which was also faced in other countries in remote areas.[31] However, at that time, and until recently, there was no comprehension of the importance of structuring sustainable criteria to prioritise renewable forms of generation in the region.

The CCC destination proposed by Law n. 12.111 provided an isonomy between the National Interconnected System and the isolated systems costs. This change was relevant to reducing the tariffs experienced by SISOL consumers, which were high due to the fuel consumed by the generating plants in the region. However, the levelling made by the new standard did not bring incentives or criteria to promote sustainable sources for the transfer of the CCC.

Even with the establishment of competitive procedures for selecting supply solutions for isolated systems, the contracting continued primarily of thermal power plants. Thus, despite the competition mechanisms, the contractors continued to have fossil fuel power plants, with the new model for passing on costs to the system. Therefore, the generation costs and the emissions were still high after the new rules.

According to current auction rules, there are no mechanisms that price externalities associated with GGE emissions. Instead, the costs associated with operating expensive thermal projects are subsidised. In this way, competitive procedures led to the contracting of oil-fired thermal plants whose fuel cost is subsidised and whose emissions are not priced, making them artificially competitive, considering their actual social costs. Thus, despite the good intention of the Brazilian legislator, the cost neutrality between SIN and SISOL ended up making the DSO neutral as to the expenses related to its purchase of electricity.

From the above, it is verified that it is necessary to seek ways to encourage the adoption of renewable solutions, also considering the guarantee of the reliability of isolated systems, which is a relevant factor and of supreme importance for consumers. Also, there must be a redesign of the SISOL subsidies and incentives. It is necessary to review the total transfer costs of contracting expensive and polluting sources to the system. Mechanisms to address this issue may result from the proper pricing of emissions, enhancing the competitiveness of renewable sources.

iii. Possible Alternatives

1. DISTRIBUTED GENERATION

In addition to the division between National Interconnected System and isolated systems, the Brazilian electricity sector is also divided into a free contracting environment (in Portuguese *Ambiente de Contratação Livre* – ACL) and a regulated contracting environment (in Portuguese *Ambiente de Contratação Regulada* – ACR). At ACL, large consumers can purchase their electricity directly from suppliers. At ACR, the consumers are called 'captive consumers' and must obtain their energy from the DSO responsible for hiring the electricity supplier.

Since 2012, through ANEEL Resolution No. 482, the prosumer figure was created, which allowed the captive consumer to produce their energy until 5 MW (in this case, the consumer would be called prosumer). This alternative is the so-called distributed generation (in Portuguese *Geração Distribuída* – GD). According to the 482 Resolution, the prosumer can generate via qualified cogeneration or renewable sources, always connected to the Distribution System Operators network. In 2015, Energy Agency Resolution No. 687 expanded the arrangements that configured GD, allowing the prosumer to share the renewable generation installed with other consumers if organised in consortium or cooperative (kinds of legal purchase group organisations).

In a distributed generation arrangement, consumers use the distributor's system as a battery, injecting any surplus portion generated, which becomes a credit that can be compensated when it demands more energy than produced. Most distributed generation consuming units opted for photovoltaic generation because it is a source of easy installation due to its interface with buildings in urban areas.

GD brings multiple benefits to the system, such as (i) low environmental impact and diversification of the energy matrix; (ii) the possibility of postponing transmission and distribution system expansion; and (iii) the chance of minimising the losses of the distribution system. However, after studies carried out by ANEEL in 2019, the Agency concluded that it would be necessary to restructure the incentives provided to distributed generation. The proposed amendment balances the costs of using the distribution network and energy charges by the prosumers, who do not pay the electric consumer charges entirely. ANEEL showed that there is still the adoption of the DSO network in the distributed generation arrangements, but due to the current incentive system, the costs related to this use are not fully recovered from these users and are transferred for non-distributed generation consumers. For this reason, the Agency proposed a transitional period for the withdrawal of the benefits granted to the arrangement not to become excessively burdensome to other consumers, who incorporate the costs of GD.

ANEEL did not implement its transition proposal because the Federal Executive Branch requested not to, stating that it was a public policy decision over which the Agency would not have competence. After that interference, the discussion is ongoing on the Federal Legislative Branch, which intends to create a law to decide how long the GD incentives will be allowed.

Despite discussions about the sustainability of distributed generation in the way it currently occurs, its rapid adoption by consumers is notorious. Since distributed generation began to be implemented in 2012, it grew quickly and right now represents around 5.9 GW of installed power in the country (ANEEL, 2021). The states with the highest installed capacity of consumers are Minas Gerais, followed by São Paulo and Rio Grande do Sul.

There are still no consistent studies that explain the spreading of distributed generation in Brazil; however, tax incentives may explain the large concentration of this type of investment in Minas Gerais. Minas Gerais is the state with more beneficial taxation for PV acquisition and surplus energy. Moreover, the purchasing power of the population living in these locations can also explain more prosumers. Despite the constant reduction in the value of solar panels, this type of project still requires high investment.

Thus, and considering that: (i) the average annual energy yield of the Northern region of the country is 1,354 kWh/kW per year,[32] which would allow the exploitation of photovoltaic sources in the area; (ii) that such a source is sustainable and would allow the gradual replacement of oil thermoelectric plants, it would be an excellent initiative to have incentives for the development of these sources of supply in isolated systems. This type of incentive can also come from the redesign of public policy discussed in the Legislative Branch,[33] allowing consumers who live in isolated systems to keep subsidies currently in force for longer to enable distributed generation development in these regions.

It is also important to mention that a pilot project will allow implementing a micro-grid system in the country. This model may serve as a support for similar solutions in isolated systems, which Distribution System Operators already operate.

This initiative is a regulatory sandbox authorised by ANEEL in September 2020, through ANEEL Authorisation No. 9,224, which allowed Paranaense Energy Company (in Portuguese *Companhia Paranaense de Energia* – Copel) to open a public call for the acquisition of energy from distributed generation. Copel is the DSO of Paraná, a state in the country's southern region, and to date, it has not opened this public call.

The authorisation allows Copel to create micro-grids based on Decree No. 5.163/2004 to improve the reliability and continuity of supply to consumers that are part of the electrical sets involved. The contracts signed may have a duration of up to 5 years, and the contracted amount may vary between 1 MW and 30 MW.[34] The agreements signed must be presented to ANEEL, demonstrating that they represent the lowest effective cost to the consumers served, which the National Agency of Electric Energy will verify.

Although this proposal is for Distribution System Operators located in the country's southern region, this model can be implemented by DSO located in isolated systems to ensure more reliability and lower emissions to generation based on distributed generation. This alternative can even make the monetisation of this project feasible and encourage its development in isolated systems. It is relevant to see how Copel's public call will unfold and to what extent this milestone will inspire similar initiatives in other locations in the country, notably in SISOL.

2. ENERGY RESEARCH OFFICE AND NATIONAL AGENCY OF ELECTRIC ENERGY PROPOSAL: HYBRID SYSTEMS

As explained by SJ 466, the connection of an isolated system to the National Interconnected System is economically justified 'when the operating and expansion cost curves reach levels that justify investments in transmission for interconnection'. Thus, for the interconnection to occur, it must be verified that the economic viability of a specific SISOL has been economically saturated, offsetting the system that costs the expansion of the National Interconnected System.

Therefore, and due to its legal competencies, the Energy Research Office constantly evaluates the feasibility of connecting an isolated system to instruct the Ministry of Mines and Energy and supports in the planning of the transmission grid expansion. In addition to its competence to help the Granting Power in the planning of the Brazilian electrical system, EPE also develops studies to promote the use of renewables in the isolated system. This effort resulted in three studies:

i) Technical Note in 2018 to analyse the *'Energy Potential of Forest Waste of Sustainable Management and Waste of Wood Industrialization* (EPE-DEA-NT-17/2018-r0)'.[35]
ii) Technical Note in 2014 to analyse *'Hybrid Systems with photovoltaic energy for Lot III of the Eletrobras Acre Distribution Reference Project* (EPE-DEE-NT-027/2014-r0)'.[36]
iii) Technical Note in 2016, to analyse *'Solar Energy for Supply of Isolated Systems of the Amazon* (EPE-DEE-NT091/2016-r0)'.[37]

The three Energy Research Office studies concluded that the combination of thermoelectric fossil fuels projects and renewable sources, mainly photovoltaic (PV), would be the most promising solution. The results of studies exploring such solutions showed more significant benefits to the system, so ANEEL, through CP 67 mentioned above, seeks to address this type of solution.

The 2014 and 2016 EPE studies concluded that hybrid systems could be a relevant solution for SISOL, as they could reduce fuel consumption by up to 26% and reduce the energy price by 9% on average (EPE, 2016). Those studies emphasised that if fossil fuel costs keep increasing and photovoltaic equipment prices decline in the future, the price difference could be more significant (EPE, 2014 and EPE, 2016), as the Energy Research Office considered conservative scenarios in its analysis.

Thus, with the regulation that will emerge from CP 67 proposed by ANEEL, it will be more transparent and safer for current PPA generators in the isolated system to understand the benefits of adding renewable solutions. The increase of hybrid enterprises can be very beneficial to the isolated systems because (i) it can reduce GGE by reducing or partially replacing fossil fuel generation; and (ii) it can reduce generation operating costs by reducing fuel consumption.

11.4 Conclusion

Three lessons can be learned from all the elements analysed in this section. From each one of the lessons, a recommendation will be made.

First, it is undeniable that the subsidies proposed had a vital role in electricity accessibility for isolated system consumers. Nevertheless, there must be a redesign of the subsidies granted to price the emissions of gases that cause the greenhouse effect. Even if it is relevant to compare the price of energy between SIN and SISOL consumers, negative externalities must be measured in the contracting of isolated systems.

Second, and as already mapped by the public stakeholders, there should be more explicit rules for hybrid solutions. The addition of a renewable power plant could represent a relevant reduction in GGEs. The proper encouragement of this 'extra' renewable generation off-grid could be a good solution for the short term, as the current generators can implement those solutions without waiting for new bids. This aspect is already under discussion at ANEEL.

Still, it is essential to pay attention to the strategy adopted by the regulator, as this could be an excellent opportunity to replace part of the energy already produced by operating fossil fuel power plants. Some specialists believe that even if the generator receives 100% of the benefit from the renewable added unit, it would positively benefit the system (Ponte, 2019). Therefore, the recommendation is that stakeholders pay attention to how ANEEL will regulate this aspect, ensuring that the proper incentives result from the public discussions.

The third lesson is that distributed generation could be a solution to be explored more fully at the isolated systems. Distributed generation is still in its infancy in the isolated system, but it could be a low-cost, clean, and less complex way of implementing off-grid renewable sources in Brazil. It is not known for sure the reason for the low existence of prosumers in SISOL, but the low purchasing power of the population and few incentives at the local level may be possible reasons.

As the generated distribution could represent a reduction of greenhouse gas emissions and allow more capacity variety available for isolated system consumers, the system should consider the possibility of maintaining the subsidies of generated distribution for off-grid consumers. Another alternative would be the implementation of special taxation as an incentive for consumers to become prosumers.

In conclusion, Brazilian OGRE is inexpressive, representing 4% of the isolated system that signifies only 0.6% of the country's energy consumption. The location of the isolated systems explain its predominant thermal electric energy composition, but the implementation of legislation without consistent incentives for renewable sources demonstrates why isolated system kept this way. There are relatively simple solutions to make Brazil have a more renewable off-grid system. However, it demands the attention of policymakers in all power spheres: federal and state levels.

Notes

1 Brazil Decree No. 41,019/1957, article 149. (República Federativa do Brasil, 'Decree n. 41019' <http://www.planalto.gov.br/ccivil_03/decreto/antigos/d41019.htm> accessed 13 June 2021).

2 "*The most suitable models will probably depend much on the specific circumstances. When delivered through an established grid, the cost per MWh is cheaper than of mini-grids or off-grid solutions, but the cost of extending the grid to sparsely populated, remote or mountainous areas can be very high and long distance networks can have high technical losses. This results in grid extension being the most suitable option for all urban zones and for around 30 % of the rural areas, but it is not cost-effective in more remote rural areas. IEA estimates that 70 % of rural areas should be connected*

either with mini-grids (65 % of that 70 %) or with small, stand-alone off-grid solutions (the remaining 35 %). These stand-alone systems have no transmission and distribution costs, but higher costs per MWh." Pérez-Arriaga IJ, Regulation of the Power Sector, vol 61 (2013).

3 'Emissões Por Atividade Econômica - Sankey | SEEG - Sistema de Estimativa de Emissão de Gases' <https://plataforma.seeg.eco.br/sankey> accessed 13 June 2021.

4 'Emissões Por Atividade Econômica - Sankey | SEEG - Sistema de Estimativa de Emissão de Gases' <https://plataforma.seeg.eco.br/sankey> accessed 13 June 2021.

5 "*The need for the development of off-grid renewable electricity (OGRE) is anchored on the twin pillars of energy security and climate change.*" (Ole NC, 'The Nigerian Electricity Regulatory Framework: Hotspots and Challenges for Off-Grid Renewable Electricity Development' (2020) 38 Journal of Energy & Natural Resources Law 367)

6 "The term OGRE refers to the production of renewable electricity from smaller electric plants at or near the point of sale, and its distribution is direct to users without connecting to a national grid". (Ole NC, 'The Nigerian Electricity Regulatory Framework: Hotspots and Challenges for Off-Grid Renewable Electricity Development' (2020) 38 Journal of Energy & Natural Resources Law 367)

7 Gustavo Kaercher Loureiro, Instituições de Direito Da Energia Elétrica: ProPedêutica e Fundamentos, Volume I (Quartier Latin 2020).

8 This idea of introducing competition in the electricity sector preceded legal attempts to enable the participation of private agents in the provision of public services and public interest services by creating the General Concessions Law in 1995 (Law n. 8987/1995) and the creation of the independent producer of electricity in 1996 (Law n. 9074/96).

9 Article 13. The activities of coordination and control of the operation of the generation and transmission of electricity in the interconnected systems will be carried out by the National Operator of the Electric System, a legal entity of private law, with authorization of ANEEL, to be integrated by concession holders, permission or authorization and consumers to which the arts. 15 and 16 of Law No. 9,074 of 1995.

10 ANEEL was created by Law No. 9,427/1998.

11 Mercedes SSP, Rico JAP and Pozzo LDY, 'Uma Revisão Histórica Do Planejamento Do Setor Elétrico Brasileiro' [2015] Revista USP 13.

12 Empresa de Pesquisa Energética, 'Planejamento Do Atendimento Aos Sistemas Isolados - Horizonte 2024 - Ciclo 2019' (Empresa de Pesquisa Energética 2020) <http://www.epe.gov.br> accessed 13 June 2021.

13 'III - Isolated Systems: electrical systems of public service distribution of electricity that, in their normal configuration, are not electrically connected to the National Interconnected System - SIN, for technical or economic reasons'. (Brazil Decree No. 41,019/1957 on electric energy regulation, art 2nd).

14 According to the National Energy Balance, produced by EPE in 2020, the Brazilian energy matrix in 2019 was composed of: 64.9% hydraulic; 9.3% natural gas; 8.6% wind; 8.4% biomass; 3.3% coal and derivatives; 2.5% nuclear; 2% oil derivatives; 1% solar. (Empresa de Pesquisa Energética, 'Brazilian Energy Balance' (Empresa de Pesquisa Energética 2020)).

15 The authors better explain that "*When asked about the lack of projects based on renewable sources, the distribution companies frequently complain about a possible lack of reliability of new technologies in such far away communities. Another stated reason was an article of Decree n° 7246 saying that in case of an auction without bids the DSO should execute its own project. As these companies only had experience with Diesel, they preferred not to take risks*". Ponte GP da; and others, 'Electricity Generation in Isolated Systems: Challenges and Suggestions for Increasing the Share of Renewable Sources' (2018) <https://proceedings.science/p/85545>.

16 da Ponte GP, Calili RF and Souza RC, 'Energy Generation in Brazilian Isolated Systems: Challenges and Proposals for Increasing the Share of Renewables Based on a Multicriteria Analysis' (2021) 61 Energy for Sustainable Development 74.

17 *Art. 2"Minister of State for Mines and Energy is responsible for the formulation, direction and implementation of national policy in matters related to mines and energy"* (República Federativa do Brasil, 'Law n. 4904' <http://www.planalto.gov.br/ccivil_03/leis/1950-1969/l4904.htm> accessed 13 June 2021).

18 Public company, subordinated to the Ministry of Mines and Energy, with the competence to "*provide services in the area of studies and research aimed at supporting the planning of the energy sector, such as electricity, oil and natural gas and its derivatives, coal, renewable energy sources and energy efficiency, among others*", 2nd article (República Federativa do Brasil, 'Law n. 10.847' <http://www.planalto.gov.br/ccivil_03/_ato2004-2006/2004/lei/l10.847.htm> accessed 13 June 2021).

19 "*Article 3° Until June 30 of each year, distribution agents must submit to the Ministry of Mines and Energy, through the Energy Research Company - EPE, proposal to plan the service of their respective consumer markets located in Isolated Systems for the five-year horizon, from the following year.*

§ 2 - The proposal for planning the service of consumer markets in Isolated Systems shall contain at least the following information:

XII - the prediction of interconnections with other Isolated Systems or with the SIN;XIV - the demonstration of the technical, economic or environmental infeasibility of the interconnection of isolated systems to the SIN; and". (Ministério de Minas e Energia, 'Portaria No 67, de 1o de Março de 2018' <https://www.in.gov.br/web/guest/materia/-/asset_publisher/Kujrw0TZC2Mb/content/id/5093329/do1-2018-03-02-portaria-n-67-de-1-de-marco-de-2018-5093325> accessed 13 June 2021.)

20 Article 4. "*VII - prepare studies necessary for the development of plans for the expansion of the generation, and transmission of electricity in the short, medium and long term.*" (República Federativa do Brasil, 'Law n. 10.847' <http://www.planalto.gov.br/ccivil_03/_ato2004-2006/2004/lei/l10.847.htm> accessed 13 June 2021).

21 Subside created by Law 10.438/2002.

22 This law "*provide rules for electricity services in Isolated Systems; amends Laws 9,991, July 24, 2000, 9,074, July 7, 1995, 9,427, December 26, 1996 10,848, of March 15, 2004; revokes provisions of Laws Nos. 8631, of March 4, 1993, 9,648, of May 27, 1998, and 10,833, of December 29, 2003; and makes other arrangements.*"

23 "7. It should also be emphasized that isolated systems did not receive specific treatment, even when the recent regulation of the electricity sector carried out by Law nº 10,848, of March 15, 2004, and decree no 5,163 of July 30, 2004. In this way, modern and obsolete institutional relations coexist, such as contracts for the supply of electricity in the mold of Law no. 8,631, of March 4, 1993, the energy of Independent Producers of Electric Energy - PIE, according to Law no 9,074 of July 7, 1995, or individuals exploring small generators and micro-power plants."

24 Created by Law 5,899/1973, CCC was structured to allow the apportionment of fuel costs used, not SIN, see: 'Article 13. The main objective of the operational coordination referred to in the previous article shall be to use the rational use of existing generating and transmission facilities that come into existence in the interconnected systems of the South-East and South regions, while also ensuring: III - that the burdens and advantages arising from the consumption of fossil fuels, to meet the needs of interconnected systems or by the imposition of national interest, be prorated among all concessionaires of those systems, according to criteria that the Executive Branch will establish'. (República Federativa do Brasil, 'Law No. 5,899/1973' <http://www.planalto.gov.br/ccivil_03/leis/l5899.htm#:~:text=L5899&text=LEI%20N%C2%BA%205.899%2C%20DE%205%20DE%20JULHO%20DE%201973.&text=Disp%C3%B5e%20sobre%20a%20aquisi%C3%A7%C3%A3o%20dos,ITAIPU%20e%20d%C3%A1%20outras%20provid%C3%AAncias.> acessed 13 June 2021).

25 About the competition between renewable and fossil fuel sources at China, the author also that: "*A factor that contributes to the lower capital cost of fossil fuel electricity projects is their "accumulation of benefits" over years of experience with usage and subsidies. Given the priority provision, existing public funds for the support of the electricity sector are to be applied to OGRE projects first of all before consideration is given to fossil fuel electricity options.*" (Ole NC, 'The Role of Renewable Energy Law 2005 in Supporting the Development of Off-Grid Renewable Electricity in China The Financial Securities for Decommissioning of Offshore Installations in Nigeria: A Review of The Legal and Contractual Regime View Project' [2019] International Energy Law Review).

26 See: "Article 2. I - Reference Project: description of electricity supply solution to serve consumers of isolated systems proposed by the local distribution agent, to be prepared according to guidelines of the Ministry of Mines and Energy;"

27 "*Article 4° Until December 1 of each year, Distributors with Isolated Systems should forward to EPE the planning of the service of their consumer markets in these Systems, for the minimum horizon of five years, from the year subsequent. (Writing by PRT MME 493, of 08.23.2011)*

§ 2 - The EPE will analyze the planning of customer market service in Isolated Systems, and may, to do so, request additional information and documents from distributors.

§ 3 - The planning of the service of consumer markets in Isolated Systems approved by the MME, based on the technical analysis of the EPE, will be forwarded to ANEEL for the preparation of the schedule of the necessary Auctions.

§ 4 - To ensure the service of the expansion of consumer markets, the distributor's planning should explain:

I - the amounts of electricity and associated power to be incorporated into each consumer market of isolated systems; and

II - the corresponding to the electricity supply solutions proposed by the distribution agent.

III - the need to contract capacity reserves that deal with the sole paragraph of Article 3".

28 Those rules became more meticulous with the regulation from 2018:

Article 3° Until June 30 of each year, distribution agents must submit to the Ministry of Mines and Energy, through the Energy Research Company - EPE, proposal of planning to serve their respective consumer markets located in Isolated Systems for the horizon of five years, from the following year.

§ 2 - The proposal for planning the service of consumer markets in Isolated Systems shall contain at least the following information:

I - a brief description of the geographical aspects of the localities, including coordinates, population, subordination political administrative, forms of access;

II - the historical values of the last three years and projections of consumption, losses, energy load, and demand, in the planning horizon foreseen in the caput;

III - the typical load curves and maximum demands year by year, on the planning horizon provided for in the caput;

IV - a description of the current supply of electricity generation, as well as other available supply solutions;

V - o maturity of existing contracts for the purchase of energy and power and rental of generating units;

WE - a self-generation deactivation programming;

VII - any desired replacement of existing supply;

VIII - the needs of contracting supply solution for supply expansion;

IX - as any needs to contract the generation capacity reserve that deals with Article 6, § 3, of this Ordinance, with the respective justifications;

X - a proposal for the division of lots, if the need for contracting is identified;

XI - the conditions of the distribution network, as well as the detailing of the needs of reinforcements and expansions;

XII - the prediction of interconnections with other Isolated Systems or with the SIN;

XIII - the schedule of implementation of determining distribution works;

XIV - the demonstration of the technical, economic, or environmental infeasibility of the interconnection of isolated systems to the SIN; and

XV - energy savings due to energy efficiency programs." (Ministério de Minas e Energia, 'Portaria No 600, de 30 de Junho de 2010' <www.epe.gov.br.> accessed 13 June 2021).

29 For more information, consult the 1733 ANEEL Ordinance attachment, available here: http://www2.aneel.gov.br/cedoc/areh20141733_2.pdf

30 The current rules approved have a more complex methodology that more appropriately considers other generation sources.

31 "It is a well-known problem, since almost every country has had to create and subsidise some kind of rural electrification programme to reach segments of the population that live in remote areas, are widely scattered in a territory, have low potential electricity demand needs and also low economic purchasing power". Pérez-Arriaga IJ, Regulation of the Power Sector, vol 61 (2013).

32 Information provided by ANEEL based on the Brazilian Atlas of Solar Energy (Pereira E and others, Atlas Brasileiro de Energia Solar (Universidade Federal de São Paulo 2017).

33 Brazil Legislative Bill No. 5.829/2019 that institutes the legal framework for distributed microgeneration and minigeneration, the Electric Energy Compensation System (SCEE), and makes other provisions. (Chamber of Deputies, Legislative Bill No. 5.829/2019 2019).

34 Despite the maximum amount allowed for the acquisition (30MW), DG projects under Resolution 482/2012 from 1MW up to 5MW may participate in Copel's public call. This point is expressly explained in the leading vote of ANEEL Authorization No. 9,224/2020, as transcribed below:

> "63. Thus, Copel-D's initial proposal for contracting plants between 5 MW and 30 MW should be expanded. This expansion also contributes so that generation plants that would eventually be classified as distributed mini-generation (under Normative Resolution No. 482, of 2012) can assess the possibility of participating in the call, contributing to the network, and selling energy." (National Agency of Electric Energy Vote for Authorization [2020])

35 Empresa de Pesquisa Energética, 'Potencial Energético de Resíduos Florestais Do Manejo Sustentável e de Resíduos Da Industrialização Da Madeira' (Empresa de Pesquisa Energética 2018).

36 Empresa de Pesquisa Energética, 'Avaliação de Sistemas Híbridos com energia fotovoltaica para o Lote III do Projeto de Referência da Eletrobras Distribuição Acre' (Empresa de Pesquisa Energética 2014).

37 Empresa de Pesquisa Energética, 'Avaliação Da Atratividade Econômica de Solução Híbrida Em Sistemas Do Grupo B Do Projeto de Referência Da Eletrobras Distribuição Amazonas'. (Empresa de Pesquisa Energética 2016).

References

Agência Nacional de Energia Elétrica, 'Resolução Normativa no 482, de 17 de abril de 2012'.

Chamber of Deputies, Legislative Bill No. 5.829/2019 2019.

'Emissões Por Atividade Econômica - Sankey | SEEG - Sistema de Estimativa de Emissão de Gases' <https://plataforma.seeg.eco.br/sankey> accessed 13 June 2021.

Empresa de Pesquisa Energética, *Avaliação Da Atratividade Econômica de Solução Híbrida Em Sistemas Do Grupo B Do Projeto de Referência Da Eletrobras Distribuição Amazonas* (Empresa de Pesquisa Energética 2016).

Empresa de Pesquisa Energética, *Avaliação de Sistemas Híbridos com energia fotovoltaica para o Lote III do Projeto de Referência da Eletrobras Distribuição Acre* (Empresa de Pesquisa Energética 2014).

Empresa de Pesquisa Energética, *Potencial Energético de Resíduos Florestais Do Manejo Sustentável e de Resíduos Da Industrialização Da Madeira* (Empresa de Pesquisa Energética 2018).

Empresa de Pesquisa Energética, *Planejamento Do Atendimento Aos Sistemas Isolados - Horizonte 2024 - Ciclo 2019* (Empresa de Pesquisa Energética 2020) <http://www.epe.gov.br> accessed 13 June 2021.

Empresa de Pesquisa Energética, *Brazilian Energy Balance* (Empresa de Pesquisa Energética 2021).

Kaercher Loureiro G, *Instituições de Direito Da Energia Elétrica: ProPedêutica e Fundamentos*, Volume I (Quartier Latin 2020).

Mercedes S S P, Rico J A P and Pozzo L D Y, 'Uma Revisão Histórica Do Planejamento Do Setor Elétrico Brasileiro' (2015) Revista USP 13.

Ministério de Minas e Energia, 'Portaria No 600, de 30 de Junho de 2010' <www.epe.gov.br.> accessed 13 June 2021.

Ministério de Minas e Energia, 'Portaria N° 67, de 1° de Março de 2018' <https://www.in.gov.br/web/guest/materia/-/asset_publisher/Kujrw0TZC2Mb/content/id/5093329/do1-2018-03-02-portaria-n-67-de-1-de-marco-de-2018-5093325> accessed 13 June 2021.

Ole N C, 'The Role of Renewable Energy Law 2005 in Supporting the Development of Off-grid Renewable Electricity in China' (2019) International Energy Law Review.

Ole N C, 'The Nigerian Electricity Regulatory Framework: Hotspots and Challenges for Off-grid Renewable Electricity Development' (2020) 38 Journal of Energy & Natural Resources Law 367.

Pereira E and others, *Atlas Brasileiro de Energia Solar* (Universidade Federal de São Paulo 2017).

Pérez-Arriaga I J, *Regulation of the Power Sector*, vol 61 (2013).

da Ponte G P and others, 'Electricity Generation in Isolated Systems: Challenges and Suggestions for Increasing the Share of Renewable Sources' (2018) <https://proceedings.science/p/85545>.

da Ponte G P, Calili R F and Souza R C, 'Energy Generation in Brazilian Isolated Systems: Challenges and Proposals for Increasing the Share of Renewables Based on a Multicriteria Analysis' (2021) 61 Energy for Sustainable Development 74.

República Federativa do Brasil, 'Decree n. 41019' <http://www.planalto.gov.br/ccivil_03/decreto/antigos/d41019.htm> accessed 13 June 2021.

República Federativa do Brasil, 'Law n. 4904' <http://www.planalto.gov.br/ccivil_03/leis/1950-1969/l4904.htm> accessed 13 June 2021.

República Federativa do Brasil, 'Law n. 10.847' <http://www.planalto.gov.br/ccivil_03/_ato2004-2006/2004/lei/l10.847.htm> accessed 13 June 2021.

República Federativa do Brasil, 'Law No. 5,899/1973' <http://www.planalto.gov.br/ccivil_03/leis/l5899.htm#:~:text=L5899&text=LEI%20N%C2%BA%205.899%2C%20DE%205%20DE%20JULHO%20DE%201973.&text=Disp%C3%B5e%20sobre%20a%20aquisi%C3%A7%C3%A3o%20dos,ITAIPU%20e%20d%C3%A1%20outras%20provid%C3%AAncias> acessed 13 June 2021.

12 Mexico

María Serna, Ngozi Chinwa Ole, and Izuoma Egeruoh-Adindu

12.1 Introduction

The Secretariat of Energy (*Secretaría de Energía*, 'SENER', for its initials in Spanish) reports that in December 2016, the Mexican electrification rate was 98.58%.[1] The situation has currently improved to 99.4% with just about 1.8 million people lacking access to electricity.[2] Providing access to these people has been challenging for the government because they are resident in remote rural areas characterised by the type of topography and sparse population that makes grid electrification impracticable.[3] The impracticability is partly because such extension would involve prohibitive expenses which translate to high electricity tariffs which most remote rural residents cannot afford.[4] It is estimated that the percentage of lack of access to electricity will increase to over 20% given the projected population growth if measures are not adopted to address the problem.[5] Access to electricity is the sturdy basis for the development of such remote rural areas.[6] In the words of Bezerre et al., 'once a community has access to electricity, it can also have access to safe potable water, better health conditions, food security, as well as lighting and information'.[7] Conversely, the lack of access to electricity undermines the development of the areas without it and, the residents cannot access basic essential facilities. Considering that the population that lacks electricity access is also in isolated places with difficulty accessing the grid, the Federal government has proposed to install isolated supply systems (off-grid renewable electricity – OGRE) consistent in solar panels and storage.[8]

The population that has access to electricity often experience disrupted power supply occasioned by the nature of the Mexican electricity grid which makes it susceptible to damage arising from extreme weather conditions such as thunderstorms and windstorms.[9] The inter-connectivity of the grid makes its infrastructure harder to be maintained, and a breakdown in a small unit will invariably affect a larger part. Regrettably, given the wide spread of the national grid, blackouts and power disruption have become frequent in Mexico. Sometimes, it is occasioned by a shortage in gas supplies and transmission losses.[10] Such power blackouts normally range from a few seconds to days.[11] Unsurprisingly, decentralised electricity technologies including OGRE are growing in popularity as backups during such periods of blackout.[12]

DOI: 10.4324/9781003178088-16

Furthermore, Mexico is a signatory to the Paris Agreement 2015 and thus has the commitment to adopt measures to mitigate climate change including in the energy sector. The electricity sector in Mexico is thermal based being founded on coal and gas electricity technologies. Addressing climate change entails a shift from fossil fuel electricity strategies to the development of clean energy sources including OGRE. However, the development of OGRE in Mexico has been seriously constrained by several factors. While there are laws adopted by the government in the area of electricity and climate change (i.e., the Electric Industry Law (*Ley de la Industria Eléctrica*) as amended in 2021; the General Law on Climate Change 2012, and the Energy Transition Law 2015. There is no literature analysing the extent to which these laws are suited to adequately support the development of OGRE in Mexico. But existing literatures evidence that the status of renewable energy including OGRE is well below the need for it,[13] Thus indications that perhaps the law is not adequate in supporting its development.

In the light of the above, this chapter analyses the extent to which relevant laws, as mentioned above, support the development of OGRE in Mexico. It argues that while the relevant laws and policies support the development of OGRE in some respects, several other factors whittle down their effectiveness in this context. In the first instance, it identifies that the barriers to its development include lack of affordability and accessibility of initial capital, opposition from local communities, and lack of political will. It argues that while the government has adopted targets for the development of clean energy, the absence of a national target for renewable energy, and OGRE, inhibits the development of the latter. The requirement for payment of application permit fees under the *Ley de la Industria Eléctrica* was argued to heighten the problem of funding for OGRE projects. The Environmental Impact Assessment process was argued to work against the development of renewable electricity projects including OGRE given the lack of non-priority for it, lengthy process, and its exclusion in arable lands. This chapter bolsters confidence in the existence of some support schemes for OGRE projects under the law including the Universal Electric Fund and bank loans, which will help investors overcome the financial barriers in the sector. However, such support has been whittled down by uncertainties introduced by inconsistencies in the pecuniary support given and lack of political support for renewable electricity. Such lack of political will to implement the relevant laws has also undermined the loans and grants issued by some banks for the purpose of supporting OGRE projects. The chapter is concluded with recommendations on how the law can be strengthened to support the optimal development of OGRE.

The chapter is divided into several sections including the first section which is the introduction. Section 12.2 contains the imperatives and barriers to OGRE development. Section 12.3 and 12.4 discuss the policy targets and regulatory environment for OGRE, respectively. Section 12.5 is the conclusion.

12.2 Imperatives and Barriers to OGRE Development

The point has been made that OGRE is needed in the context of addressing energy security in Mexico. While a chunk of the population have access to electricity, about 1.8 million do not.[14] Even for the ones that have, power blackouts and disruption imply that residents, households, and businesses could spend days without electricity.[15] The situation is exacerbated by high electricity prices and low household income which reduces the electricity that the population can access at any one time.[16] It is for this reason that research has pointed out that about 13.5 million people in Mexico are wallowing in electricity poverty.[17] Most of this population are in remote rural areas, where OGRE remains the most viable option of electrification with regards to cost and suitability.[18] Even in urban areas, OGRE technologies are used by some businesses and households as backup in such periods of power blackouts.[19]

The electricity sector in Mexico is characterised by thermal gas/coal plants which constitutes 76.9% of the current mix, with hydro-plants having the remaining percentage.[20] Mexico is the world's tenth-largest oil and natural gas producer, thus justifying the heavy reliance on fossil fuels for energy generation.[21] However, the population continues to expand exceeding the available fossil fuel. As a result, Mexico now imports natural gas with antecedent volatility of such supply. In the words of Gonzalez-Lopez, 'natural gas imported mostly from the United States had volatile prices, considerably increasing its cost, and putting the combined cycle generation at risk'.[22] In contrast, renewable energy is abundant in every nook and cranny of the country.[23] It has also become cost competitive with fossil fuel electricity. Bloomberg opines that 'the cost of producing an energy unit with solar or wind systems in Mexico today is equal to that of a combined cycle gas plant'.[24] The United States Department of Energy National Renewable Energy Laboratory (NREL) concludes 'that, in light of this potential and the low cost of renewable energy generation, Mexico is ideally poised to become a clean energy powerhouse'.[25] In particular, OGRE has the potential to facilitate universal access to electricity in Mexico.[26]

Similarly, climate change is another imperative for the development of OGRE. Mexico is the second largest emitter of greenhouse gases (GHGs) in Latin America and the Caribbean.[27] As stated, Mexico is a signatory to the Paris Climate Change Agreement 2015.[28] Pursuant to this Agreement, it has adopted a National Determined Contribution (NDC) which commits to achieving a 25% reduction in the emission of greenhouse gases below the business-as-usual level in the country by 2030.[29] The power sector contributes significantly to the emission of GHGs in Mexico.[30] Renewable energy, including OGRE, offers the greatest potential for mitigating climate change in the country. As such, the optimal development of OGRE in the context of where it is needed remains the way forward. The country is also extremely vulnerable to the adverse effects of climate change and renewable energy, including OGRE, can help in cushioning it.[31] Regardless of this, the NDC does not mention renewable energy including OGRE as an option for reducing the emission levels of the country. It is for this reason that Rennkamp opines that 'Renewable energy plays a relatively small role in the Mexican NDC'.[32] Thus, the government of Mexico cannot afford to pay lip service to its pledges under the Paris Agreement 2015. The Mexican Law on Climate Change 2012 (amended in 2020) contains plans for the general development of clean energy.[33] The latter should serve as the basis for the development of OGRE.

The Mexican electricity sector including OGRE is privatised implying that private investment is crucial for the development of the sector.[34] As such, the sector must be investor friendly without which the optimal development of the sector may not be attainable. Regrettably, the full potential of renewable energy including OGRE has not been realised, especially in the context of climate change and energy security.[35] The limited development of renewable energy including OGRE is a result of the existence of barriers, including having decades of public support for fossil fuel which continually puts it first over and above renewable energy.[36] There is also the problem of accessing the needed capital for OGRE development.[37] Government is also struggling with the needed political will to support the development of renewable energy including OGRE.[38] These barriers whittled down the development of OGRE and will not provide an opportunity for an energy mix that will provide a future to electrify the unelectrified population in Mexico.

Fossil fuel electricity has enjoyed, and continues to enjoy, government support and experience with usage which makes it a preferred option for investors. The Mexican electricity sector is founded on fossil fuel options, i.e., thermoelectric plant powered by coal and natural gas in 1879.[39] Since then, residents and government have become conversant with its use even as the technologies have evolved progressively.[40] Besides the experience with usage, the mentioned fossil fuel technologies continually benefit from government grants and subsidies which translates to cheaper cost.[41] While the government attempts to support renewable energy, the prolonged support for fossil fuel electricity options will continually privilege the latter over the former. It is for this reason that some law makers in Mexico have 'argue [*sic*] for a review of the current subsidy scheme that promotes fossil fueled energy rather than renewable energy'.[42]

Another problem is the affordability or accessibility of the capital needed for OGRE projects. Most of the renewable electricity projects in Mexico are on-grid with a few OGRE projects, together constituting 23.5% of the electricity mix.[43] Irrespective of the type of renewable electricity, it is capital intensive and not easily affordable without recourse to some sort of external funding.[44] The two methods of funding electricity projects are debt with recourse to shareholders and debt without recourse (project financing).[45] While accessing funds through the two methods is not a problem for major actors, it is for medium and small investors.[46] Regrettably, such major actors are often more interested in fossil fuel or on-grid renewable electricity projects which have more profit and lower investment risks than off-grid. It is the medium and small investors that are interested in OGRE but they often do not have the necessary requirements to access the debts.[47] In the words of the International Renewable Energy Agency (IRENA), 'Some non-utility and small-scale power projects experience difficulties in qualifying for or negotiating reasonable finance despite the fact that installing these systems creates major energy savings'.[48]

'Equally lacking is the public and political drive to comprehensively develop a national renewable power industry'.[49] Prior to the current administration, Mexico was renowned for being proactive and ambitious in adopting climate change mitigation measures including the development of renewable energy. For instance, the country was the first to submit their climate change plan ahead of the Paris Climate Change Agreement 2015. However, the present administration has adopted a pro-fossil fuel stand while antagonising the development of renewable energy. State electricity companies own and run fossil fuel electricity projects while private investors are the major stakeholders in the renewable energy sector.[50] The current government has adopted a policy that urges the Mexican State Productive Enterprise, *Comisión Federal de Electricidad* (CFE) to prioritise the purchase of electricity from the state-owned companies that produce fossil fuel electricity over and above cheaper renewable electricity alternatives.[51] The latter is because the administration sees renewable energy sources as a competition with the CFE. This led to proposals for a constitutional amendment to prevent the use of renewables like OGRE in other areas in order to promote a monopoly in the power sector.[52] Again, the government is investing in developing gasoline refineries instead of investing in renewable energy including OGRE.[53] In the words of Flannery,

> In summary, Mexico is quickly moving in the wrong direction on climate action by not seeking to further decarbonize its power sector and enhance deployment of ever-cheaper sources of domestic, renewable energy. It is not just bucking but almost completely ignoring global trends.[54]

12.3 Policy Targets

It is apt to mention that the Electric Industry Law (*Ley de la Industria Eléctrica*, 'LIE' for its initials in Spanish) does not explicitly contain a target for the OGRE sector. However, it provides that Federal government will promote the electrification of rural communities and marginated urban zones.[55] LIE did not say how this can be attained. Given the imperatives for the development of OGRE in remote rural areas and urban area as backup, one can argue that it is implied that the government has the obligation to promote it.[56] However, such argument is watered down greatly by the presence of other known electricity alternatives such as natural gas and coal. In the absence of an express target for the development of renewable electricity including OGRE, there was no impetus for the government to abandon such fossil fuel electricity option for a riskier one.

However, the Law on Climate Change 2012 and the Energy Transition Law 2015 remedies the gap in the LIE by providing for clean energy targets including renewable electricity. The Climate Change Law sets a target for the integration of clean energy in the electricity mix to be 35% by 2024 and 50% by 2050, respectively.[57] The latter was reiterated in the Energy Transition Law 2015.[58] While the Climate

Change Law did not define what is clean energy, the Energy Transition Law 2015 defines it to include fossil fuel technologies that have carbon capture and sequestration technology, nuclear energy, renewable energy, efficient cogeneration plant.[59] Unlike the Electric Industry Act, the Energy Transition Law recognises expressly that renewable energy is an option in the context of climate change mitigation.

However, the absence of an express target for renewable electricity and OGRE undermines the extent to which the provisions of the Climate Change Law and Energy Transition Law will positively drive their development. IRENA recognise that the adoption of targets is a precursor for effectively promoting the development of renewable energy including OGRE.[60] For such targets to be effective, 'it must be specific, measurable, achievable, realistic, and time-bound (SMART), containing the numerics and deadline for achieving such a target'.[61] On the basis of such a target, measures are adopted and assessed for the development of OGRE.[62] In the words of Ole, 'the absence of a national target for renewable energy and sectoral targets for OGRE … whittles down the possibilities of adopting effective legal measures for addressing the identified … barriers'.[63] In the context of Mexico, there is no such benchmark for adopting and assessing legal measures for the development of OGRE or renewable electricity more generally. Thus, the development of the sector will continue to be affected until the situation is reversed. It is apt to mention that investor's confidence in the sector is undermined by a lack of serious commitment by the government to develop the sector which is evidenced by absence of sector targets for renewable electricity and OGRE.[64]

The situation is worsened by the inclusion of fossil fuel in the definition of clean energy. While renewable electricity including OGRE is cheaper than fossil fuel with carbon capture device, the accumulated decade of experience with the latter's usage and government's pro-nationalisation stance in supporting it, naturally makes it a more preferable option. Therefore, it is not surprising that the government has mandated CFE to prioritise fossil fuel electricity from the state company over renewable electricity including OGRE. Notably, carbon capture and sequestration at its best captures at most 90% of GHG emissions.[65] However, there are also some carbon capture technologies that are not efficient for this purpose.[66] As such, there is no provision that aims to guarantee that any of such carbon capture technologies to be used by state-owned enterprises would be fit for purpose.

Irrespective of the abovementioned absence of targets, commentaries and policy documents emphasise that achieving universal access to electricity, and achieving the targeted level of mitigation of climate change, entails the optimal development of renewable electricity including OGRE in the context where it is needed.[67] While their position may not augment for the lack of sectoral targets for the duo, it may ultimately persuade government to adopt such targets.

12.4 Regulatory Environment

12.4.1 Definition of Off-Grid

Off-grid generation is outlined in the LIE as 'isolated supply' (*abasto aislado* in Spanish) with the following definition: 'the generation or import of electricity to satisfy own needs or for export, without transmitting the energy through the National Transmission Network or the General Distribution Networks'.[68] In that sense, we identify two main characteristics of Mexican off-grid generation: (a) the energy generated is for own/private consumption and sale and (b) the energy cannot be transmitted through the transmission or distribution grids.[69]

The Energy Regulatory Commission published Agreement No. A/049/2017[70] ('Own needs Agreement') to interpret the definition of 'own needs' that applies to off-grid generation. The Agreement defines the concept of *own needs* as 'the energy consumed by the load centers of the *same natural or legal person*, or, of a group of these that belong to the same economic interest group'.[71] For the off-grid generator, this implies that the permit holder must be (a) the natural or legal person who consumes the energy; (b) one of the people who make up the economic interest group; or, (c) the generator if it belongs to the same economic interest group. For the energy consumers, this means that they must be incorporated into the same legal person as the generator, or into other companies that belong to the same economic interest group. This requirement is difficult to execute in rural or marginated zones, where it is difficult for the population to coordinate themselves or have the requirements to be incorporated into legal entities or companies. This is proof that the current regulation for off-grid renewable energy generation permits is not fit for addressing energy poverty in rural or marginated zones in their entirety.

12.4.2 Licensing Regime for Electricity Projects

The regulatory framework consists of (a) Electric Industry Law known as *Ley de la Industria Eléctrica* (LIE) in Spanish; (b) Electric Industry Law known as *Ley de la Industria Eléctrica* (LIE) Regulations; and (c) Own Needs Agreement. Additionally, depending on the nature of the project, social and environmental regulations will apply. As mentioned before, the own-needs restriction for isolated supply generation permits, inhibit the development of off-grid renewable energy for the population's benefit and it appears to be designed to consider business and industrial uses. If the generation is over 0.5 MW, it requires a permit from the Energy Regulatory Commission (*Comisión Reguladora de Energería*) (CRE). Such permits are issued upon the payment of an application fee. Below that 0.5 MW capacity, generators are called 'exempted generators' since they do not require a permit. Additionally, if isolated supply is generated via renewable energies, the CRE may establish metering and reporting requirements.[72] In general, electricity projects are subject to the allocation of permits/authorisations.

While the exemption may benefit stand-alone renewable electricity projects, it does not confer any added benefits as per privileging OGRE over competing fossil fuel electricity options. Exempted generators include all categories of off-grid generation. Fossil fuel electricity options of 0.5 MW and below are exempted.[73] Such exempted fossil fuel electricity options is already preferable to OGRE because of accumulated benefits of experience with usage and decades of subsidies that make its initial capital outlay cheaper than the latter. So, an exemption that benefits both genres of off-grid electricity will ultimately not be beneficial to OGRE if it does not level the playing field between it and fossil fuel electricity technologies.

What is more, from an investor's perspective, a mini-grid renewable electricity project is more profitable than a stand-alone project because it uses economies of scale and has more customers. On the other hand, stand-alone electricity projects which are 0.5 MW and below are mostly executed by consumers who are interested in generating electricity for their personal consumption.[74] As earlier stated, most rural dwellers are unable to afford the initial cost of such stand-alone projects. In which case, private and public investors are anticipated to assist by developing OGRE projects for sale to consumers.[75] Given that such investors are driven by profit, they are naturally attracted to mini-grid renewable electricity as opposed to stand-alone because of its cheaper cost and greater profit. For instance, mini-grid renewable electricity projects compose of a larger chunk of OGRE capacity in Mexico.[76] Regrettably, most mini-grid electricity projects are more than 0.5 MW in capacity and unable to benefit from this exemption.[77] Conversely put, investors in mini-grid renewable electricity projects would still have to obtain a permit after payment of application fees. The additional cost of permits will heighten the problem of affordability and accessibility of initial capital cost of OGRE projects.

12.4.3 Environmental Impact Statement

Off-grid projects with a generation over 3 MW[78] require the submission of an Environmental Impact Statement (EIS) before the Secretariat of Environment and Natural Resources (*Secretaría de Medio Ambiente y Recursos Naturales*, 'SEMARNAT' for its initials in Spanish) in terms of the Regulations of the General Law of Environmental Balance and Environmental Protection in matters of Environmental Impact Assessment (*Reglamento de la Ley General del Equilibrio Ecológico y la Protección al Ambiente en Materia de Evaluación de Impacto Ambiental*, 'LGEEPA Regulations'). The Environmental Impact Assessment (EIA) process normally starts with the renewable energy developer submitting an EIS usually alongside a permit application in the context of where same is needed. In response, the Secretariat of Environment and Natural Resources (SEMARNAT) may accept such EIS unconditionally, partly, or reject.

While it is recognised that renewable energy has a crucial role to play in environmental protection and preservation, there is no privilege accorded to it under the EIA process. In fact, renewable energy project developers are expected to pass through the same process of EIA as would a fossil fuel electricity project.[79] It is reported that renewable energy projects including OGRE go through a lengthier process of EIA in comparison to fossil fuel because of limited experience with it by SEMARNAT personnel.[80] The situation has become exacerbated by SEMARNAT's restrictive attitude towards some renewable energy projects.[81] For instance, off-grid solar electricity projects of certain magnitude are not allowed in arable land. It is apt to mention that Mexico is filled with arable land.[82]

12.4.4 Social Impact Assessment

Depending on the technology used and the location of the off-grid systems, a social impact assessment may have to be submitted to the Secretariat of Energy (*Secretaría de Energía*, 'SENER' for its initials in Spanish), in terms of the General Guidelines regarding Social Impact Assessment in the Energy Sector (*Disposiciones Administrativas de Carácter General sobre la Evaluación de Impacto Social en el Sector Energético*). There have been cases where the population is against renewable energy projects, mostly due to the over-promising of energy companies that state that the population will benefit from the project, only to find out that most of the benefits are given to the investors and their related companies.[83] To prevent these discomforts, SENER should closely follow all social communications vis-à-vis the committed actions in favour of the population. However, under the current regulatory framework, SENER only foresees that the SIA is fulfilled, with no regulations as to what the companies may promise outside the SIA.

At the beginning of the current Mexican political administration, the Social Impact Assessment was the first permit to be halted. The offices to file for its approval were closed indefinitely. Since CRE requires to attach to the generation permit filing, the proof of the corresponding SIA submission before SENER, energy projects, those involving renewable energies, had their first impasse.

12.4.5 Pricing Mechanisms

OGRE power purchase is not regulated, so service providers may offer different mechanisms. However, given the case that there were one or few suppliers of off-grid electricity in each project, the Energy Regulatory Commission has the power to impose tariff regulations to avoid abuses, in case of complaints. This has not happened yet, but CRE has the authority to promote competition, regulate tariffs, and protect consumers' rights. Given that the own-needs definition requires that consumers form part of the same legal entity as the generator, or of another legal entity from the same economic interest group, projects should be careful when choosing the legal scheme for off-grid consumers, to protect consumers and to ensure that they do not lose their nature.

12.4.6 Support Schemes

The Universal Electric Service Fund (*Fondo de Servicio Universal de Energía*, 'FSUE') is a public trust fund created under Mexico's Energy Reform of 2013 to address electrification in rural communities and marginated urban zones.[84] Though the fund was developed with specific targets, with provisions for use of solar panels and storage equipment, it gave room for both grid extension and off-grid, stand-alone electrification projects to be funded by the FSUE. Thus, research has shown that between 2016 and 2019, three rounds of tenders were issued for grid extension and three for off-grid schemes.[85] However, these records were solely for solar and not other renewable energy sources. The restriction to OGRE solar alone is not helpful in the context of achieving energy security because a more diversified option would have implied a greater chance of attaining sustainable universal electrification, especially in the context of the abundance of other renewable energy sources in Mexico like bioenergy, hydro, wind, etc. Given that most rural areas are isolated and off from the electricity grid, it would have been pertinent to promote OGRE in the electrification of these rural communities and marginated urban zones.[86] However, the FSUE has a specified target which is on-grid and use of solar panels which are also not affordable to the rural dwellers and the fund may not be channelled to another course other than the one for which it was instituted. It is apt to mention that the FSUE does not support OGRE in all urban areas irrespective of the need for it as a backup to on-grid electricity.[87]

Besides the abovementioned, the FSUE has been severely affected by the lack of political will by the government to support renewable energy.[88] Thus, the fund has been suspended intermittently. For instance, throughout the year 2019, it was suspended.[89] Such suspension makes the fund unattractive to investors given the obvious lack of uncertainty in continuity.

Additionally, there are some financing programmes targeted at supporting renewable energy implementation towards the energy transition, mostly financed by *Nacional Financiera* (NAFIN), a state-owned development bank.[90] These financing programmes are of diverse nature and scale, for example, there is a specific programme to finance the purchase of photovoltaic systems with certain suppliers.[91] NAFIN is also available for large scale renewable energy projects, to support the energy transition in terms of energy reform. The programmes range from loans to green bonds. There are also other banks supporting the development of renewable energy in their own right.[92] However, such support has slowed down in recent times given government policy bearing towards intensifying support for fossil fuel electricity.

Accordingly, the Energy Transition Act provides support schemes that promote renewable energy sources including OGRE.[93] The law does this by regulating the obligations of power companies related to the mandatory share of clean energies and promotes the push for the reduction of polluting emissions of the electric power industry.[94] This is done to ensure the competitiveness of the clean power sector within the new wholesale electricity market. The law empowers the Secretariat of Energy (SENER) to promote the generation of clean power to reach the levels set forth in the Climate Change Act of 2012 (as amended in 2020) for the electric power industry, including a minimum share of clean energies in power generation of 35% by 2024.

SENER is responsible for establishing obligations for the mandatory acquisition of clean energy certificates that the suppliers and qualified users participating in the wholesale power market and the holders of interconnection agreements must be complied with on an annual basis. The issuance of clean energy certificates (CECs) is to serve a dual capacity which is both a green incentive and a regulatory mechanism to meet the country's clean energy targets.[95] A transitory provision, valid for four years allows companies to defer 50% of their CEC obligation for two years. The law also establishes the National Commission for Efficient Use of Energy (CONUEE), tasked with drafting the Energy Strategy that sets medium- and long-term goals (15 and 30 years horizon, respectively), as well as the National Programme for Sustainable Use of Energy (PRONASE) which are provided for in chapter V of the Act. The CONUEE and the PRONASE serve as the two main implementation tools of the law and support schemes.[96]

Similar to most efforts made to support OGRE, the implementation of the relevant provisions of the Energy Transition Act and Climate Change Act has been halted due to a lack of political will by the current administration to implement it. Assuming, without conceding that it is not, the support provided by the Transition Act and Climate Change Act are generally aimed at boosting the development of renewable energy, it is not tailored towards the peculiar challenges of OGRE. For instance, the obligation on CEC can be fulfilled by sourcing electricity from renewable energy. On-grid renewable electricity, though presented with its peculiar challenges, is still more profitable than OGRE from an investor perspective because it enjoys economies of scale. Thus, investors will naturally tilt towards on-grid especially as they do not have any social obligation to advance sustainability over profit.

12.4.7 Purchasing Arrangements

There are no regulations in terms of purchasing arrangements for off-grid electricity consumers. As stated before, isolated supply is only contemplated for own needs. Given the case, CRE could intervene to protect consumers' rights, provided that the legal schemes do not denaturalise their consumer position. Purchasing arrangements for photovoltaic systems are of a private nature. To address the initial capital costs, suppliers may offer financing mechanisms. However, some suppliers have stated that lease-to-own mechanisms of the photovoltaic systems proved to be ineffective. Some reasons were (a) customers did not give maintenance once the payment period was over; (b) customers had to save on their own to buy the batteries that need to be replaced regularly; and (c) once the payment is complete, customers have no incentive to move up to a larger system that could be of greater benefit to them.[97] In response, some suppliers have started offering solar-as-a-service schemes, where suppliers retain the ownership of the photovoltaic system and sell the electricity at tiered rates. These rates are not regulated yet but are offered in a range from USD 8–13 per month, and cover maintenance and battery upgrades.[98]

12.5 Conclusion

An analysis of the relevant legal framework for OGRE shows that Mexico has seriously derailed from its initial efforts to support its development and that of renewable energy in the light of energy security and climate change. In the words of Herrera, 'overall Mexico's contribution does not go far enough nor reflect sufficient urgency in the face of the climate emergency'.[99] The current government commits to energy sovereignty and development as its priority.[100] Thus, its decision to support government-owned electricity companies which are sadly focused in the development of fossil fuel options. It is apt to mention that directing the attention of such state companies to the development of renewable energy may cost in the immediate future but will be more beneficial as per energy security and climate change. Thus, efforts should be made by concerned individuals and non-state actors to sensitise the government to introduce a gradual time-bound direction of the focus of state electricity companies towards renewable electricity development. As stated, the benefits of renewable electricity including OGRE far outweigh that of fossil fuel. Developing OGRE and other renewable electricity options will be of immense benefit to the country and can also be embedded in the energy sovereignty vision of the current administration.

Relevant laws should be amended to thin out areas that are not favourable to the development of renewable electricity including OGRE. The Climate Change Law and Energy Transition Law should be amended to explicitly provide for national and sector-specific targets for renewable electricity and OGRE, respectively. Such a target is exigent in the context of having a standard for adopting and assessing measures for the development of the sector. Provisions of LIE on exempted generation should be reviewed upwardly to accommodate and privilege most OGRE projects. The framework for EIA should be reviewed to extensively favour renewable electricity projects eliminating every exemption. It was clearly shown that there are measures provided for the support of OGRE but implementation remains the critical challenge. Sensitisation of government by private actors is key to garnering the needed political will to steer action towards such implementation. What is more, existing financial support for fossil fuel will continue to undermine efforts in developing OGRE. Thus, support for fossil fuel electricity options should be gradually eliminated and support for OGRE intensified to drown the effect of previous support for the former. The scope of FSUE should be extended to support OGRE in urban areas especially in the context where the beneficiaries are unable to afford it.

Renewable electricity including OGRE is growing in popularity as a preferred option of electrification because of the benefit it has economically, socially, and environmentally. Thus, it is trendy for a country to prioritise the development of renewable electricity. Mexico cannot afford to retrogress from this moving trend especially if it wants to remain economically and environmentally relevant. Joining the trend entails strengthening the role of law for the optimal development of OGRE through amending the relevant laws in the light of the recommendations made in this chapter.

Notes

1 Electrificación rural para comunidades fuera de la red utilizando generación de energía renovable con sistemas híbridos <https://base.energia.gob.mx/ER1218/LB_FSUE.PDF> accessed 8 May 2022.

2 World Bank, 'Access to Electricity (% of Population)-Mexico' (2022) <https://data.worldbank.org/indicator/EG.ELC.ACCS.ZS?locations=MX> accessed 18th May 2022.

3 Raul Jimenez, 'Rural Electricity Access Penalty in Latin America: Income and Location' (2016) 10, <https://publications.iadb.org/publications/english/document/Rural-Electricity-Access-Penalty-in-Latin-America-Income-and-Location.pdf> accessed 18th May 2022. For a definition of remote rural áreas, see I M Bugaje, 'Remote Area Power Supply in Nigeria: The Prospects of Solar Energy' (1999) 18 Renewable Energy 491.

4 Claudia Salgado, 'Electricity Sector' (2021) <https://www.trade.gov/country-commercial-guides/mexico-electricity-sector> accessed 18th May 2022.

5 World Bank, 'Switching on Remote Communities through Electricity Access in Mexico' (2017) <https://www.worldbank.org/en/results/2017/11/01/switching-on-remote-communities-through-electricity-access-in-mexico> accessed 18th May 2022.

6 David I Stern and others, 'The Impact of Electricity on Economic Development: A Macroeconomic Perspective' (2019) <https://escholarship.org/uc/item/7jb0015q> accessed 18th May 2022.

7 Paula Borges da Silveira Bezerra and others, 'The Power of Light: Socio-Economic and Environmental Implications of a Rural Electrification Program in Brazil' (2017) 12 Environ. Res. Lett. 095004.

8 Electrificación rural para comunidades fuera de la red utilizando generación de energía renovable con sistemas híbridos <https://base.energia.gob.mx/ER1218/LB_FSUE.PDF >accessed 8th May 2022.

9 Mexpereince, 'Dealing with Electrical Power Cuts in Mexico' (2022) <https://www.mexperience.com/when-the-lights-go-out/> accessed 18th May 2022.

10 Rebecca Conan, 'Northern Mexico Loses Power as Cold Dents US Gas Flow' (2022) <https://www.argusmedia.com/en/news/2187069-northern-mexico-loses-power-as-cold-dents-us-gas-flow> accessed 18th May 2022.

11 Ibid.

12 Ngozi Chinwa Ole, 'The Nigerian Electricity Regulatory Framework: Hotspots and Challenges for Off-Grid Renewable Electricity Development' (2020) 26(4) Journal of Energy and Natural Resource Law 394.

13 Meghan L O Sullivan, 'Mexico's Energy Reforms: A Blow to Realizing the Most Competitive and Dynamic Region in the World' (2022) <https://www.brookings.edu/blog/up-front/2022/02/28/mexicos-energy-reforms-a-blow-to-realizing-the-most-competitive-and-dynamic-region-in-the-world/> accessed 11th May 2022.

14 World Bank, 'Access to Electricity (% of Population)-Mexico' (n 2).

15 Alexa Spenca and others, 'Sustainability Following Adversity: Power Outage Experiences are Related to Greater Energy Saving Intentions in the United Kingdom and Mexico' (2021) 79 Energy Research and Social Science 102143.

16 Business Standard, 'Mexico Suffers Another Day of Rolling Blackouts Due to Storm in Texas' (2022) <https://www.business-standard.com/article/international/mexico-suffers-another-day-of-rolling-blackouts-due-to-storm-in-texas-121021700077_1.html> accessed 20th May 2022.

17 María José Goytia, 'Millions of Mexicans Live in Energy Poverty' (2022) <https://mexicobusiness.news/energy/news/millions-mexicans-live-energy-poverty> accessed 20th May 2022.

18 Mc Kinsey &Company, *How Mexico can Harnass its Superior Energy Abundance* <https://www.mckinsey.com> accessed 9th May 2022.

19 Ibid.

20 Adán L Martínez-Cruz and Héctor M Núñez, 'Tension in Mexico's Energy Transition: Are Urban Residential Consumers in Aguascalientes Willing to Pay for Renewable Energy and Green Jobs?' (2021) 150 Energy Policy 112145.

21 Victor G Carreon-Rodriguez, 'Mexico' in Beni Trojbicz (ed) *Oil Wealth and Federal Conflict in American Petrofederations* (Elsevier 2022) 123–174.
22 Rafael Gonzalez-Lopez and Natalie Ortiz-Guerrero, 'Integrated Analysis of the Mexican Electricity Sector: Changes during the Covid-19 Pandemic' (2022) 35(6) Electronic Journal 107142.
23 Andresde Jesus Fernandez and Jim Watson, 'Mexico's Renewable Energy Innovation System: Geothermal and Solar Photovoltaics Case Study' (2022) 43 Environmental Innovation and Societal Transitions 200.
24 Rodrigo Osorio, 'Energy Poverty: The Incessant Struggle' (2020) <https://mexicobusiness.news/energy/news/energy-poverty-incessant-struggle> accessed 20th May 2022.
25 NREL, 'News Release: NREL Identifies Abundant Renewable Energy Resources as Key to Mexico's Clean Energy Ambitions' (2022) <https://www.nrel.gov/news/press/2022/nrel-identifies-abundant-renewable-energy-resources-as-key-to-mexicos-clean-energy-ambitions.html> accessed 20th May 2022.
26 Ibid.
27 USAID, 'Mexico-Climate Change Fact Sheet' (2022) <https://www.usaid.gov/sites/default/files/documents/USAID-Climate-Change-Fact-Sheet-Mexico.pdf> accessed 20th May 2022.
28 Amanda Maxwell, 'Mexico Ratifies the Paris Agreement' (2022) <https://www.nrdc.org/experts/amanda-maxwell/mexico-ratifies-paris-agreement> accessed 20th May 2022.
29 UNFCCC, 'Mexico Intended Nationally Determined Contribution' (2015) <https://www4.unfccc.int/sites/submissions/INDC/Published%20Documents/Mexico/1/MEXICO%20INDC%2003.30.2015.pdf> accessed 20th May 2022.
30 USAID, 'Greenhouse Gas Emissions Factsheet: Mexico' (2022) <https://www.climatelinks.org/resources/greenhouse-gas-emissions-factsheet-mexico> accessed 20th May 2022.
31 Maria Fernanda Mac Gregor-Gaona and others, 'Assessing Climate Change Risk: An Index Proposal for Mexico City' (2021) 65 International Journal of Disaster Risk Reduction 102549.
32 Britta Rennkamp and others, 'Competing Coalitions: The Politics of Renewable Energy and Fossil Fuels in Mexico, South Africa and Thailand' (2017) 34 Energy Research and Social Science 214.
33 The Mexican General Law on Climate Change 2012 (amended in 2020), Article 33(3).
34 Flory Anette Dieck-Assad, 'Private Vs. Public Investment in the Mexican Utility Company: A Case Study' (2016) 12(1) Journal of International Energy Research 27.
35 Gibran S Aleman-Nava and others, 'Renewable Energy Research Progress in Mexico: A Review' (2014) 32 Renewable and Sustainable Energy Reviews 140.
36 Raúl Camba and others, 'How Mexico can Harness its Superior Energy Abundance' (2019) <https://www.mckinsey.com/industries/oil-and-gas/our-insights/how-mexico-can-harness-its-superior-energy-abundance> accessed 21st May 2022.
37 Bruno Bernal Trujillo, 'Barriers to Developing Solar Energy in Mexico and Egypt' (2021) Ch. 2, <https://dadun.unav.edu/bitstream/10171/61018/1/Tesis_BernalTrujillo21.pdf> accessed 21st May 2022.
38 Energy Safety Nets: Mexico Case Study <https://www.seforall.org> accessed 11th May 2022.
39 Victor Carreón and Armando Jimenez, 'The Mexican Electricity Sector: Economic, Legal and Political Issues' (2019) <https://fsi-live.s3.us-west-1.amazonaws.com/s3fs-public/evnts/media/Mexico.pdf> accessed 21st May 2022.
40 Ibid.
41 Britta Rennkamp and others, 'Competing Coalitions: The Politics of Renewable Energy and Fossil Fuels in Mexico, South Africa and Thailand' (n 32) 214.

42 Ibid.
43 Daniela van Schagen Mendoza, 'Opportunities in the Mexican Renewable Energy Sector' (2018) 10, <https://www.rvo.nl/sites/default/files/2019/04/opportunities-in-the-mexican-renewable-energy-sector.pdf> accessed 21st May 2022.
44 Elizabeth Lokey, 'Barriers to Clean Development Mechanism Renewable Energy Projects in Mexico' (2009) 34 Renewable Energy 504.
45 Bruno Bernal Trujillo, 'Barriers to Developing Solar Energy in Mexico and Egypt' (n 37).
46 Ibid.
47 Ibid.
48 IRENA, *Renewable Energy Prospects: Mexico* (IRENA 2015) 70.
49 Ibid, 70–71.
50 Mark Stevenson, 'Mexican Court Blocks Law Favoring State Electricity Plants' (2021) <https://www.seattletimes.com/business/mexican-court-blocks-law-favoring-state-electricity-plants/> accessed 21st May 2022.
51 Nathanial Parish Flannery, 'Political Risk Analysis: Is Mexico Declaring War Against Clean Energy?' (2022) <https://www.forbes.com/sites/nathanielparishflannery/2021/04/22/political-risk-analysis-is-mexico-declaring-war-against-clean-energy/> accessed 21st May 2022.
52 Ibid.
53 Ibid.
54 Ibid.
55 The Electric Industry Law (*Ley de la Industria Eléctrica)* 1992, Article 13.
56 Ngozi Chinwa Ole, 'The Paris Agreement 2015 as a Primer for Developing Nigerian Off-Grid Solar Electricity' (2018) 26(3) African Journal of International and Comparative Law 426.
57 The General Law on Climate Change 2012, Transitory Articles, Article 4.
58 The Energy Transition Law 2015, Article 4.
59 IRENA, 'Energy Transition Law (ETL)' (2015) <https://www.irena.org//media/Files/IRENA/Agency/Webinars/LTES-Campaign---15-Nov--Mexico.pdf?la=en&hash=0B23554E9B91D51D350646ACCEAB46BD4DDD877A> accessed 26th May 2022.
60 IRENA, *Renewable Energy Target Setting* (IRENA 2015) 27, 33, 34.
61 Ibid, see also Sadie Cox and others, 'Solar Power Policy Overview and Good Practices' (2015) NREL/TP-6A20–64178, 4.
62 Ngozi Chinwa Ole, 'The Role of Renewable Energy Law 2005 in Supporting the Development of Off-Grid Renewable Electricity Law in China' (2019) 5 International Energy Law Review 132.
63 Ngozi Chinwa Ole, 'The Nigerian Electricity Regulatory Framework: Hotspots and Challenges for Off-Grid Renewable Electricity Development' (n 12) 394.
64 Ibid.
65 Andrew Moseman, 'How Efficient is Carbon Capture and Storage?' (2022) <https://climate.mit.edu/ask-mit/how-efficient-carbon-capture-and-storage> accessed 27th May 2022.
66 Ibid.
67 Heiner Von Lupke and Mareike Well, 'Analyzing Climate and Energy Policy Integration: The Case of the Mexican Energy Transition' (2020) 20 Climate Policy 832; Jorge Alejandro Silva and Maria Antonieta Andrade, 'Solar Energy Analysis in Use and Implementation in Mexico: A Review' (2020) 14(6) International Journal of Energy Sector Management 1333.
68 Article 22 LIE. Se entiende por abasto aislado la generación o importación de energía eléctrica para la satisfacción de necesidades propias o para la exportación, sin transmitir dicha energía por la Red Nacional de Transmisión o por las Redes Generales de Distribución....

69 Nevertheless, the LIE also contemplates that if the off-grid generator wants to sell surpluses or purchase energy shortages, they will do it through the Wholesale Electricity Market. The latter implies that the generator would be interconnected to the grid and transmit the energy. Article 23 LIE.

70 Acuerdo de la Comisión Reguladora de Energía por el que se emite el criterio de interpretación del concepto "necesidades propias", establecido en el artículo 22 de la Ley de la Industria Eléctrica, y por el que se describen los aspectos generales aplicables a la actividad de Abasto Aislado published in the Federal Registrar on November 21st, 2017.

71 Agreement, 2.1

72 Article 126, f. V. LIE. *La CRE podrá establecer requerimientos de medición y reporte relacionados con la generación de Energías Limpias mediante el abasto aislado.*

73 Ernst & Young, 'Solar Energy Investment in Mexico' (2018) <https://energypedia.info/images/archive/0/00/20180903193554%21Investment_Opportunities_Solar_2018.pdf> accessed 7th June 2022.

74 Ibid.

75 Gibran S Aleman-Nava and others, 'Renewable Energy Research Progress in Mexico: A Review' (2014) 32 Renewable and Sustainable Energy Reviews 140.

76 Ibid.

77 Ibid.

78 Article 5, K) LGEEPA Regulations.

79 Bruno Bernal Trujillo, 'Barriers to Developing Solar Energy in Mexico and Egypt' (n 37) 42.

80 Ibid. See also Miguel Saldivia Olave and Sofia Vargas-Payera, 'Chapter 12 - Environmental Impact Assessment and Public Participation of Geothermal Energy Projects: The Cases of Chile, Costa Rica, Colombia, and Mexico' in Lucas Noura Guimaraes (ed), *The Regulation and Policy of Latin American Energy Transitions* (Elsevier 2020) 209–220.

81 Ibid.

82 UNCTAD, *Mexico's Agricultural Development: Perspectives and Outlook* (UNCTAD 2013) 2–5.

83 Ibid.

84 Conecc, 'Improving and Focusing Electricity Subisidies: Option for Optimisation in Mexico' (2018) <https://www.energypartnership.mx/fileadmin/user_upload/mexico/media_elements/reports/ElectricitySubsidies-MEX.pdf> accessed 9th June 2022.

85 Energy Safety Nets: Mexico Case Study <https://www.seforall.org> accessed 11th May 2022.

86 Ibid.

87 For an extended literature on the use of OGRE as back-up, see Ngozi Chinwa Ole, 'The Role of Renewable Energy Law 2005 in Supporting the Development of Off-Grid Renewable Electricity in China' (2019) 5 International Energy Law Review 132–133.

88 Juan C Percino-Picazo and others, 'Analysis of Restructuring the Mexican Electricity Sector to Operate in a Wholesale Energy Market' (2021) 14 Energies 3331, 3346.

89 Green Climate Fund, 'Energy Efficiency Promotion Programme for MSMEs' (2021) <https://www.nafin.com/portalnf/files/secciones/emisiones-relaciones-internacionales/fondo-verde-clima/documentos/NAFIN_CN_Eficiencia_energe_769_tica_en_PyMEs.pdf> accessed 9th June 2022.

90 Ibid.

91 Ibid.

92 Inter-American Development Bank, 'The-Role-of-National-Development-Banks-in-Intermediating-International-Climate-Finance-to-Scale-Up-Private-Sector-Investments' (2012) <https://publications.iadb.org/publications/english/document/The-Role-of-National-Development-Banks-in-Intermediating-International-Climate-Finance-to-Scale-Up-Private-Sector-Investments.pdf> accessed 9th June 2022.

93 Energy Transition Act 2015, Article 16.
94 Ibid.
95 Fernando Perez-Lozado, 'Change of Clean Energy Rules in Mexico: Potential Impact for Investors' (2020) <http://arbitrationblog.kluwerarbitration.com/2020/01/13/change-of-clean-energy-rules-in-mexico-potential-impact-for investors/#:~:text=Subsequently%2C%20Certificates%20of%20Clean%20Energy,only%20from%20renewable%20sources%2C%2012> accessed 9th June 2022.
96 Grantham Research Institute on Climate Change and Environmental Law, 'Climate Change Laws of the World-Energy Transition Law of Mexico' <https://climate-laws.org/> accessed 11th May 2022.
97 J Deign, 'Lifting the Lid on Mexico's Off-Grid Opportunity' (2019) <https://www.greentechmedia.com/articles/read/lifting-the-lid-on-mexicos-off-grid-opportunity> accessed 10th May 2022.
98 Ibid.
99 Caroline Herrera, 'Mexico Publishes Unambitious Updated NDC' (2021) <https://www.nrdc.org/experts/carolina-herrera/mexico-publishes-unambitious-updated-ndc> accessed 10th June 2022.
100 Ibid.

References

Alejandro Silva Jorge and Maria Antonieta Andrade, 'Solar Energy Analysis in Use and Implementation in Mexico: A Review' (2020) 14(6) International Journal of Energy Sector Management 1333

Aleman-Nava Gibran S and others, 'Renewable Energy Research Progress in Mexico: A Review' (2014) 32 Renewable and Sustainable Energy Reviews 140

Bernal Trujillo Bruno, 'Barriers to Developing Solar Energy in Mexico and Egypt' Ch. 2. <https://dadun.unav.edu/bitstream/10171/61018/1/Tesis_BernalTrujillo21.pdf> accessed 21 May 2022

Carreon-Rodriguez Victor G, 'Mexico' in Beni Trojbicz (eds), *Oil Wealth and Federal Conflict in American Petrofederations* (Elsevier 2022), 123–174

Dieck-Assad Flory Anette, 'Private Vs. Public Investment in the Mexican Utility Company: A Case Study' (2016) 12(1) Journal of International Energy Research 27

'Electrificación rural para comunidades fuera de la red utilizando generación de energía renovable con sistemas híbridos' <https://base.energia.gob.mx/ER1218/LB_FSUE.PDF> accessed 8 May 2022

Fernanda Mac Gregor-Gaona Maria and others, 'Assessing Climate Change Risk: An Index Proposal for Mexico City' (2021) 65 International Journal of Disaster Risk Reduction 102, 549

Fernandez Andresde Jesus and Watson Jim, 'Mexico's Renewable Energy Innovation System: Geothermal and Solar Photovoltaics Case Study' (2022) 43 Environmental Innovation and Societal Transitions 200

Gonzalez-Lopez Rafael and Natalie Ortiz-Guerrero, 'Integrated Analysis of the Mexican Electricity Sector: Changes During the Covid-19 Pandemic' (2022) 35(6) Electronic Journal 107, 142

Heiner Von Lupke and Mareike Well, 'Analyzing Climate and Energy Policy Integration: The Case of the Mexican Energy Transition' (2020) 20 Climate Policy 832

I Stern David and others, 'The Impact of Electricity on Economic Development: A Macroeconomic Perspective' <https://escholarship.org/uc/item/7jb0015q> accessed 18 May 2022

IRENA, 'Energy Transition Law (2015) ETL' <https://www.irena.org//media/Files/IRENA/Agency/Webinars/LTES-Campaign---15-Nov--Mexico.pdf?la=en&hash=0B23554E9B91D51D350646ACCEAB46BD4DDD877A> accessed 26 May 2022

IRENA, 'Renewable Energy Prospects: Mexico' (2015) IRENA 70

Jimenez Raul, 'Rural Electricity Access Penalty in Latin America: Income and Location' <https://publications.iadb.org/publications/english/document/Rural-Electricity-Access-Penalty-in-Latin-America-Income-and-Location.pdf> accessed 18 May 2022

L Martínez-Cruz Adán and Héctor M Núñez, 'Tension in Mexico's Energy Transition: Are Urban Residential Consumers in Aguascalientes willing to Pay for Renewable Energy and Green Jobs?' (2021) 150 Energy Policy 112, 145

M I Bugaje, 'Remote Area Power Supply in Nigeria: The Prospects of Solar Energy' (1999) RE 18, 491

Mexpereince, 'Dealing with Electrical Power Cuts in Mexico' <https://www.mexperience.com/when-the-lights-go-out/> accessed 18 May 2022

Moseman Andrew, 'How Efficient is Carbon Capture and Storage?'<https://climate.mit.edu/ask-mit/how-efficient-carbon-capture-and-storage> accessed 27 May 2022

NREL, 'News Release: NREL Identifies Abundant Renewable Energy Resources as Key to Mexico's Clean Energy Ambitions' <https://www.nrel.gov/news/press/2022/nrel-identifies-abundant-renewable-energy-resources-as-key-to-mexicos-clean-energy-ambitions.html> accessed 20 May 2022

Ole Ngozi Chinwe, ' The Role of Renewable Energy Law 2005 in Supporting the Development of Off-grid Renewable Electricity Law in China' (2019) 5 International Energy Law Review 132

Ole Ngozi Chinwe, 'The Nigerian Electricity Regulatory Framework: Hotspots and Challenges for Off-grid Renewable Electricity Development' (2020) 26(4) Journal of Energy and Natural Resource Law, 394

Ole Ngozi Chinwe, 'The Paris Agreement 2015 as a Primer for Developing Nigerian Off-grid Solar Electricity' (2018) 26(3) African Journal of International and Comparative Law 426

Parish Flannery Nathanial, 'Political Risk Analysis: Is Mexico Declaring War Against Clean Energy?'<https://www.forbes.com/sites/nathanielparishflannery/2021/04/22/political-risk-analysis-is-mexico-declaring-war-against-clean-energy/> accessed 21 May 2022

Paula Borges da Silveira Bezerra and others, 'The Power of Light: Socio-Economic and Environmental Implications of a Rural Electrification Program in Brazil' (2017) Environmental Research Letters 12

Raúl Camba and others, 'How Mexico Can Harness Its Superior Energy Abundance' <https://www.mckinsey.com/industries/oil-and-gas/our-insights/how-mexico-can-harness-its-superior-energy-abundance> accessed 21 May 2022

Rebecca Conan, 'Northern Mexico Loses Power as Cold Dents US Gas Flow' (2022) <https://www.argusmedia.com/en/news/2187069-northern-mexico-loses-power-as-cold-dents-us-gas-flow> accessed 18 May 2022

Rennkamp Britta and others, 'Competing Coalitions: The Politics of Renewable Energy and Fossil Fuels in Mexico, South Africa and Thailand' (2017) 34 Energy Research and Social Science 214

Sadie Cox and others, 'Solar Power Policy Overview and Good Practices' (2015) NREL/TP-6A20-64178, 4

Salgado Claudia, 'Electricity Sector'<https://www.trade.gov/country-commercial-guides/mexico-electricity-sector> accessed 18 May 2022

Spenca Alexa and others, 'Sustainability Following Adversity: Power Outage Experiences are Related to Greater Energy Saving Intentions in the United Kingdom and Mexico' (2021) 79(102) Energy Research and Social Science 143

Stevenson Mark, 'Mexican Court Blocks Law Favoring State Electricity Plants' <https://www.seattletimes.com/business/mexican-court-blocks-law-favoring-state-electricity-plants/> accessed 21 May 2022

The Electric Industry Law (Ley de la Industria Eléctrica) 1992, Art. 13

The Mexican General Law on Climate Change 2012 (amended in 2020), Art 33(3)

USAID, 'Greenhouse Gas Emissions Factsheet: Mexico' <https://www.climatelinks.org/resources/greenhouse-gas-emissions-factsheet-mexico> accessed 20 May 2022

World Bank, 'Access to Electricity (% Of Population)-Mexico 2022' <https://data.worldbank.org/indicator/EG.ELC.ACCS.ZS?locations=MX> accessed 18 May 2022

World Bank, 'Switching on Remote Communities Through Electricity Access in Mexico' <https://www.worldbank.org/en/results/2017/11/01/switching-on-remote-communities-through-electricity-access-in-mexico> accessed 18 May 2022

13 Trinidad and Tobago

Alicia Elias-Roberts

13.1 Introduction

The twin island Republic of Trinidad and Tobago is located between the Southern Caribbean and North Atlantic Ocean, northeast of Venezuela. Covering an area of 5,128 sq. km, the country has a population of 1.4 million people, with Trinidad being the larger and more populous island.[1] Approximately 95% of the population resides in Trinidad, spread over the country's area of about 4,768 sq. km.[2] Trinidad and Tobago are classified as a small island developing state given its social, economic, and environmental vulnerabilities. The country heavily depends on its oil and gas sector to support its economy and society. Like other small island developing states, Trinidad and Tobago faces the challenge of climate change,[3] small economies of scale, and increased economic, social, and environmental vulnerability.

Trinidad and Tobago has a very unique renewable energy case compared to other Commonwealth Caribbean nations because of its robust oil and gas industry. Trinidad and Tobago is unique in the sense of low energy prices, high per-capita energy consumption, lack of energy efficiency standards, and low awareness regarding energy efficiency.[4] According to an Inter-American Development Bank Final Report in 2012,[5] Trinidad and Tobago generated 8.5 terawatt-hour (TWh) of electricity with CO_2 emissions of 700g for every kilowatt-hour (kWh) generated in 2010. When compared to international benchmarks for 2010, these figures demonstrate high electricity consumption and CO_2 emissions per capita from energy-related activities at nearly 2.5 times the world average.[6]

The use of renewable resources and off-grid renewable electricity in Trinidad and Tobago has been negligible. Additionally, given the country's highly energy-intensive industry, Trinidad and Tobago has the second largest per capita carbon dioxide (CO_2) emissions that is about six times that of the world average, producing 34 metric tons CO_2 per capita in 2014.[7] Trinidad and Tobago has consistently ranked as a top ten country for high energy consumption over the past ten years.[8] The petrochemical and heavy industries in Trinidad and Tobago are responsible for 56% of the country's emissions.[9] Trinidad and Tobago has a large carbon footprint and hence emission reduction is critical for the country's future. In light of the above, the carbon intensity pathway which characterises the country's

DOI: 10.4324/9781003178088-17

electricity industry provides a strong impetus for developing clean energy, including off-grid renewable electricity.

The current status of electricity generation in Trinidad and Tobago is almost 100% based on natural gas. The contribution of renewable energy generation is small, with only a few residential and commercial micro-scale systems connected to the grid or operating as standalone systems.[10] The Government of Trinidad and Tobago has reported that they have plans to improve the efficiency of the existing electricity generating plants as well as an intention to incorporate a small percentage of renewable electricity-based generation over the coming years. Considering the above it is important to develop new laws, policies, and strategies to ensure long-term sustainable development. Specifically, there is also the need to support the development of off-grid renewable electricity. A key feature of this development will be greater levels of energy efficiency, which would allow for energy security in the long term, a reduction in greenhouse gas emissions, and increased revenue and cost savings. Further, an important factor to support OGRE in Trinidad and Tobago is the fact that it remains the most viable option for the electrification of remote rural areas in the twin islands. Also, OGRE can be used as a backup to support the power supply in urban areas.

13.2 Drivers for Off-Grid Renewable Electricity Development

Trinidad and Tobago's National Energy Policy Framework[11] can be seen as part of the nation's continuously evolving domestic energy policy framework and is guided by several factors. The current policies and strategies concerning the promotion of sustainable and renewable energy in the country can be found in several policy documents including: the Green Paper on Minerals Policy,[12] the Report on a Framework for Development of a Renewable Energy Policy,[13] and the National Energy Efficiency Monitoring Report of Trinidad and Tobago.[14] From these various policy documents, the development of renewable energy resources and off-grid renewable electricity is guided by the following main factors:

- a recognition that the country's hydrocarbon resources are finite and depleting;
- achieving long-term security of energy supply;
- contributing to the country's efforts as a signatory to the Kyoto Protocol to address the climate change issue; recognising the threat faced as a Small Island State;
- informing and educating the public on alternative clean energy and energy efficiency measures;
- ensuring that a sustainable balance is maintained with the environment; and
- providing a sufficient and affordable energy supply for societal needs.[15]

Renewable Energy development is identified as a strategy to achieve sustainable growth and development in Trinidad and Tobago. Notwithstanding the country's current favourable energy situation, Trinidad and Tobago needs to continually focus on safeguarding the energy security of the state while addressing the global threat of climate change, to which small island states are particularly vulnerable. Some of the of key drivers for decreasing the use of non-renewable fossil fuel and developing renewable energy and off-grid renewable electricity systems are summarised below.

Trinidad and Tobago's Intended Nationally Determined Contribution (INDC) is based on its Carbon Reduction Strategy developed for three major sectors: its power generation, transportation, and industrial sectors. These sectors are responsible for most emissions that impacts the country's National Climate Change Policy.[16] Trinidad and Tobago's aim is to achieve a reduction objective in overall emissions from the three sectors of 15% by 2030 from BAU, which in absolute terms is an equivalent of 103 million tonnes (103,000,000) of CO_2e.[17] The approximate cost of meeting this objective is US$2 billion, which is expected to be met partly through domestic funding and conditional on international financing including through the Green Climate Fund. Trinidad and Tobago indicated in its NDC that it will commit to unconditionally reduce its public transportation emissions by 30% or 1,700,000 tonnes CO_2e compared to 2013 levels by December 31, 2030.[18]

While the NDC states that mitigation options were identified and will include options such as clean technology, fuel switching, and renewable energy and energy efficiency technologies, the submission failed to specify OGRE targets. The impact of not setting clear national targets for renewable energy and sectoral targets for OGRE weakens the possibilities of adopting effective legal measures to address the barriers for OGRE. The International Renewable Energy Agency (IRENA) recommends that a requirement for fine-tuning the role of law in efficiently supporting the development of renewable energy, including OGRE, is the adoption of national and ensuing sector-specific targets.[19] These targets must be SMART; that is, Specific, Measurable, Achievable, Realistic, and Time-bound, and contain the numeric and deadline for achieving such a target. The Government of Trinidad and Tobago ought to adopt the SMART targets, which would help in adopting legal measures to address the various identified barriers.[20]

A major driver for the introduction of renewable energy systems and, more specifically, the development of off-grid renewable electricity in Trinidad and Tobago is the mitigation of climate change; through reduction of greenhouse gas (GHG) emissions.[21] In 1994, the United Nations Framework Convention on Climate Change (UNFCCC) entered into force with the ultimate objective being the stabilisation of greenhouse gas concentrations in the atmosphere at a level that will prevent dangerous human interference with the climate system.[22] The UNFCCC was later complemented by the 1997 Kyoto Protocol under which member countries are committed to reducing their GHG emissions by an average of 5% by 2012 against 1990 levels.[23] More recently the 2015 Paris Agreement to the United Nations Framework Convention on Climate Change which entered into force in November 2016[24] has its goal to limit global warming to well below 2, preferably to 1.5 degrees Celsius, compared to pre-industrial levels. To achieve this long-term temperature goal, parties to the Agreement aim to reach global peaking of greenhouse gas emissions as soon as possible to achieve a climate neutral world by mid-century. The Paris Agreement is a landmark multilateral environmental treaty because, for the first time, a binding agreement brings all nations, regardless of their territorial size, population, economy, etc. into a common cause to undertake ambitious efforts to combat climate change and adapt to its effects.

Trinidad and Tobago is a ratified signatory to the UNFCCC, the Kyoto Protocol, and the Paris Agreement. Renewable Energy development is one of the strategies being adopted to mitigate the impact of climate change and investments have been significant and growing over the last decade. The threat of climate change is very real for small island states like those of the Caribbean where serious negative socio-economic implications could result from an event such as rising sea level (depending on its severity). This is because of their limited size, geographical location, high exposure of infrastructure, and limited adaptive capacity. Renewable energy development would promote energy security and the sustainable development of these states.

Another driver to support the development of off-grid renewable electricity in Trinidad and Tobago is energy security. The supply of fossil fuels is subject to significant changes and fluctuations as revisions are made to proven resources and reserves. The introduction of power generation from renewable energy sources provides an alternative energy resource, the availability of which is not subject to the cost fluctuations as is the case with fossil fuels. This is linked to a third major driver, which is fossil fuel depletion. Reducing the demand for conventional energy sources will extend the lifetime of the country's limited fossil fuel reserves and renewable energy could be developed as an alternative energy source.

Fourth, the country's national development is another major driver. Through research, development, and demonstration renewable energy systems are becoming more affordable and accessible. This also creates investment opportunities and is attractive to new investors, both locally and internationally. Off-grid renewable electricity in Trinidad and Tobago can also lead towards foreign direct investment, the creation of several new jobs, and the development of the economy. This will also foster local innovation and entrepreneurship and expand opportunities for participation of all citizens.

Additionally, there can be an increase in domestic export of petroleum products. The reduction in domestic demand for petroleum products through the use of off-grid renewable electricity in Trinidad and Tobago would make greater quantities of petroleum products available for exports. Finally, there will be hundreds of new employment opportunities. The introduction of off-grid renewable electricity will open up new job markets for the local development, manufacture, installation, and maintenance of these systems.

The various drivers discussed above are all linked and cannot be properly considered in isolation. For instance, the size of crude and natural gas reserves and depletion rates, are linked to monetisation of reserves, and several other issues including energy security and affordability, local content and participation, value maximisation, pricing and market characteristics, economic sustainability, environmental considerations, and the international environment. The development of off-grid renewable electricity and energy efficiency policy measures will therefore require a careful and studied appreciation of the several other factors which are a critical aspect of the nation's continued development.

13.3 Barriers to Off-Grid Renewable Electricity Development

This section discusses some of the key barriers that has hindered the development of renewable energy in Trinidad and Tobago. There are several practical limitations and market barriers to the implementation of renewable energy sources and off-grid renewable electricity. It is important that these barriers be taken into consideration when planning for the introduction of renewable energy systems locally.

Petroleum product prices are highly subsidised in Trinidad and Tobago, and this has resulted in the country having the lowest diesel fuel and commercial and industrial electricity prices in most of the Latin American and Caribbean region.[25] This acts as a major barrier which hinders the development of renewable energy in Trinidad and Tobago. The government's policy to subsidise electricity rates and transportation fuel prices was adopted and implemented as an indirect means of ensuring that citizens share in the wealth generated from the country's crude oil and natural gas resources. However, this government policy poses a challenge for renewable energy to compete with traditional energy sources and serves as a disincentive to consumers to conserve energy. Subsidies significantly lower the final energy prices, putting renewable energy at a competitive disadvantage. One option would be to develop an enabling tax incentive regime to make the market viable for renewables.

To develop off-grid renewable electricity in Trinidad and Tobago requires a higher initial capital cost when compared with conventional energy sources, providing less installed capacity per initial dollar invested. As a result of this, renewable energy investments generally require higher amounts of financing for the same capacity as compared to conventional fossil fuel systems which may result in higher lending rates. The higher initial capital costs for renewable energy are reflected in the cost of generation. According to a study performed by the University of Trinidad and Tobago,[26] when comparing the costs of producing electricity with different renewables in Trinidad and Tobago, the actual cost of electricity to local consumers is approximately US4 cents/kWh on average, which makes renewables very uncompetitive economically, except for biomass, wind, and small hydro.[27] Another barrier to consider is the current lack of fiscal incentives. In Trinidad and Tobago limited fiscal incentives have been introduced for the importation, manufacturing, and use of renewable energy technologies and products. These incentives need to be expanded. Additionally, there may be added transaction costs with off-grid renewable electricity projects when compared to other types of projects. This may be because off-grid renewable electricity projects usually require additional information not readily available and additional time for planning and financing compared to conventional energy projects. Also, there is usually unfamiliarity with the technologies and uncertainties about their performance.

There will also be commercialisation barriers that can hinder the development of renewable energy projects and off-grid renewable electricity in Trinidad and Tobago. As a new technology, off-grid renewable electricity systems will encounter a number of challenges to commercialisation. These can include undeveloped infrastructure, lack of economies of scale in production of renewable energy products, identification of appropriate sites with accessibility to transmission lines, development of new renewable energy industry standards, determination of competitive tariffs for renewable energy usage, and training of local personnel in all aspects of the renewable energy industry.[28]

Other barriers to consider are the limited development of the local market and lack of education and sensitisation about off-grid renewable electricity. The local off-grid renewable electricity market in Trinidad and Tobago is still in an infant stage and, as a result, the penetration of renewable energy technologies faces inadequate information, lack of access to capital, and high transaction costs as barriers. This is also linked to the lack of education and awareness about this new product and industry. The general population of Trinidad and Tobago are not informed about the several issues that are necessary to facilitate renewable energy development. For instance, the importance of energy, the real cost of producing it, the need to conserve it, and the relationships between energy and critical issues such as global climate change; the long-term cost savings to the consumer over the lifetime of renewable energy technologies, which can offset the high cost of installation; and capacity in the renewable energy field is minimal locally and needs to be built further through increased training and skills development.

It should be noted that Trinidad and Tobago does not have a strategy to collect local renewable energy data which can then be easily accessed by the general public. Such data should be developed and include national wind, solar, and wave resource maps; average local costs of installation and operation of small-scale renewable systems; basic information about local needs for installation and operation of small-scale renewable energy systems and local suppliers of renewable systems.[29]

The development of renewable energy in Trinidad and Tobago should consider that there may be environmental concerns with these projects. Research has indicated that hazardous materials used in the manufacture of photovoltaic modules (PVs) could pose a health and safety hazard.[30] These materials include cadmium compounds, arsenic compounds, carbon tetrachloride, tellurium compounds, and a number of flammable gases. However, proper area and personal monitoring, as well as established proactive programmes in industrial hygiene and environmental control can diminish these risks. On the other hand, the emissions of particulates, heavy metals, carbon dioxide, sulphides, and nitrides produced during the entire life cycle of the modules (thin film) are extremely low, that is, 100–360 times lower than that of coal and oil power plants and as a result, the associated risks are minimal. In the case of wind technologies potential adverse environmental impacts may be associated with the noise of operating turbines and erection of turbines in the migratory path of birds.

Competing land use issues may be a potential barrier for large-scale power generation as wind farms require more land area than conventional gas-fired power plants of equivalent capacity; albeit this land can also be used for agricultural purposes. In the case of current solar technologies, 5–10 acres per megawatt of power are required; however, small-scale applications are feasible for residential and commercial areas.[31] With respect to bio-fuel production, the dedication of crops and land is required, which can impact negatively on food security.

13.4 Policy Targets and Framework for Off-Grid Renewable Electricity

According to a 2019 presentation by the Ministry of Energy and Energy Industries[32] in Trinidad and Tobago the government has the following targets for renewable energy:

- Carbon Reduction Strategy 2040: focused on three main emitting sectors: power generation, transport, and industry;
- Nationally determined contributions under the Paris Agreement:
  - Conditional: 15% reduction in emissions from BAU by 2030 from the three main emitting sectors (power generation, transport, and industry);
  - Unconditional: 30% reduction in emissions by 2030 in the transport sector based on 2013 levels; and
- National Renewable Energy Target: 10% electricity generation from renewable energy by 2021.

It is commendable that the Government of Trinidad and Tobago has established a politically approved and time-bound target for their Renewable Energy target, especially for the contribution of renewable energies in the electricity generation and other sectors. The target set by the Government of Trinidad and Tobago is adequate for the short term and will have to be evaluated and renewed by the end of 2021. This sort of commitment has been done in other countries of the Caribbean. For instance, Grenada has officially declared that by 2020, 20% of all domestic energy usage for electricity and transport will originate from renewable energy sources.[33]

As explained above, Trinidad and Tobago's electricity rates are some of the lowest in the Caribbean at approximately US$0.04 per kilowatt-hour (kWh), well below the regional average of US$0.33/kWh. Unlike many Caribbean Island nations, Trinidad and Tobago has significant oil and natural gas reserves and is a net exporter of these fuels.[34] However, subsidies for domestic energy consumption have created a fiscal burden on the government that could be reduced through the expansion of alternative transportation fuels, energy efficiency, and renewable energy.

Some of the initiatives that the Government of Trinidad and Tobago has planned to meet the targets identified above include the following:[35]

- Utility scale renewable energy: 2 MW to 130 MW from wind and/or solar.
- Waste to energy: up to 10 MW at the Beetham Landfill.

- UAE CREF US$3 million grant funding to support a carport mounted solar installation at the Queen's Park Savannah and EC Chargers.
- Feed-In Tariff Policy – this was approved in 2015 but is currently under review for re-submission and implementation. The focus of this energy supply policy is to support the development of new renewable power generation by setting the right price to drive renewable energy deployment. This mechanism places an obligation on electricity utilities to purchase electricity generated by renewable sources at a calculated percent of the retail price or its 'avoided cost'.
- Energy Efficiency Action Plan – developed by a multi-stakeholder committee led by the Ministry of Public Utilities.
- Renewable Energy and Energy Efficiency Education Project.
- Solar LED lighting in play parks.

While the initiative that the Government of Trinidad and Tobago has planned to meet the targets identified are adequate, there are a few challenges. Lack of resource assessments and coordinating institution to monitor the initiatives remain a challenge. Another major challenge remains the legislative barriers, which are discussed below.

13.5 Legal and Regulatory Environment for Off-Grid Renewable Electricity

This section provides a short overview on the legal and regulatory framework for the off-grid renewable electricity system in Trinidad and Tobago. The laws are analysed to evaluate their alignment with the relevant policies and their general effectiveness in addressing the identified barriers to off-grid renewable electricity development.

Trinidad and Tobago does not have a legislative framework to deal with off-grid renewable electricity development. The current laws in place have a gap in this area. There is also the lack of a facilitative environment for potential renewable energy businesses, investors, and independent power producers to assess their potential risks to be able to plan for long-term investments.[36] The current legal and regulatory framework does not include a licencing regime for electricity projects; pricing mechanisms; support schemes; purchasing arrangements; distribution framework, etc. However, the Environmental Management Act[37] does provide for environmental regulations like Environmental Impact Assessment.

The Environmental Management Act requires a Certificate of Environmental Clearance under the Certificate of Environmental Clearance (Designated Activities) Order, 2001 for a number of activities listed in the Schedule as designated activities. The Certificate of Environmental Clearance Rules, 2001 (CEC Rules) sets out the application procedure for a CEC, the requirements and standards for the preparation of an Environmental Impact Assessment (EIA), public access information, etc. Certificates of Environmental Clearance must be issued in compliance with the CEC Rules.[38] These Rules, together with the supporting Certificate of Environmental Clearance (Designated Activities) Order, 2001 define the activities that require such a Certificate. With regard to off-grid renewable electricity, the project activities will require a Certificate of Environmental Clearance. Note that the off-grid renewable electricity projects may also fall under the Minerals Act of Trinidad and Tobago, 2000 (the 'Minerals Act'), where EIAs and CECs are mandatory.[39] The Minerals Act provides for EIAs as well as rehabilitation and a rehabilitation bond. It states that 'a licensee who by virtue of his actions, either advertently or inadvertently, causes damage to public or private property shall be liable to restore or rehabilitate the said property to its former state'.[40]

The Regulated Industry Commission (RIC) is established under the *Regulated Industries Commission Act* No. 26 of 1998,[41] and is responsible for regulating supply of electricity in Trinidad and Tobago. The *RIC Act* specifies that the RIC is to

> Review the principles for determining rates and charges for services every five years or, where the license issued to the service provider prescribes otherwise, at such shorter interval as it may be determined.

Also, under the current Act, electricity generation is not possible without prior licencing through the Ministry of Energy and Energy Industries, unless an exemption is granted (*RIC Act No. 26 of 1998*, Part IV).

The *Trinidad and Tobago Electricity Commission (T&TEC) Act*[42] does not include any regulations that makes provision for interconnection by potential renewable energy power generators to the national electrical grid. The current laws do not provide guidelines to set feed-in tariffs for potential independent renewable energy producers feeding into the electric power grid.[43] The *Trinidad and Tobago Electricity and Commission Act* was originally enacted in 1945 and was subsequently amended several times. As the name suggests, the objective of the Act was the establishment of the state-owned utility enterprise T&TEC, which was created to combine the services of generation, transmission, and supply of electricity for the whole country under one body. Currently, T&TEC remains the sole buyer of all generated electricity and has a country-wide monopoly for the distribution of electric power.

The *T&TEC Act* states in Part VI, section 31. (2) that the Commission may 'with the consent of the Minister, purchase energy from an approved generator of electricity'. It further says in paragraph (3) of the same section that

> the right to generate energy in any part of Trinidad and Tobago for the public or any member thereof is vested in the Commission who may, subject to subsection (5) enter into a license agreement with an **approved generator** of electricity permitting the approved generator of electricity the non-exclusive right to generate electricity.
>
> (bold for emphasis)

It further outlines and clarifies in the same section 31:

> (3A) 'Subject to this Act, the right to supply energy in any part of Trinidad of Tobago to the public or any member thereof, either directly or indirectly, is vested in the Commission.'
>
> (3C) 'The Commission may, with the approval of the Minister, by Order declare a body corporate or firm to be an approved generator of electricity.'
>
> (5) 'The right of an approved generator of electricity to generate energy is subject to such terms and conditions as the Minister may determine.'

Based on the existing legislation, the 'approved generator' is the classical Independent Power Producer who delivers all its electricity to the sole buyer under a negotiated Power Purchase Agreement (PPA). Currently, the *T&TEC Act* neither allows for smaller operators to feed electricity into the grid nor is self-generation or wheeling of electricity between power source and demand sink permitted. The same accounts for the *RIC Act*.[44]

It will therefore be necessary to amend both the *Trinidad and Tobago Electricity Commission Act* and the *RIC Act* accordingly and allow access to the grid. The Regulated Industries Commission (RIC) should continue to play the regulatory body role to control the extent and scope of any access to the grid in line with the government's overall laws and policy objectives. The regulatory power should also be exercised in a discretionary manner dependent on the capacity demand and technical circumstances. In addition, it is recommended to simplify the permission process for operators of small-distributed generation and lessen the licencing procedure. A simple request for registration and control by the electrical inspectorate should be adequate.

In order to encourage further growth and development of renewable energy, additional pieces of legislation need to be examined, noting where industry incentives can be included and in compliance with the renewable energy policy measures being considered. Below is a list of some legislation which should be reviewed and amended where necessary to incorporate renewable electricity.

1. *Trinidad and Tobago Electricity Commission*, Chap 54:70, Order No.42 of 1945.
2. *The Regulated Industries Commission*, Chap 54:73, Act No.26 of 1998.
3. *The Fiscal Incentives Act*, Chap 85:01, Act No.22 of 1979.
4. *The Environmental Management Act*, Chap 35:05, Act No.3 of 2000.
5. *The Petroleum Act*, Chap 62:01, Act No.46 of 1969.
6. *The Petroleum Production Levy and Subsidy Act*, Chap 62:02, Act No.14 of 1974.
7. *The Tourism Development Act*, Chap 87:22, Act No.9 of 2000.
8. *Income Tax (In Aid of Industry) Act*, Chap 85:04, Act No.12 of 1950.
9. *The Town and Country Planning Act*, Chap 35:01, Act No.29 of 1960.
10. *The Public Transport Service Act*, Chap 48:02, Act No.11 of 1965; and
11. *The Motor Vehicle and Road Traffic Act*, Chap 48:50.

13.6 Conclusion

The importance of creating the appropriate legal and regulatory environment cannot be underscored enough. Trinidad and Tobago needs to give priority to reviewing the legal and regulatory framework to support the development of off-grid renewable electricity in the country. Notwithstanding the continued importance of Trinidad and Tobago's petroleum resources, the government recognises that renewable energy, clean energy production, and the maximisation of energy efficiency are critical elements of the drive for sustainable development. As discussed above, the use of renewable energy resources is occurring at a very low rate in Trinidad and Tobago. Further, over the past ten years the country has consistently ranked as a top ten country per capita for carbon dioxide (CO_2) emissions and high energy consumption. In light of the country's very large carbon footprint, reduction in greenhouse gas emissions is critical for the country's future. The development of clean energy, and off-grid renewable electricity in particular, is one path that could place the country towards taking measures to deal with climate change and global warming. It is critical that the country embark on a long-term political and societal commitment to a framework of policies and ideas that would propel local action in these areas, and having the necessary legal and regulatory provisions to support off-grid renewable electricity in Trinidad and Tobago is one such start.

Notes

1 Information taken from the CIA, The World Factbook (2022), available at https://www.cia.gov/the-world-factbook/countries/trinidad-and-tobago/, accessed 24 June 2022.
2 Ibid.
3 See *United Nations Framework Convention on Climate Change* (UNFCCC), New York: United Nations, General Assembly (1992) A/RES/48/189.
4 See Personal Financial, 'Energy: The Countries with the Highest Per Capita Energy Consumption' 19 October 2020, available at https://personal-financial.com/2020/10/19/energythe-countries-with-the-highest-per-capita-energy-consumption/, accessed 24 June 2022.
5 Natacha C. Marzolf and others, *A Unique Approach for Sustainable Energy in Trinidad and Tobago*, Inter-American Development Bank Final Report, Energy Division (2012).
6 Ibid. *See also* the International Energy Agency, *CO2 Emissions from Fuel Combustion* (2012); and International Energy Agency, *Greenhouse Gas Emissions from Fuel Combustion: Overview* (2021), available at https://www.iea.org/reports/co2-emissions-from-fuel-combustion-overview, accessed 06/28/2021.
7 World Bank, *CO2 Emissions (Metric Tons Per Capita from 1960–2016)* (2022), available at https://data.worldbank.org/indicator/EN.ATM.CO2E.PC?locations=TT, accessed 24 June 2022.
8 See n. 4 Personal Financial.
9 UNEP, ACP-MEA & UNFCCC, *Profile of Emissions Reduction Potentials in Developing Countries* (2013), available at https://backend.orbit.dtu.dk/ws/portalfiles/portal/60555273/FINAL_SUMMARY_REPORT.pdf accessed 24 June 2022.
10 See n. 5. Natacha C. Marzolf, et al.
11 See Government of Trinidad and Tobago, Ministry of Energy and Energy Affairs, A Report of the Renewable Energy Committee, *Framework for the Development of a*

Renewable Energy Policy for Trinidad and Tobago (January 2011), available at https://www.energy.gov.tt/wp-content/uploads/2014/01/Framework-for-the-development-of-a-renewable-energy-policy-for-TT-January-2011.pdf, accessed 24 June 2022.

12 Government of Trinidad and Tobago, Ministry of Energy and Energy Affairs, *Green Paper on Minerals Policy* (2014), available at https://www.energy.gov.tt/wp-content/uploads/2014/01/Green-Paper-on-Minerals-Policy.pdf, accessed 24 June 2022.

13 See n. 11, *Framework for the Development of a Renewable Energy Policy for Trinidad and Tobago.*

14 Delena Indar, *National Energy Efficiency Monitoring Report of Trinidad and Tobago* (Economic Commission for Latin America and the Caribbean (ECLAC) Report, United Nations, 2019).

15 See n. 11, *Framework for the Development of a Renewable Energy Policy for Trinidad and Tobago*, at 2.

16 Trinidad and Tobago, *Intended Nationally Determined Contribution (NDC) under the United Nations Framework Convention on Climate Change* (2018), available at https://www4.unfccc.int/sites/ndcstaging/PublishedDocuments /Trinidad%20and%20Tobago%20First/Trinidad%20and%20Tobago%20Final%20INDC.pdf, accessed 24 June 2022.

17 Ibid.

18 Ibid.

19 See, IRENA, *Vision and Mission*, available at https://www.irena.org/statutevisionmission, accessed 07 November 2021.

20 See similar arguments made about energy supply by Ngozi Chinwa Ole, 'The Nigerian Electricity Regulatory Framework: Hotspots and Challenges for Off-Grid Renewable Electricity Development' [2020] Journal of Energy & Natural Resource Law 1.

21 See *UNFCCC supra* n. 3 and *Paris Agreement to the United Nations Framework Convention on Climate Change*, December 12, 2015, T.I.A.S. No. 16-1104.

22 Ibid.

23 R K Pachauri and A Reisinger (eds.), *IPCC, 2007: Climate Change 2007: Synthesis Report. Contribution of Working Groups I, II and III to the Fourth Assessment Report of the Intergovernmental Panel on Climate Change* (IPCC, Geneva, Switzerland, 2008), available at https://www.ipcc.ch/report/ar4/syr/, accessed 24 June 2022.

24 *Paris Agreement*, see n. 21.

25 *See* n. 5, Natacha C. Marzolf, et al. at xiii.

26 Ibid. Note that the information from this source and data was provided by University of Trinidad and Tobago. Cost estimates ranges are based on current commercial and industrial long-term Turn-key Energy Buy-back Contracts within Trinidad and Tobago.

27 Ibid.

28 See n. 11, *Framework for the Development of a Renewable Energy Policy for Trinidad and Tobago*, at p. 22.

29 Ibid at 23.

30 Dávid Strachala, Josef Hylský, Jiří Vaněk, Günter Fafilek and Kristýna Jandová, 'Methods for Recycling Photovoltaic Modules and Their Impact on Environment and Raw Material Extraction' (2017), 22(3) Acta Montanistica Slovaca 257–269.

31 D Fernández-Bellon, M W Wilson, S Irwin and J J C B O'Halloran, 'Effects of Development of Wind Energy and Associated Changes in Land Use on Bird Densities in Upland Areas' (2019) 33(2) Conservation Biology 413–442.

32 Zahra Cielto-Bowrin, Sustainable Energy Development Analyst, Government of Trinidad and Tobago, Ministry of Energy and Energy Affairs, *Country Presentation: Trinidad and Tobago- Caribbean Renewable Energy Statistics Training*, Barbados, Nov. 27-29, 2019. Report of the Renewable Energy Committee, available at https://www.irena.org/-/media/Images/IRENA/Agency/Data-Statistics/Caribbean-Renewable-Energy-Statistics-Training/Country-Presentations/Trinidad-and-Tobago.pdf?la=en&hash=423FB360F4FDBE394EC5855B3629FBDD015F9443 accessed 24 June 2022.

33 Government of Grenada, *The National Energy Policy of Grenada,* November 2011.
34 The National Renewable Energy Laboratory (NREL), US Department of Energy, Office of Energy Efficiency and Renewable Energy, *Energy Snapshot: Trinidad and Tobago* (May 2015).
35 See n. 27, Zahra Cielto-Bowrin.
36 See n. 11, *Framework for the Development of a Renewable Energy Policy for Trinidad and Tobago.*
37 The Environmental Management Act, Chap 35:05, Act No.3 of 2000, Laws of Trinidad and Tobago.
38 The CEC rules were made under s, 26 of the EMA, See https://www.ema.co.tt/images/Files/pdf/certiticate_of_environmental_clearance_rules.pdf, accessed 24 June 2022.
39 Minerals Act, 2000 (no. 61 of 2000). Chap. 61:03, Law of Trinidad and Tobago, s 43.
40 Ibid, Minerals Act, s 43 (4).
41 The Regulated Industries Commission, Chap. 54:73, Act No.26 of 1998, Laws of Trinidad and Tobago.
42 Trinidad and Tobago Electricity and Commission Act, Chap. 54:70, Laws of Trinidad and Tobago.
43 See ibid.
44 Inter-American Development Bank: IDB Report, *Sustainable Energy for Trinidad and Tobago* (23 August 2011).

Bibliography

Treaties

Paris Agreement to the United Nations Framework Convention on Climate Change, 12 December 2015, T.I.A.S. No. 16-1104.

United Nations Framework Convention on Climate Change (UNFCCC), New York: United Nations, General Assembly, 1992.

Statutes and Statutory instruments

Minerals Act, 2000 (No. 61 of 2000). Cap. 61:03, Law of Trinidad and Tobago, s 43.

The CEC rules, made under s, 26 of the EMA, https://www.ema.co.tt/images/Files/pdf/certiticate_of_environmental_clearance_rules.pdf accessed 24 June 2022.

The Environmental Management Act, Chap 35:05, Act No.3 of 2000, Laws of Trinidad and Tobago.

The Regulated Industries Commission, Chap. 54:73, Act No.26 of 1998, Laws of Trinidad and Tobago.

Trinidad and Tobago Electricity and Commission Act, Chap. 54:70, Laws of Trinidad and Tobago.

Journal Articles

Ole N C, 'The Nigerian Electricity Regulatory Framework: Hotspots and Challenges for Off-grid Renewable Electricity Development' (2020) 38 (4) Journal of Energy & Natural Resource Law, 367–390

Fernández-Bellon D, Wilson M W, Irwin S and O'Halloran J J C B, 'Effects of Development of Wind Energy and Associated Changes in Land Use on Bird Densities in Upland Areas' (2019) 33(2) Conservation Biology 413–442.

Strachala D, Hylský J, et al, 'Methods for Recycling Photovoltaic Modules and Their Impact on Environment and Raw Material Extraction' (2017) 22(3) Acta Montanistica Slovaca 257–269.

Reports

Cielto-Bowrin Z, 'Sustainable Energy Development Analyst, Government of Trinidad and Tobago, Ministry of Energy and Energy Affairs', *Country Presentation: Trinidad and Tobago- Caribbean Renewable Energy Statistics Training*, Barbados, 27–29 November 2019. Report of the Renewable Energy Committee, https://www.irena.org/-/media /Images/IRENA/Agency/Data-Statistics/Caribbean-Renewable-Energy-Statistics Training/ Country-Presentations/Trinidad-and Tobago.pdf?la=en&hash=423FB360F4 FDBE39 4EC5855 B3629FBDD015F9443 accessed 24 June 2022.

Government of Grenada, *The National Energy Policy of Grenada*, November 2011.

Government of Trinidad and Tobago, Ministry of Energy and Energy Affairs, *A Report of the Renewable Energy Committee: Framework for the Development of a Renewable Energy Policy for Trinidad and Tobago*, (January 2011), https://www.energy.gov.tt/wp-content/uploads/2014/01/Framework-for-the-development-of-a-renewable-energy-policy-for-TT-January-2011.pdf accessed 24 June 2022.

Government of Trinidad and Tobago, Ministry of Energy and Energy Affairs, *Green Paper on Minerals Policy* (2014), https://www.energy.gov.tt/wp-content/uploads/2014/01/Green-Paper-on-Minerals-Policy.pdf accessed 24 June 2022.

Indar D, *National Energy Efficiency Monitoring Report of Trinidad and Tobago*, Economic Commission for Latin America and the Caribbean (ECLAC) Report, United Nations, 2019.

Information taken from the CIA World Factbook,https://www.cia.gov/the-world-factbook/countries/trinidad-and-tobago/ accessed 24 June 2022.

Inter-American Development Bank: IDB Report, *Sustainable Energy for Trinidad and Tobago* (23 August 2011).

International Energy Agency: CO_2 Emissions from Fuel Combustion (2012) https://www.iea.org/reports/co2-emissions-from-fuel-combustion-overview accessed 24 June 2022.

Natasha Marzolf, et al, *A Unique Approach for Sustainable Energy in Trinidad and Tobago*, Inter-American Development Bank Final Report, Energy Division (2012).

Pachauri R K and Reisinger A (eds), *IPCC, 2007: Climate Change 2007: Synthesis Report. Contribution of Working Groups I, II and III to the Fourth Assessment Report of the Intergovernmental Panel on Climate Change* (IPCC, Geneva, Switzerland, 2008), https://www.ipcc.ch/report/ar4/syr/ accessed 24 June 2022.

Personal Financial, 'Energy: The Countries with the Highest Per Capita Energy Consumption' (19 October 2020), https://personal-financial.com/2020/10/19/energythe-countries-with-the-highest-per-capita-energy-consumption/ accessed 24 June 2022.

The National Renewable Energy Laboratory (NREL), US Department of Energy, Office of Energy Efficiency and Renewable Energy, *Energy Snapshot: Trinidad and Tobago* (May 2015)

Trinidad and Tobago Intended Nationally Determined Contribution (NDC) under the United Nations Framework Convention on Climate Change (2018), https://www4.unfccc.int/sites/ndcstaging/PublishedDocuments/Trinidad%20and%20Tobago%20First/Trinidad%20and%20Tobago%20Final%20INDC.pdf accessed 24 June 2022.

UNEP, ACP-MEA & UNFCCC (2013), *Profile of Emissions Reduction Potentials in Developing Countries*, https://backend.orbit.dtu.dk/ws/portalfiles /portal/60555273/FINAL_ SUMMARY_REPORT.pdf accessed 24 June 2022.
World Bank (1960–2016), *Co2 emissions* (Metric tons per capita), https://data.worldbank.org/indicator/EN.ATM.CO2E.PC?locations=TT accessed 24 June 2022.

Part 5

Towards a Supportive Regulatory Framework for Off-Grid Renewable Electricity

14 Reflections and Conclusion

Gustavo Kaercher Louriero and Eduardo G. Pereira

14.1 Introduction

OGRE is a decentralised form of renewable energy access directed at consumers not connected to the main (national or regional) grid. As seen throughout the book, national authorities tend to regulate energy matters – OGRE included – balancing different criteria and goals. Economic growth, energy security, environmental concerns, and the provision of fundamental services – such as electricity – for users located in isolated rural areas and poor urban communities are the most relevant criteria identified in that regard.

In particular, countries that are advanced in the development of grid development stage, where the grid is already substantially developed, OGRE tends not to play as remarkable a role as compared to countries with specific characteristics or limitations – or with different priorities. That is assessed by comparing the role of OGRE in Brazil, Mexico, and China with other countries such as Trinidad and Tobago in the Americas, the Philippines and Indonesia in Asia, or the African countries of Uganda, Kenya, and Nigeria.

Notwithstanding the above, Brazil, Mexico, and China are prone to implementing OGRE to replace the current diesel-sourced electricity plants off the grid already existent, which is an important milestone for addressing environmental concerns (especially in China, where many coal-sourced plants are (still) being commissioned at a steady pace). Furthermore, there are clear synergies from OGRE that could produce other beneficial outcomes. Reducing China's problems with network stabilization is one example, among others, of OGRE's beneficial 'by-products'.

On the other hand, Trinidad and Tobago, the Philippines, Indonesia, Uganda, Kenya, and Nigeria are still struggling to put in place initial OGRE measures to provide its most vulnerable citizens with access to clean electricity. Some of the major barriers generally appointed are the lack of: (i) prioritisation of renewable energy and environmental consciousness amidst the population, (ii) concrete regulatory targets and proper control of emissions, (iii) financing instruments, and (iv) infrastructure and a developed domestic market. Some of these countries pose specific problems that require a more thorough analysis, such as: (i) the political challenge in Trinidad and Tobago, given the abundance of oil and gas and, at

DOI: 10.4324/9781003178088-19

the same time, the island's seawater inflows, (ii) the infrastructure challenge in Indonesia and the Philippines, presented by their geographic configuration, an archipelago of thousands of islands, and (iii) regulatory and financing challenges faced by the African countries of Uganda, Kenya, and Nigeria.

14.2 Summary of OGRE

In order to summarise the current pace of OGRE worldwide, this section is divided into three parts: (i) natural resources, (ii) OGRE regulation, and (iii) financial aspects. Each of the parts are divided by continent, much like the structure of the book. We start with Africa, then Asia and the Americas.

A) Africa

As the chapters of this book suggest, Africa is the continent where OGRE would be most valuable to put in place, since electricity has not reached a considerable number of inhabitants, being mainly located in rural areas. Therefore, despite the fact that to this date the grid itself needs to be improved and expanded, off-grid new connections based on renewable sources would be a reasonable way to match an energy transitioning world with the population needs of the many existing isolated rural areas.

Even though a quick first approach to providing electricity for the population would be based exclusively on financial criteria, given that electricity is in fact a fundamental product for people and there is a scarce level of affordability in this case, renewable energy has to be harnessed by local government as much as possible, marrying the countries' needs with the worlds.

There is also an additional global cooperation aspect in Africa, which may foster the implementation of renewable-sourced energy projects in the continent. Climate change is, after all, a global phenomenon and implementing renewable sources of energy globally plays a vital role in dealing with it.

Kenya

As pointed out by the authors of the Kenyan chapter,[1] the country has both an abundant natural resource inventory and a robust regulatory framework regarding energy policy. Reportedly, OGRE policies such as the traditional *Kenyan Rural Electrification Programme* (KREP) were boosted by the more recent *Energy Acts* of 2006 and 2019. Also, renewable energy currently represents 65% of total energy investments and approximately half of Kenya's electricity output comes from geothermal sources – a specific, non reproductible, characteristic of the Kenyan territory.

On the other hand, the main challenges still pending to be addressed relate to financial affordability of projects, further improvements in governance in the energy sector, and both political and social enrolment aimed at finding new solutions. Reportedly, the vast majority of financial investment in renewables in

Kenya comes from foreign institutions such as the World Bank – a circumstance that needs to be balanced with local solutions.

There are also relevant gaps identified in law, such as the lack of regulatory mechanisms that foster public involvement and contribute to the spread of educational information between rural dwellers. Finally, the duplicity of functions between public administration's own bodies creates conflicts of authority.

Nigeria

Nigeria presents financial, regulatory, political, and social challenges, all combined with an extensive use of fossil fuels. Even though the *Nigerian National Petroleum Policy* (NNPP) has acknowledged the need for more balance and diversification between energy sources, natural gas still is the source of approximately 80% of the electricity generated in the country. The *National Energy Policy* (NEP) of 2003 and the *National Renewable Energy and Energy Efficiency Policy* (NREEEP) of 2015 were also meant to boost renewable energy, including OGRE, but have not been able to implement energy transition as desired. [2]

Regulation enacted by the Nigerian Electricity Regulatory Commission (NERC) also constitutes the framework for OGRE. A few examples of such regulation include the following subjects: *Permits for Captive Power Generation* of 2008, *Embedded Generation* of 2012, *Independent Electricity Distribution Network* of 2012, and *Mini-Grid* of 2016. As stated by the World Bank through *The World Bank's Regulatory Indicators for Sustainable Energy Report* of 2016, Nigeria has been under-utilising its regulatory framework already in force to expand OGRE and provide electricity to consumers located in isolated areas. That shows how regulation by itself is not sufficient to push countries into a more sustainable future.

Additionally, even though NERC has enacted The *Regulation for Feed-in Tariff of Renewable Energy Sourced Electricity* (REFIT) in 2015, it expressly excludes OGRE from its potential beneficiaries. So, there are also insufficient subsidies or financial aid for people who simply cannot afford electricity.

Uganda

Like Nigeria, Uganda's feed-in tariffs programme does not include OGRE. However, it has a more favourable scenario to renewables compared to Nigeria. Approximately 80% of the electricity generated in Uganda comes from hydropower. Like Kenya, the country is gifted with natural resources.[3]

On the other hand, only 28% of the population has access to electricity, while almost 5% is covered by OGRE. The third phase of the *Energy for Rural Transformation* (ERT), a World Bank-supported programme, has targeted OGRE as a key strategy for the population's electricity access. The current *Electricity Act* of 1999 has proved to be a complex and costly regulatory framework for investors in OGRE, mainly in matters related to approval and licensing proceedings. It is expected to be created as a one-stop shop for mini-grid developers, which would launch new investments.

Finally, the *Statutory Instrument No. 75* of 2002 created the Rural Electrification Agency, which assists the Rural Electrification Board in providing access for rural consumers of electricity through the Rural Electrification Fund. The Fund is not specifically aimed at OGRE, but for energy projects in rural communities. It is important to point out in this topic that OGRE is not exclusively aimed at rural communities, but rather for some urban areas as well.

Overall

In general terms, it can be said that the African countries surveyed in this book rely heavily on technical and financial programmes put in place by multilateral institutions such as the World Bank. These institutions also play an important role in organising special initiatives that materially provide access to electricity in isolated poor areas, such as the ERT in Uganda.

When it comes to natural resources, each country has a particular characteristic: Kenya relies on its geothermal sources, Nigeria on oil and gas, and Uganda on hydropower. Even though there is a huge potential for increase in power generation from renewable sources, investments still struggle with regulatory and financial issues, keeping away a vast number of people from electricity access. Even though there is a regulatory framework in most countries, regulation has not so far been sufficient by itself to provide access to the majority of the population in isolated areas. The absence of clear targets in regulation are also mentioned as a regulatory deficiency that needs to be addressed in the relevant countries analysed in Africa.

B) Asia

The three cases analysed in Asia are particularly relevant to realise that oil, gas, and coal still play a major role in the world energy economy. As it emerges from the chapters dedicated to the Asian countries, coal-fuelled power plants continue to this day to be the preferential targets for new investment and coal-fuelled electricity's stake is consequently growing rapidly in these countries nowadays.

It is relevant to point out that China, Indonesia, and the Philippines provide an extensive coverage of their territory via the regular grid. However, isolated communities (especially in remote islands) face more challenges to get access to electricity and this is where OGRE could provide a relevant contribution.

China

China has had an extraordinary performance in economic growth through the years and has grown to become the leading economy in the world in terms of gross domestic product (GDP). In this context, almost the entire population – which is also the biggest in the world – has been connected to grid electricity. In terms of natural resources, China has not installed sufficient technology so far, e.g., photovoltaic and wind turbines that meets the country's massive needs.[4]

Even though China's five-year plan for 2021–2025 reportedly acknowledges the importance of renewable energy, there is a combination of regulatory and political issues hindering further development. Just like the Nigerian regulation case, the country has no clear targets or a benchmark to follow in this area. Consequently, OGRE becomes an eventful topic that does not generate lasting or focused interest in political agents.

The *Electricity Law* of 1996 combined with the *Renewable Energy Law* of 2005 is the regulatory ground for OGRE, but they have also not been able to implement a cohesive renewables agenda. It is true that China is leading a major transformation in implementing renewables given the necessity of adapting its current highly carbonised energy matrix. However, that effort is not running in parallel with the implementation of OGRE solutions.

The Philippines

In the Philippines, the scenario for OGRE, and renewables in general, is somewhat similar, but aggravated by financial and natural restrictions not existent in China.

The archipelago of the Philippines is formed by approximately 7,000 islands, which hinder the implantation of renewables projects and favours transportable fuels such as diesel as a source of electricity. Also, the country does not have robust natural resources for energy use, so it has always been dependent to a great extent on imported oil and gas, making the Philippines vulnerable to price (and stock) fluctuations.[5]

On the financial side, the Philippines is still a country under development with significant restrictions. Even though the country has created the Renewable Energy Trust Fund (RETF), targeting investments in renewable energy, those investments have been having difficulties in materialising, as reported by the OECD. The *Electric Power Act* of 2001, the *Renewable Energy Act* of 2008, and the *Philippines Act* of 2008 that created the RETF are the regulatory grounds for OGRE.

Some of the regulatory gaps identified include the lack of prioritisation of renewables – particularly OGRE – over the increasing use of fossil fuels, no clear benchmarks for OGRE, and provisions that further engage the government to energy transition implementation initiatives.

Indonesia

In Indonesia, the context for OGRE is partially similar to the Philippines as it is a developing country with an extensive number of islands. However, Indonesia's economic development and the great amount of natural resources in place provides a better situation for Indonesia to overcome their challenges (especially access to electricity in the rural eastern islands).[6]

The total installed capacity of renewable energy is fairly low in Indonesia (circa 13%) compared to its potential as highlighted by the Ministry of Energy and Mineral Resources. Therefore, more investment is needed to increase the output of renewable energy (including OGRE) as they need to address the common challenge to balance the so-called energy transition, energy security, and energy justice.

Furthermore, Indonesia is lacking in two critical areas: (i) dedicated regulation to incentivise off-grid renewable electricity systems, and (ii) the demand for such electricity is low as over 99% of Indonesia's electricity comes from the main grid.

Overall

These cases in Asia show a large dependence on coal and a tendency of increasing levels of disparity between a renewable energy matrix and the current configuration of energy sources, requiring more attention and urgent political measures to counterbalance this trend. In fact, the concern in these countries even transcends OGRE, since it is more fundamentally a (general) question of energy transition (more renewables sources, connected or not connected to the grid).

However, the situation in the Philippines and Indonesia is far more complex than in China as they do not possess the same economic power and their territory contains thousands of remote islands which require a more dedicated system to address their concerns.

C) Latin America

In Brazil and Mexico, the vast majority of the population is, as in China, connected to the electricity grid. Trinidad and Tobago is a particular case of an American island dependent on oil and gas, where OGRE would fit well for isolated areas. Even though Latin America is a developing region of the world, its historical and geographical characteristics are particularly different from Africa and Asia, while naturally more susceptible to North American and European political influence but also to the incoming of new technologies.

Brazil, for instance, is one of the world's largest economies and plays a notably key geopolitical role in energy and environmental matters, while also a leader in the run for energy transition. Mexico, on the other hand, is a strategic hub in Latin America and has a rather expanded electricity grid. Last, but not least, Trinidad and Tobago is a Caribbean island with one very abundant natural resource: oil and gas. Brazil and Mexico are also home to private investments in Latin America and the world, including the energy sector. In the case of OGRE, there is a necessary cooperation element between private and public initiatives capable of creating efficient solutions, provided that they are backed by a robust regulatory framework. In Trinidad and Tobago, government plays a much larger role, but could – for instance – promote renewable energy and OGRE through a classic energy transition process directly funded by oil and gas. Climate change effects have enormous impact and should be immediately addressed by Trinidad and Tobago, urging well-fitting solutions such as OGRE.

Brazil

Brazil is highly reliant on hydropower due to its natural resources, with approximately 60% as the source of electricity, but has consistently been making efforts in diversifying its energy matrix with both renewable energy and natural gas as alternatives because of the intermittency of hydropower.[7]

Law No. 12,111 of 2009 and Decree No. 7,246 of 2010, which regulate electricity access for isolated areas ('*Sistemas Isolados*' or SISOL), are applicable for OGRE, even though the majority of these systems are still fuelled by fossil products such as diesel. The National Agency of Electricity (ANEEL) also has an important role in conducting bids for the SISOL.

The country put in place economic subsidies such as the Fuel Consumption Account ('*Conta Consumo de Combustível*' or CCC) that incentivise the use of fossil fuels in SISOL, which is as important barrier for renewables. Regulation has prioritised access to electricity in SISOL, regardless of the source of energy, rather than renewable energy. A possible solution proposed in the mid-term is the use of hybrid systems, combining fossil fuels and renewables while energy transition takes place.

Even though Brazil has a clean energy matrix overall, SISOL and incentives such as CCC remain as a regulatory gap to be addressed for the benefit of OGRE. One of the reasons for that is the highlighted relevance of energy security over environmental gains for off-grid communities, whereas the majority of the country's energy resources are decarbonised already.

Even though diesel has historically been competitive in public bids, more recently that trend might be reversing, given the geopolitical challenges in the context of oil and gas, while the cost of solar panels are reduced drastically and photovoltaic regulated alternatives such as distributed generation are expanding rapidly.

Mexico

Although the vast majority of the Mexican population possess access to electricity (i.e., over 99%), renewables play a secondary role, if compared to Brazil. Approximately 75% of the electricity is generated by oil, gas, and coal.[8] Mexico possesses a similar challenge as indicated for Indonesia and Nigeria as their economy and energy mix are heavily reliant on the abundant amount of oil, gas, and coal resources. So, they do need to address their climate goals but at the same time deal with their energy security and justice issues.

Furthermore, it is important to highlight, as indicated by the authors of the Mexican chapter, there is no specific regulation regarding OGRE, while affordability in rural areas remains a barrier for access to electricity. But the challenges are somehow similar to other countries which have low demand in those areas and lack of infrastructure in place.

Trinidad and Tobago

Trinidad and Tobago is a very particular case of Caribbean islands where oil and gas are the source of practically 100% of their electricity. The country is a large exporter of oil and gas. There is a clear contradictory situation in Trinidad and Tobago's case because climate change must be a national priority, since the island is in real danger due to the increase in sea levels. But, at the same time, its economy was largely structured on an oil and gas basis, and is now largely dependent on such products.[9]

Just like Mexico, as indicated by the authors of Trinidad and Tobago, there is no specific regulation regarding OGRE, nor feed-in tariffs. Jointly with financial limitations, no concrete political measures were implemented to truly shift the country towards energy transition, which is necessary in the context of communities with low income and low scale for infrastructure projects. In other words, a successful energy transition in Trinidad and Tobago relies on providing an affordable energy supply, as stated in the authors' text.

In the regulatory context, the establishment of clear targets for addressing urgent issues such as climate change and energy access by reducing emissions gradually is key. Trinidad and Tobago has effectively pushed itself towards that direction, as indicated, but continuous evaluation has to be put in place, in parallel with new legislation covering OGRE.

Overall

Brazil and Mexico are well positioned in terms of reach of their electricity grid, but there is a clear distinction between them when it comes to OGRE. Even though Brazil has not accomplished a significant level of renewable projects for isolated areas, the subject has a well-established regulation, while Mexico has not produced a piece of regulation regarding OGRE nor has been engaged in renewables.

Trinidad and Tobago has no regulation regarding OGRE nor has been able to release itself from oil and gas dependence through a planned schedule. The islands should promote immediate measures such as pro-renewables in the near future because climate change is a threat to the islands' future existence.

14.3 Reflections and Recommendation

Given the overall summary above, the following sections focus on four core themes in the context of OGRE: (1) key aspects regarding climate change and how it relates to OGRE, (2) a big picture of the regulatory framework on OGRE and its importance, (3) financial matters pending solution that limit OGRE, and finally (4) how the concept of energy justice matches with OGRE. Since (1), (2), (3), and (4) are intertwined topics, the final remarks propose a liaison between them that puts OGRE at centre stage.

14.3.1 Climate Change

A lot has been discussed about climate change over the last decades, associated with the energy transition process from fossil fuels to renewables. In fact, more recently, even oil and gas corporations have shown commitment to clean energy targets and roadmaps for said transition. Robust rules were enacted and judicial decisions enforced, especially in European countries such as the Netherlands.

Therefore, it is not unfounded to argue that climatic concerns have effectively guided decisions made by regulators, investors, and society at large, because of its irreversible and meaningful effects.

For that matter, the *United Nations Framework Convention on Climate Change* (UNFCCC) of 1992, the *Kyoto Protocol* of 1997 which represented a whole new paradigm in environmental legislation throughout the world, and the *Paris Climate Change Agreement* of 2015 are the most important pieces of international agreement concerning climate change. The signatory parties of the *Paris Climate Change Agreement* agreed to implement internal policies that restrain the levels of climate change to below 2°C compared to pre-industrial levels. More recently, the Glasgow COP 2021 showed how difficult it is to harmonise such different interests, but a common word was the expansion of renewables including OGRE.

When the energy matrices of countries are compared, there are clear disparities arising out of different characteristics, such as natural resources, geographic location and size, economic development, and political issues. From the nine cases presented, six are still struggling with significant obstacles to address climate change and energy transition (Nigeria, China, Indonesia, the Philippines, Mexico, and Trinidad and Tobago); three of them are reasonably adjusted (Kenya, Uganda, and Brazil).

As it seems, no sensitive and substantial political action has been practised in Trinidad and Tobago, since the shares of oil and gas have been practically the same in the electricity matrix for the last decade. Nigeria and Mexico also fit into that trend.

Despite the new investments and huge pressure from the international community, the share of coal in the electricity matrix of China is decreasing slowly, while in Indonesia and the Philippines is actually growing quite rapidly. Electricity in Brazil and Uganda is based on hydropower, and in Kenya on geothermal power, with a small share of fossil fuels. Both Kenya and Uganda have also been able to significantly reduce their use of oil for electricity purposes compared to ten years ago, which is an extraordinary milestone for containing climate change. Oil has been substituted mainly by more hydropower in Uganda and more geothermal power in Kenya.

From the above, it is noticeable that commitment to climate change mitigation has to be closely monitored in many countries, especially the ones without evident renewable natural resources, such as hydropower. It has been claimed that solar and wind power have, only recently, become more affordable and that intermittency is an additional problem to their use. Political measures are urgently required to address such problems.

14.3.2 Regulatory Framework

The chapters of this book demonstrate that there are important regulatory provisions to be created, implemented, and put into practice. Regulation is a proper mechanism for fostering renewable energy and incentivising corporations to act for the benefit of isolated rural areas. Nonetheless, on its own, it is clearly not capable of reaching practical results.

It has been evidenced, through the case studies, different levels of lack of political engagement reflected into an insufficient regulatory framework regarding OGRE. In fact, many countries' policies show they are not fully committed to renewables or OGRE, possibly because of not foreseeing an overall positive outcome, particularly under a financial perspective.

As anticipated before this book, besides lack of specific regulation concerning renewables or OGRE, even where regulation exists, some of the regulatory barriers are the absence of clear targets or unclear governance (as in the case of Nigeria and Kenya, respectively). Consequently, the policies implemented do not seem to be enforceable unless there are administrative watchdogs with sufficient authority and coordination powers to evaluate those policies.

In China, the greatest developing country in the world, despite the fact that the *Renewables Energy Law* was enacted in 2005 and access to the grid was extremely well-enacted, coal investments have grown substantially over the last few years. This clearly shows that regulation is insufficient by itself to promote concrete measures in favour of renewable energy or OGRE. The case of the Philippines is also indicative that there will be a critical problem in Southeast Asia on that matter, one of the most dynamic regions and an economic hub of the world.

There are other challenges regarding off-grid access for instance in Brazil, where there is a specific regulation for isolated areas, but at the same time there is a current incentive for fossil fuelled plants, which is preventing further development of OGRE. Subsidies for fossil fuels, even in isolated systems, manifest an environmental harmful sign, but cannot be instantly eliminated for obvious reasons. It becomes necessary to implement a planned programme that makes room for the elimination of fossil fuel subsidies while prioritising renewables. The lack of specific regulation addressing OGRE is a very sensitive matter for new investments, since financial opportunities will not emerge without legal certainty. Even where there is specific regulation, new investments will also not emerge without proper enforcement and political engagement towards them.

14.3.3 Funding and Financial Aid

When it comes to developing countries, financial limitations are key to understanding low access to fundamental services – and more specifically OGRE. Such countries are generally characterised by macroeconomic unbalance, high inflation, high interest rates, and oscillations in the exchange rate of its currency, all circumstances that disincentivise the development of an internal robust financial market. National government, normally the primary source for (local) investors, lack the means to finance OGRE-related projects, relying exclusively on its own economic capacity and quite often lack of internal market (especially in isolated areas). These are the reasons why African countries have received financial aid from multilateral institutions as funding for their OGRE projects on an ordinary basis, which is indeed a rational way to face this issue. Without further support from these institutions, which are largely funded by developed countries, the financial gap for OGRE can hardly be addressed. The Paris Agreement also prescribed funding of renewables projects for developing countries.

Feed-in tariffs are an ordinary financial mechanism to promote the growth of renewable energy share in power production. Consumers connected to the grid are charged and compensate for investments in renewable energy by paying an additional fee in their electricity bills. That is a successful formula practised in Europe. However, as seen in the cases of Nigeria and Uganda, feed-in tariffs expressly exclude OGRE. Instead, OGRE is to be incentivised through tariffs and other subsidies, since consumers in isolated areas and poor communities are more vulnerable than consumers connected to the grid.

There are studies analysing the implementation of feed-in tariffs for mini-grids in African countries, such as Tanzania. The authors conclude that such a programme is financially viable based on photovoltaic and hydropower provided that feed-in tariffs are used as an incentive, while benefits outweigh costs in the long term.[10] There are more recent studies on that matter for Tanzania.[11] This could be an extrapolating financial model for another countries, such as Kenya, Nigeria, and Uganda, or even for other parts of the world.

Finally, local funds such as the ones existent in China (Renewable Energy Development Fund – REDF) and the Philippines (Renewable Energy Trust Fund – RETF) may also serve as a relevant source of funding, even though they require proper governance and clear set up rules to adequately fulfil their original purpose.

14.3.4 Energy Justice

The concept of energy justice is relevant to the understanding of OGRE in its political and social perspective. It has emerged from the application of justice principles to global energy systems.

In this context, Jenkins et al. (2016) offers a methodology for understanding and acknowledging injustice based on three tenets: distribution, recognition, and procedure.[12] The key dimensions sought by researchers in this field are evaluative and normative, in order to assess injustices and formulate recommendations on how to solve such problems.

From the distribution tenet, it is possible to identify where injustices are located, and then analyse how they can be solved. The example given by the author is of feed-in tariffs charged by the German government in favour of wind renewable energy new projects in the context of the *Energiewende* programme, which would be subsidised by all consumers. Those with lower income are more substantially affected by such a subsidy.

This *prima facie* injustice is shown to be sound public policy if it is taken into account that renewable energy replacing nuclear energy diminishes the risk of a nuclear incident as a by-product. Or, as a decentralised infrastructure solution, the burdens of the programme are distributed more evenly across the country, for a wider number of consumers, than a centralised solution charged against local communities (including low-income individuals).

The recognition tenet 'emphasizes the need to understand different types of vulnerability and specific needs associated with energy services among social groups (especially marginalized communities[13])'. Finally, the procedure tenet refers to the assessment of a fair process, and then analyses which processes shall be implemented. In short, these tenets are closely related to social enrolment in energy projects, in the sense that the interested parties affected by a public policy participate and contribute somehow to a better outcome.

Energy justice is an important analytic tool for assessing and remediating conflicts and elaborating better solutions concerning energy projects. Further engagement of both political and social agents is one of the critical top missing characteristics of the case studies for further development regarding renewables in general and OGRE in particular.

14.4 Final Remarks

The case studies from this book have shown that OGRE is a topic that needs further engagement by the international community. OGRE is definitely a multi-beneficial solution that both provides access to electricity for consumers in isolated rural areas and poor communities, and transforms the energy matrix of countries towards low-carbonised transition.

Climate change is an issue that needs urgent addressing, especially when it becomes clear that concrete measures towards energy transition are not under way in many places of the world, not only poor countries but also countries under an advanced development stage (especially by the competing interest of energy security as highlighted in the current crises in eastern Europe). But climate change is a global issue that needs to be dealt with by each and every country. If this picture is not quickly reversed, consequences for more vulnerable countries may be radical.

Regulation is a key element for fostering OGRE, but it is not enough. Many countries have enacted renewable energy regulations, including off-grid solutions for isolated areas, and nonetheless have not developed their OGRE programmes. The lack of plain targets and governance matters are relevant in explaining why regulation has been failing.

Financial aid by multilateral institutions are also important funding sources for developing countries in order to overcome financial restrictions imposed by macroeconomic deficiencies. Another example of a financial alternative is the implementation of feed-in tariffs to compensate OGRE costs by charging the whole net of consumers. There are extrapolating models from African countries that could be used for that matter.

The concept of energy justice contributes for more rational decision-making based on political and social enrolment. There are new trends regarding access to off-grid electricity by consumers in isolated areas, such as hybrid off-grid power plants.[14,15] Also, plants that utilise hydrogen to surpass intermittency limitations of renewable energy and logistical difficulties.[16]

Notes

1 Part Two, Chapter 3.
2 Part Two, Chapter 1.
3 Part Two, Chapter 2.
4 Part Three, Chapter 3.
5 Part Three, Chapter 1.
6 Part Three, Chapter 4.
7 Part Four, Chapter 1.
8 Part Four, Chapter 3.
9 Part Four, Chapter 4.
10 Moner-Girona, Magda, et al. (2016). Adaptation of feed-in tariff for remote mini-grids: Tanzania as an illustrative case. *Renewable and Sustainable Energy Reviews, 53.* Available on: https://www.sciencedirect.com/science/article/pii/S1364032115009181.
11 Simpson, Nicholas Philip, et al. (2021). Adoption rationales and effects of off-grid renewable energy access for African youth: A case study from Tanzania. *Renewable and Sustainable Energy Reviews, 141.* Available on: https://www.sciencedirect.com/science/article/abs/pii/S1364032121000885.
12 Jenkins, K., McCauley, D. A., Heffron, R., Stephan, H., & Rehner, R. (2016). 11 Energy justice: A Conceptual Review. *Energy Research & Social Science* 174-182.
13 Lee, Joohee, & Byrne, John. (2019). Expanding the conceptual and analytical basis of energy justice: Beyond the three-tenet framework. *Frontiers in Energy Research, 7.* Available on: https://www.frontiersin.org/articles/10.3389/fenrg.2019.00099/full
14 Zebra, Emília Inês Come, Windt, Henny J. van der, Nhumaio, Geraldo, & Faaji, André P. C. (2021). A review of hybrid renewable energy systems in mini-grids for off-grid electrification in developing countries. *Renewable and Sustainable Energy Reviews, 144.* Available on: https://www.sciencedirect.com/science/article/pii/S1364032121003269.
15 Suresh, Vendoti, & Muralidhar, Mahankali. (2020). Modelling and optimization of an off-grid hybrid renewable energy system for electrification in a rural areas. *Energy Reports, 6.* Available on: https://www.sciencedirect.com/science/article/pii/S2352484718304499.
16 Marocco, Paolo, Ferrero, Domenico, Lanzini, Andrea, & Santarelli, Massimo. (2022). The role of hydrogen in the optimal design of off-grid hybrid renewable energy systems. *Journal of Energy Storage, 46.* Available on: https://www.sciencedirect.com/science/article/pii/S2352152X21015589.

Index

access to reliable energy 73
Administrative License Law of the People's Republic of China 139, 149, 153
Africa 260
Africa Growth Together Fund 102
African Development Bank 57, 84, 95, 102
Agenda for Sustainable Development 13, 29, 35, 38
anthropogenic interference 19, 53
Asia 1, 3, 4, 36–37, 134, 259, 262

bankability 174, 177, 181
Basic Electricity Generation Cost 176
Bio-Electricity *4*
Biofuel Power Plant 179, 185
Biogas for Electricity *4*
biomass plants 2, 40
bio materials 1
Brazil 264, 268
budget cuts 174
build-own-operate-transfer 181

capacity-building actions 59
capital intensive: projects 176, 194; venture 116
captive consumer 213
carbon emission(s) 139; reduction 85
carbon lock-ins 25
Carbon Reduction Strategy 243, 247
China 262, 268–269; RE Law 2005 144; Renewable Energy Law 2005 136
China's 14th five-year plan 2021–2025 144
clean energy development 174
climate change 266; action 59; mitigation 20, 55, 136, 206, 226–227, 267; objectives 163
climate conditions 94; targets 44
climate-sensitive sectors 75
coal production 175
cogeneration plant 227
commercialisation: of RE 157; renewable energy 162
concrete regulatory targets 259
Conference of (the) Parties 53–54
consequences of climate change 18
Constitution of: Kenya 78, 81, 91; Nigeria 98; the Republic of Indonesia 177; Uganda 112
construction of OGRE technologies 42
contractual vehicles 3
conventional: energy 98; fossil fuel 176, 245; sources 2, 40
COP *see* Conference of (the) Parties
COP Meeting 2021 58
corporate income tax 189, 190
cost-reflective tariff 121, 155, 161
Criminal Code (of Nigeria) 100

deforestation 39
Department of Energy (of the Philippines) 158
deployment of mobile technologies 116
depreciation and amortisation 189
Development Bank of the Philippines 162
differentiated responsibility 18
Directorate General of: Climate Change Control (of Indonesia) 175; New Energy Renewable Energy and Conservation of Energy (of Indonesia) 180
Directorate of Water Resources Management (of Uganda) 123
dissemination of RE benefits 160
distributed generation 213–214, 216, 250, 265
diversification of energy sources 75
DOE *see* Department of Energy (of the Philippines)
DOE National Renewable Energy Plan 2011 158

domestic: energy usage 247; legal measures 6, 27; legislation 21, 75; mitigation measures 54

economic: liberalisation 112; policy environment 43; sustainability 244
Economic Community for West Africa 26, 97
economic growth and: development 39–41, 83, 157; diversification 40
Economic Sustainability Plan (of Nigeria) 95
economies of scale 177, 232, 241, 246
ECOWAS *see* Economic Community for West Africa
ECT *see* Energy Charter Treaty
electricity: connection costs 113; demand 111–113, 155; energy tariffs 82; generation licence 99; plants 1; Power Sector Reforms Act (of Nigeria) 96; pricing scheme 100; purchase price 190; quality 206; sector transformation 7; subsidy 176; transmission 81, 98
electricity access: and consumption (in Uganda) 113; in rural areas 117, 119
Electric Power Industry Reform Act 2001 159, 162
Electric Power Industry Reform Act of 2001 (of Philippines) 155
electrification: of rural communities 226; rate 4, 222; subsidies 102
emission of GHGs 224
employment generation 40, 41
end-user subsidies 112, 117
energy: deficiency (in Nigeria) 103; demand(s) 40, 135; diversification 97; infrastructures 2; justice 266, 269, 270–271; matrix 270; poverty 39, 103, 228; security 136, 137
Energy and Petroleum: Regulatory Authority (of Kenya) 79; Tribunal (of Kenya) 79
Energy Charter Treaty 20
Energy for Rural Transformation 111–112, 114–115
Energy Information Administration 43
enforcement of regulations 77
environmental: concerns 259; degradation 82, 118; energy institutions 27; impact assessment 6, 140; preservation 121; protection 229; protection objective 75
Environmental Approval forms (in Indonesia) 184
Environmental Impact Assessment Act 1992 (of Nigeria) 99
EU *see* European Union
European Convention on Human Rights 23
European Union 26

FDI 84–85
Federative Republic of Brazil 203
financial: barriers 57, 160, 163, 223; feasibility analysis 182
financing instruments 259
fiscal and non-fiscal incentives 157
fiscal incentives 191, 245
Fossil fuel electricity 4; options 228; technologies 52, 53, 137
fossil fuels 39, 43, 52, 61, 110, 157, 224, 244; power plants 194, 208–209, 212, 216; projects 20; reserves 244
framework for sustainable energy 17–18, 22–23, 27
fund universalisation 206

gas electricity technologies 223
gasoline refineries 226
General Law on Climate Change 2012 (of Mexico) 223
generation arrangements 213
Geothermal Power Plant 179, 193
geothermal power project 76, 84
GHG emissions 227
Glasgow 26th Meetings of the COP 53
Glasgow COP 2021 55
global climate action 20, 54
Global Climate Risk Index 2021 175
global commitment to sustainable energy 25
global energy: institutions 18; transition 1, 3
global energy transition agenda 143
global environmental protection 1
Global Environment Facility 56
global sustainable: development drive 99; energy governance 25
governance practices 44
Government of Uganda 110
Green Climate Fund 56, 243
greenhouse gas emission(s) 203–204, 242–243, 251; strategies 6, 55, 56, 60
green technology 93
grid-based electrification 122
Grid-connected electricity 39
grid expansion 93, 102, 118, 120, 186, 215
grid extension projects 120
growth and sustainable development 44
growth of renewable energy 163
Guideline on the Domestic Product Utilization for Electricity Infrastructure (of Indonesia) 176

high electricity connection cost 113
high electricity tariff 75
high importation costs 74
high operating costs 211
high-polluting energy sources 38
human capital development 43
hybrid: energy generation systems 77; renewable energy options 94; solutions 210
hydrocarbon resources 242
hydroelectric dams 94
Hydro Power Plant 101, 179, **185**, 188, 192–193
hydro risk mitigation mechanisms 77

IEA 38
IEC *see* International Energy Charter
ILO *see* International Labour Organis(z)ation
Indonesia 263
Indonesia's total electricity mix 174
industrial electricity prices 245
Industry Reform Act 2001 162
initial capital 208, 223; cost(s) 5, 53, 57, 136–137, 143–144, 154–155, 229, 232, 245
installed electricity capacity 74, 112
Inter-American Development Bank Final Report in 2012 241
interconnected mini-grid 101
Intergovernmental Panel on Climate Change 2, 254
international: energy governance 17; energy governance framework 27; energy organisations 26; environmental laws 75
international climate change regime 7, 23–24, 53, 55, 60
International Energy Agency 38
International Energy Charter 21
international financial institutions: World Bank 21–22, 57, 79, 114; International Monetary Fund 21
international framework for sustainable energy 17–18; law 17
International Labour Organis(z)ation 2, 40
International Renewable Energy Agency 3, 30, 37, 55, 67, 124, 151, 225, 243; United Nations Sustainable Development Goals 3
investment climate 58, 194
IPCC *see* Intergovernmental Panel on Climate Change
IRENA 2, 22, 227, 243; *see also* International Renewable Energy Agency
IRENA Off-grid Renewable Energy Statistics 2021 4

Johannesburg Plan of action 19

Kenya 73, 75–76, 260, 269; Civil Aviation Authority Clearance 80
Kenyan Rural Electrification Programme 75
Kyoto Protocol 7, 23–24, 53–54, 242

Latin America 264
LCPD 76
Least Cost Power Development Plan 73, 91
legal and regulatory framework 248
legislative barriers 248
Licensing and Environmental Impact Assessment 3
Light Up Africa 101
limiting global warming 2
local supply deficiencies 203
long-term international financing 116
low budgetary allocation 76
low-cost electricity 113
low-cost power supply 110

marginated zones 228
maximisation of energy efficiency 251
MDGs *see* UN, Millennium Development Goals
Medium and Long-Term Development Plan for Renewable Energy 2007 138
methodological: holism 6; institutionalism 6
Mexican: Law on Climate Change 2012 225; off-grid generation 228; State Productive Enterprise 226
Mexico 264
Mexico's Energy Reform of 2013 230
micro-grid system 214
Mid Term National Development Plan for 2020–2024 (of Indonesia) 175
Millennium Development Goals 38
mini-grid: developers 122, 144, 261; electricity projects 229; infrastructure 36; regulation 117; renewable electricity 1, 162, 229; renewable energy technologies 1; technologies 39
Mini Hydro and Micro Hydro Power Plant 179
'minimum subsidy tender' 102
Ministry of Energy and Mineral Resources (of Indonesia) 173
mitigation of greenhouse gases 204
mobile telecommunication platforms 116
mobilisation of climate finance 56, 58, 60
modular geothermal technologies 77
multilateral institutions 268, 270

multiple permit systems 5
Multi Year Tariff Order 100

National: Council of Water and Electric Energy (of Brazil) 204; Determined Contributions 54; Development and Reform Commission (of China) 138; Energy Administration (of China) 139
National Energy Efficiency Monitoring Report of Trinidad and Tobago 242
National Energy Policy (of Kenya) 76
National Energy Policy (of Nigeria) 97
National Environmental Management: Authority 81; Authority (of Kenya) 80, 123
National Environmental Standards and Regulations Enforcement Agency (of Nigeria) 100
national grid 1
National Interconnected System (of Brazil) 203
nationally determined contributions 20, 54
Nationally Determined Contribution to the Paris Agreement 75
National Renewable Energy and Energy: Efficiency Policy (in Nigeria) 97; Board (of the Philippines) 159
National Transmission Network or the General Distribution Networks (of Mexico) 228
National Treasury of the Philippines National Oil Company 160
NDCs *see* nationally determined contributions
NEP *see* National Energy Policy
network interconnection study 182
Nigeria 93–94, 97, 102, 261, 268–269; Energy Access Fund 102; Infrastructure Debt Fund 102
Nigerian: Bulk Electricity Trader 101; Electricity Regulatory Commission 96, 261; energy sector 98; National Petroleum Corporation 97; National Petroleum Policy 96
non-fossil fuel sources 26
nonrenewable energy sector 44
North Atlantic Ocean 241

ocean energy 193
OECD *see* Organisation for Economic Cooperation and Development
off-grid: developers 99; development 101–102; diesel electricity projects 157; rural electrification 94; solar industry 41; utilisation 77
off-grid electricity: access 118; consumers 232; generation 94; sector 5, 155; sources 93
off-grid electrification development 119
off-grid energy development 95
off-grid financing scheme 120
off-grid fossil fuel electricity 158
off-grid power: projects 100; solutions 95; system 96
off-grid renewable: electricity 1, 73, 101; electricity projects 245; energy 2; energy electricity 95, 188; energy solutions 44; energy supply systems 118
offshore oil platforms 97
OGRE 36, 110, 227; investors 5, 116, 157, 161; photovoltaic projects 143; technologies 45, 111, 114, 117, 120, 123, 138, 155, 158, 160, 224; *see also* off-grid renewable, electricity
Oiapoque Thermal Power Plant 210
oil companies 18
on-grid: electricity 1, 99–100, 115, 173, 231; renewable electricity 155; renewable electricity options 5
open international economic system 19
operational: feasibility analysis 182; subsidies 161
optimal development of OGRE 2, 6, 135, 136, 156, 223
Organisation for Economic Cooperation and Development 22

Paris Agreement 7, 19, 25, 55, 66, 116, 146, 151, 175, 225, 243, 269; Committee on Capacity Building 59
Paris Climate Change Agreement 3, 6, 8, 53–54, 155
payment of subsidies 143
perception of renewable electricity 3
Philippines 154, 158, 160, 259, 263, 268; Amusement and Gaming Corporation 160; Clean Air Act 160; Renewable Energy Act 2008 156–158, 162, 172
photovoltaic: equipment prices 215; generation 213; modules 246; solar system(s) 75, 232
'pioneer industries' 189
policy objectives 96, 98, 119, 121, 156, 250
political measures 266
Polluter pay(s) 18
power: blackouts 135, 222, 224; purchase agreements 8, 78, 101, 181–183, 186–187, 189–191, 193, 205; utilisation per capita 178

predominance of fossil sources 208
price fluctuations 157
pricing: mechanism(s) 188, 194, 248 of electricity 100
principle of sustainable development 18
private: law entity 205; sector capital 157; sector participation 111, 159
production of renewable electricity 73, 111
profitability 100
promotion of: capacity building 77; energy efficiency 19; renewable energy 23, 116; sustainable energy 18, 19, 24
proper control of emissions 259
prosumers 5, 213, 214, 216
Protocol and the Southern African Development Community 26
public policy 206, 207, 213, 214, 269, 270
Public-Private Partnership Act (in Kenya) 84
purchasing arrangements 248

rainfed agriculture 75
RDS *see Royal Dutch Shell*
RE development benchmark 74
reduction of GHG emissions 116
regulatory: environment 94, 103, 118, 136, 139, 144, 251; framework 268; framework of renewable electricity (in Indonesia) 180; support framework 144
Regulatory Impact Analysis Report 210
reliability of: electricity supply 183; isolated systems 212
renewable electricity: generation 40, 181, 194, 196; power plants 176; projects 143; technologies 59, 137, 155, 157, 167
renewable electricity-based generation 242
renewable energy: definition of 1; developers 100–101, 103, 143; development 141
Renewable Energy Development Fund (of China) 141
renewable energy electricity generation 42, 186; generation targets 41; illiteracy 145; policy 42; potentials 94; regulation 1; solar wind geothermal bioenergy hydropower ocean 36; sources 1, 141
Renewable Energy Law of China 2009 137
Renewable Energy Sources Generated Electricity 77
renewable off-grid projects 99
renewable power generation 248
Republic of Trinidad and Tobago 241
RES-E *see* Renewable Energy Sources Generated Electricity
residential electricity tariff 113
RE technologies 79, 157, 160, 162
Rio Declaration 19, 22; on Environment and Development 19
Royal Dutch Shell 23
rural electrification 73; (in Uganda) 110; infrastructures 120; in Kenya 74; projects 101, 120, 124; strategy 136; subsidies 111
Rural Electrification Act of 1936 (of Uganda) 114
Rural Electrification Agency 102
Rural Electrification Agency (of Uganda) 123
Rural Electrification Board (of Uganda) 123
Rural Electrification Fund (of Nigeria) 101
Rural Electrification Strategy and Plan (of Uganda) 112

SADC *see* Protocol and the Southern African Development Community
SDGs 37, 39; *see also* Sustainable Development Goals
SDS 39
Secretariat of Energy (of Mexico) 231
sector-specific targets 243
self-reliance in energy generation 96
sequestration technology 227
Small Power Utilities Group of the National Power Corporation 157
small-scale renewable energy 186
smallscale renewable energy systems 246
social impact assessment 6, 230
societal progress and development 38
Soft law instruments 19
solar: insolation 114; irradiation 114; photovoltaic 173; photovoltaic cells 2, 40; photovoltaic facilities 42
solar-fed minigrids 74
Solar Power Plant 179
special taxation 216
SSA *see* sub-Saharan Africa
standalone projects 162–163
standalone systems 111–112, 242
State Electricity Regulatory Commission (of China) 139
state-owned companies 5, 191, 226
Stockholm Declaration (of 1972) 19
structures for sustainable energy governance 18
sub-Saharan Africa 36, 43
sustainable: economic growth and development 19; energy development 73, 84, 157
sustainable development 2; patterns 17; principle 18, 23; requirements 100

Sustainable Development Goals 2, 5–6, 20, 37
sustainable energy: promotion 21, 24; sources 18; systems 174
Sustainable Energy for All: Action Agenda 73; Initiative (in Uganda) 111

tariff: fixing 100; for household consumer 188; increase 207; regime 100; regulation 6; structure 3, 117, 207
tax loss 189
technological capacity 5, 53, 137, 154–155, 157, 174
technology development and transfer 58
technology-specific clearance 80
thermal electric energy 216
thermic power plant 211
thermoelectric: generation projects 209; plant 225
tidal waves 1, 79, 94
transboundary energy problems 17
Transmission Company of Nigeria 98
transmission efficiency 77
transmission lines 93, 246
transnational law 1
Trinidad and Tobago 241, 244, 264; and policy targets 247–248
Trinidad and Tobago's Intended Nationally Determined Contribution 243

Uganda 110–111, 121, 261, 269; Electricity Act of 1999 111; National Bureau of Standard 123; and OGRE 112, 114, 116; Photovoltaic Pilot Project for Rural Electrification 114
Ugandan: Cabinet 111, 119, 121; economy 110, 121; electricity regulator 117
UN: Climate Change Executive Secretary 55; General Assembly Resolution 65/151, 20; Millennium Development Goals 20; Sustainable Development Summit in 2015 38
UNDESA *see* United Nations Department of Economic and Social Affairs
UNDP *see* United Nations Development Program
UNEP *see* United Nations Environment Programme
UNFCCC 20, 53, 54; *see also* United Nations Framework Convention on Climate Change
UNGA 26
United Nations Department of Economic and Social Affairs 21
United Nations Development Program 21
United Nations Development Programme 114
United Nations Environment Programme 21, 68
United Nations Framework Convention on Climate Change 19, 53, 62, 75, 175, 243, 251, 253
United Nations Sustainable Development Goals 2, 37
universal access to: electricity 2, 4, 227; sustainable electricity 5
universal electricity access 115
Universal Electric Service Fund (of Mexico) 230
unsustainable energy utilisation 18
USAID 125, 127, 131, 132, 160
U.S. Bureau of Labor Statistics 41, 47n46
US Global Research Programme and the American Geophysical Union 52
utilisation of renewable energy 94, 111, 137, 141–142, 156, 174, 188

Vision 2040 (of Uganda) 110, 114

waste power plant 173
wind farms 41, 246
Wind Power Plant 180, 185, 193
wind power project 5, 76, 84
World Bank 21–22, 26, 57, 114
World Bank's Regulatory Indicators for Sustainable Energy Report of 2016 97
World Commission of Environment and Development 19
World Energy Outlook (report) 22
world's Gross Domestic Product 40
WTO 24, 26; *see also* World Bank

For Product Safety Concerns and Information please contact our EU
representative GPSR@taylorandfrancis.com
Taylor & Francis Verlag GmbH, Kaufingerstraße 24, 80331 München, Germany

www.ingramcontent.com/pod-product-compliance
Ingram Content Group UK Ltd.
Pitfield, Milton Keynes, MK11 3LW, UK
UKHW020151200926
475702UK00016B/2652

* 9 7 8 1 0 3 2 0 1 2 9 5 7 *